D1548665

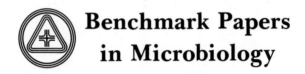

Benchmark Papers in Microbiology

Series Editor: Wayne W. Umbreit
Rutgers—The State University

Published Volumes and Volumes in Preparation

**Benchmark Papers
in Microbiology**

—— A *BENCHMARK* ® Books Series ——

MICROBIAL GROWTH

Edited by
P. S. S. DAWSON
National Research Council of Canada

**Dowden, Hutchinson
& Ross, Inc.**
Stroudsburg, Pennsylvania

Distributed by:
HALSTED PRESS
A Division of John Wiley & Sons, Inc.

Library of Congress Cataloging in Publication Data

Dawson, Peter Stephen Shevyn, 1923-   comp.
   Microbial growth.

   (Benchmark papers in microbiology ; v. 8)
   1. Microbial growth. I. Title.
[DNLM: 1. Growth--Collected works. 2. Microbiology
--Collected works. QW52 D272m]
QR86.D38   576'.1   74-26644
ISBN 0-470-19971-7

Exclusive Distributor: Halsted Press
a Division of John Wiley & Sons, Inc.

Acknowledgments
and Permissions

ACKNOWLEDGMENTS

NATIONAL ACADEMY OF SCIENCES—*Proceedings of the National Academy of Sciences (U.S.)*
   Bacterial Synchronization by Selection of Cells at Division

THE ROYAL SOCIETY OF LONDON—*Proceedings of the Royal Society of London*
   On the Biology of *Bacillus ramosus* (Fraenkel), a Schizomycete of the River Thames

PERMISSIONS

The following papers have been reprinted with the permission of the authors and the copyright holders.

ACADEMIC PRESS, INC.
   Experimental Cell Research
      Induction of Synchronous Cell Division in Mass Cultures of *Tetrahymena piriformis*
   Methods in Cell Physiology
      Continuous Synchronous Cultures of Protozoa

AMERICAN ASSOCIATION FOR THE ADVANCEMENT OF SCIENCE—*Science*
   Description of the Chemostat

AMERICAN SOCIETY FOR MICROBIOLOGY
   Bacteriological Reviews
      The Earlier Phases of the Bacterial Culture Cycle
      Microbial Growth Rates in Nature
   Journal of Bacteriology
      Kinetics of Growth of Individual Cells of *Escherichia coli* and *Azotobacter agilis*

ANNUAL REVIEWS, INC.—*Annual Review of Microbiology*
   The Growth of Bacterial Cultures

BLACKWELL SCIENTIFIC PUBLICATIONS LTD.—*Science Progress*
   A Review of *Growth, Function, and Regulation in Bacterial Cells*

CAMBRIDGE UNIVERSITY PRESS FOR THE SOCIETY FOR GENERAL MICROBIOLOGY
   Journal of General Microbiology
      The Continuous Culture of Bacteria: A Theoretical and Experimental Study
      Dependency on Medium and Temperature of Cell Size and Chemical Composition During Balanced
         Growth of *Salmonella typhimurium*
   Symposium of the Society for General Microbiology
      The Chemical Composition of Micro-organisms as a Function of Their Environment

FEDERATION OF EUROPEAN BIOCHEMICAL SOCIETIES—*European Journal of Biochemistry*
   Synchronization of *Escherichia coli* in a Chemostat by Periodic Phosphate Feeding

HER MAJESTY'S STATIONERY OFFICE—*Microbial Physiology and Continuous Culture* (Proceedings of the 3rd International Symposium)
   The Analogy Between Batch and Continuous Culture
   Balanced and Restricted Growth
   The Growth Rate of Microorganisms as a Function of Substrate Concentration
   Properties of Microorganisms Grown in Excess of Substrate at Different Dilution Rates in Continuous Multistream Culture Systems

INSTITUT PASTEUR, PARIS—*Annales de l'Institut Pasteur*
   La Technique de culture continue, théorie et applications
   for the Syndics of Newnham College, Cambridge

LONGMANS, GREEN & COMPANY LTD. —*Bacterial Metabolism*
   Growth and Nutrition

MACMILLAN JOURNALS LTD.—*Nature*
   Integration of Cell Reactions

PRINCETON UNIVERSITY PRESS—*The Chemistry and Physiology of Growth*
   The Kinetics of Growth of Microorganisms

PUBLISHING HOUSE OF THE CZECHOSLOVAK ACADEMY OF SCIENCES
   Continuous Cultivation of Microorganisms: A Symposium
      The Physiological State of Microorganisms During Continuous Culture
   Continuous Cultivation of Microorganisms (Proceedings of the 4th Symposium)
      Continuous Phased Culture—Experimental Technique

ROCKEFELLER UNIVERSITY PRESS —*Journal of General Physiology*
   Culture Conditions and the Development of the Photosynthetic Mechanism: II. An Apparatus for the Continuous Culture of Chlorella

SOCIETY FOR APPLIED BACTERIOLOGY—*Journal of Applied Bacteriology*
   Studies of the Growth of *Penicillium chrysogenum* in Continuous Flow Culture with Reference to Penicillin Production

UNIVERSITY OF CHICAGO PRESS—*Journal of Infectious Diseases*
   Life Phases in a Bacterial Culture

JOHN WILEY & SONS, INC.—*Essays in Biosynthesis and Microbial Development*
   The Regulation of Secondary Biosynthesis

Series Editor's Preface

Growth, in the culture of microorganisms, is more than cellular growth since it implies the growth of a population—that is, the growth of a group of cells from division to division, together with the reproduction and division cycles. Dr. Dawson has collected in one place a great deal of information on the culture cycle—including how a population of microorganisms grows, how it waxes and wanes, how it reaches its limit, sustains itself, and dies. The subject of population growth has been the basis of much of microbial physiology and has, of course, a variety of practical applications, implied in the answers to the following questions. Why does a population exhibit lag? Why does it stop growing? What conditions and controls the level of population eventually reached? What determines the rate at which this population level is attained? Why does it die? These have been the questions approached by microbial physiology; a number of the answers are immensely clarified by the selection of papers offered in this Benchmark volume.

Wayne W. Umbreit

Preface

In recent years the study of microbial growth has broadened and developed from the narrow confines of earlier years. The naïve simplicity of former times has disappeared, replaced by more involved patterns of thought and activity. Today microbial growth is a topic of wide and profound interest — if, sometimes, only vaguely understood.

In 1949 Monod observed that the study of bacterial growth was not in itself a specialized subject but a method basic to the discipline, an observation very true now for microbial growth as it relates to the wider field of microbiology.

Perhaps it is because very few textbooks give growth more than a cursory mention that its basic significance is not generally recognized: for, regrettably, the fact that growth is indeed basic to the discipline goes largely ignored in the general practice of microbiology and related endeavors. The omission often leads to superficial experimentation and much wasted effort, thus cluttering the literature with a lot of meaningless data.

It is hoped that this collection of papers may draw attention to the importance and basic significance of growth in microbiology.

It is inevitable that a very limited collection of papers, such as this one, in which some of the papers have had to be shortened to fit into the restricted space available, necessarily omits many other contributions equally deserving of inclusion—a circumstance that is very much regretted. It is hoped, however, that in spite of this handicap, the selections included cover the many experimental, theoretical, and philosophical developments on the broad canvas of microbial growth; and that they serve, in a roughly chronological manner, to present the perspectives necessary to show how fundamental is the study of growth for microbiology.

P. S. S. Dawson

Contents

Contents by Author

Introduction

The observations of his "animalcules" by Antonie van Leeuwenhoek (1632– 1723) with his newly discovered single-lens microscopes, and the development of pure culture methods by Koch (1847– 1910), may be singled out initially as forming the beginnings of the study of microbial growth. The serious study of microbial growth, however, begins about 1895, with the contributions of Ward, Müller, and others.

At the turn of the century, the early studies of growth were chiefly concerned with observing and determining, by use of a microscope, changes of shape, size, and number of organisms. There were obvious limitations to these procedures, and such experiments soon became linked to qualitative observations of the populations developing in the simple cultures. The excerpts (Paper 1) from Ward's report to the Royal Society of London in 1895 show what could be accomplished with the simple methods then available.

A general pattern of development in a culture, later known as the growth curve or "growth cycle," soon became matter of fact and served as a focal point for studies of microbial growth during the next 50 years, the era of the method of batch culture. The first 20 years of this period saw minor improvements of technique that permitted the pattern of the growth curve to be divided, at first, into three or four well-defined stages, and then later, more carefully, into seven. A classic paper by Buchanan (Paper 2) reviewed these developments and proposed the more-detailed, seven-stage template which had considerable influence during the next several decades.

As the pattern of develoment in a culture became familiar, more interest was taken of the many variations in form of the growth curve; closer and more critical attention to detail appeared. Total, as opposed to viable, counts of cells were distinguished and their experimental reproducibility examined; changes taking place in the morphology of the growing cells were closely studied; increasing attention was shown in the development of alternative methods for determining the crop of cells that could be harvested at different stages in the development of the culture; and the dry weight of organisms was introduced as a convenient method for estimating the amount of growth in a culture.

Microbiology in the early 1920s was not a closely knit discipline but an aggregation of different branches of bacteriology and biology; under such diverse influences two general trends appeared, the emergence of growth-cycle concepts in microbial growth studies, and recognition of physiological implications to growth. Certain phases of the growth curve became associated with characteristic activities in the cells; for instance, "physiological youth" and "young cells" with the early period, and "senility" and "old cells" with the later phase. The terminology in use was confusing; certain terms, such as "age," "growth cycle," "young," and "old," were used imprecisely and inconsistently with reference to cells and cell populations, so their varied meanings conflicted and often produced confused thinking. Some of these troubles persist and are occasionally encountered today [see D. W. Tempest, *Advan. Microbiol. Physiol.,* **4**, 225 (1970)].

Progress during the era of batch culture was largely related to gradual improvements of technique. During the early half of the period, the microscope served as the chief analytical tool, with studies largely concentrated on the cell and the study of cell numbers; later there was a distinct change to the study of cell mass and its composition. The transition was gradual and not clear cut.

In the 1930s studies of microbial growth concentrated on different phases of the growth curve, with the lag phase tending to receive most attention. The intriguing changes taking place in cells during this phase, coupled with the concept of physiological youth, had a special appeal for many workers. The review of Winslow and Walker (Paper 3) was a commentary of the contemporary progress in microbial growth studies at the time.

Improvements were not always of microbiological origin; the influence of biochemistry was becoming increasingly evident. For some time microbial materials had been used in studies of cell metabolism, as, for example, the use of "proliferating" and "resting" cells by the Cambridge School in England. Such investigations led to close observations of cell production and performance, and to new methods for examining, measuring, and reporting growth of cultures. Much better control of experimental parameters for growth, such as temperature of incubation and aeration by shaking procedures, were being used; experiments were being conducted more precisely by the introduction and use of chemically defined media, increased use of chemical assays and standardized procedures, and later by the measurement of cell density by optical methods.

The late Marjory Stephenson, who was largely responsible for the rise of chemical microbiology in Cambridge and closely associated with these developments in growth studies, gathered a distillate of this progress into a chapter on growth in her book *Bacterial Metabolism* (excerpted in Paper 4). This influential contribution documented many of the important experimental developments of that period. It is evident from this collection that growth was now seen to be closely linked with mass increase, especially in relation to metabolic parameters in the culture. The development of dry-weight determinations and then of optical methods for determining the microbial concentration in a culture were largely responsible for this change. Optical methods were convenient and rapid; being more easily calibrated against dry weight than by cell number, these methods were used increasingly, and the trend continues today.

In the mid-1940s, toward the end of the 50-year period, it was generally not very clear what constituted growth: although much was known superficially about different aspects of the problem, growth itself remained an enigma, a complex phenomenon, an unknown. The article by Van Niel (Paper 5) began to penetrate the mist and to provide some perspectives on the problem of growth; but it was Jacques Monod who cleared the mist away.

In a now-classical study entitled "Recherches sur la croissance des cultures bactériennes" published in 1942 Monod revealed how the nutrient supply available to the organisms controlled growth in a culture. At the time of its introduction, the significant part that this contribution was to play as a hinge upon which most future developments were to turn was not appreciated. Because of its size, the monograph is viewed briefly here by the contemporary glances which Stephenson took of its important parts. It is summarized by Monod's subsequent review (Paper 6) in 1949 of growth in bacterial cultures.

Stephenson's extract shows that Monod still recognized the general basic growth curve described earlier by Buchanan; but, significantly, he included another curve, which drew attention to the changes of growth rate taking place in the culture. In this curve the constancy of the growth rate in the phase of exponential growth is easily apparent and contrasts with the variations taking place in the phases preceding and following it. The distinction is important, because it reveals the fundamental difference in approach that Monod had to the problems of growth and the main reason for his ultimate success: by concentrating on the exponential phase of a constant growth rate he was able to obtain reproducible results on which he could build, an advantage lost to his contemporaries, who were concentrating on the variable inconstancies of the enigmatic lag phase. Other excerpts from Monod's "Recherches," given later in Stephenson's paper, show experimentally how the amount and rate of growth in the culture were related to the amount and concentration of substrate available in the culture. In his review Monod summarized how growth in a culture could be formulated in terms of growth constants specific for the organism and a substrate. He showed also how one limiting substrate could be followed by another, and recorded his discovery of the phenomenon of "diauxie." In this way the variability of growth curves, which had perplexed microbiologists for so long, was clarified.

Monod's work had two major influences on growth studies: a more enlightened use of the batch method of culture and the successful development of continuous culture as a technique. The batch culture, unchallenged and unquestioned for over half a century, was now seen to be limited and suitable for meaningful work in only two ways: first, when restricted to the exponential phase, because only there can reproducibility be obtained (Papers 7 and 8); and secondly, for the overall evaluation of the range of an organism's capabilities on a substrate (Paper 9). For other uses the batch method would still be useful in an empirical manner, but then expediency would appear to be the main justification for its employment.

In 1950 the technique of continuous culture brought revolutionary changes to the study of microbial growth, a new plateau of growth studies was reached, and a new era began. Monod was largely responsible for this development with the introduction of his theory of continuous culture (Paper 10), a logical extension of his earlier study of growth.

From his formulation of the relationship between growth rate and substrate concentration in a culture, Monod realized that theoretically it should be possible to continue growth at any fixed rate provided that the appropriate concentration of growth-limiting nutrient could be maintained in the growing culture. By using the simple device of constant dilution at constant volume, this was achieved experimentally, and in the publication of his theory of continuous culture, Monod recorded the successful operation of his continuous culture apparatus, the bactogen. This breakthrough was a double event, because simultaneously, and quite independently, Novick and Szilard (Paper 11) described a similar development, the chemostat. Also contemporary with these events was Bryson's introduction of the turbidostat, another variation of the idea.

A noteworthy article by Herbert, Elsworth, and Telling (Paper 12), established and consolidated the Monod theory, and by substantiating the general validity of this theory, established also the general veracity of Monod's basic formulation of growth and of the growth constants. This was done using bacteria, but Pirt and Callow (Paper 13) established that the same principles applied also for the continous culture of mycelial-producing organisms. Herbert was largely mentor and guide to the early developments of continuous culture and this paper has to be considered with several others by Herbert that appeared soon after; two of these (Papers 14 and 15) were of special significance for studies of microbial growth.

The relevance of environmental change in the culture vessel for regulating and controlling the growth and development of the culture in a cultivation system was now indicated by Herbert (Paper 14), who pointed out that the system of cultivation, by controlling the nutrient environment, regulated the growth of cells and the development of the cell population in that system. If the environment remained unchanged, as in continuous culture, a steady unchanging growth equilibrium was obtained—the steady state; but, if the environment changed, as in batch culture, the growth was manifested by changes as portrayed in the growth curve. Pointing out that the growth cycle of batch culture was a misnomer and an artifact arising from the development of a population in a closed system, Herbert demolished the idea, prevalent till then, that growth entailed a predestined and necessary sequence of change. The advent of continuous culture had quite suddenly revealed growth to be a condition that tended to be in balance with its surroundings — a surprising new orientation for studies in microbial growth.

It was soon apparent that in the absence of genetical change, succeeding generations of organisms in a continuous culture were similar — unchanged in composition, morphology, and cellular activities. Growth proceeded in a condition of equilibrium, manifested in the culture as a steady state. The properties of a steady state being characteristic of a growth rate, and therefore related to the doubling time, a fundamental property of the cell, automatically related growth on an absolute basis. Thus the growth curve and the stage and age of batch culture, all of which have no absolute meaning, could be replaced by the more significant parameter of mean doubling time, or growth rate, of the cells in continuous culture.

Herbert then proceeded to show (Paper 15) the fundamental difference between "closed" and "open" systems for growth, and of their significance for the cultivation of cells: how the many different methods employed for growing cells could be related to one or other of the two systems, and how these could become modified in multistage arrangements. Herbert extended the Monod theory to multistage continuous culture systems and indicated possibilities that existed for obtaining different results from growth carried out in the different systems.

In the early 1950s the plateau of continuous growth had been reached quickly, and directly, by the Monod and related theories; but another, meandering, ascent was also being completed. This alternative, by an empirical and qualitative, rather than quantitative approach, had a physiological instead of a mathematical basis. It was to become an increasingly important aspect in growth studies.

The rise of biochemistry and chemical microbiology in the 1930s had seen a deepening interest in cell metabolism which had encouraged close observation of cellular activities and produced attempts to prolong growth, increase cell yields, and extend various phases of the growth curve. Such activities led to sporadic attempts to cultivate microbes in a continuous manner. Most of these endeavors were seen at the time to be extensions of the batch method of cultivation and had little impact and very limited success, although one of the more successful attempts, that by Myers and Clark (Paper 16), had several significant innovations that foretold future developments.

Meanwhile, however, there were the little-known efforts of Utenkov in Russia and, later, of Málek in Czechoslovakia, who considered continuous culture to be a new method, different from batch culture, and of use for conducting physiological studies. With the recognition of continuous culture as a technique in the early 1950s, Málek was able, at last, to draw attention to such developments in his paper read at the first international symposium on continuous culture in Prague in 1958 (Paper 17), Málek outlined the empirical development of the technique, the importance of the qualitative changes taking place during cell growth and development, and introduced the concept of the "physiological state."

Málek's persistent emphasis on the physiological aspects of microbial growth now commanded increasing attention as discrepancies began to accumulate between results obtained from growth in batch and continuous systems. There appeared to be a dichotomy in growth, and in growth studies, that was reflected in both batch and continuous cultures. In batch culture there were the long-recognized differences between "proliferating" and "resting" cells — or of the "trophophase" and "idiophase" conditions of Bu'lock's more contemporary studies — and of differences between the steady states of fast- and slow-growing cells in continuous culture.

It began to appear that growth was not just a quantitative increase in mass and number but also involved concomitant qualitative changes; growth involved both multiplication and development in an intricately enmeshed relationship within the cell. The problem, seen at its simplest as the correlation of physiological state with growth rate, is a conundrum that presently has to be rationalized and solved.

A number of ways are being explored for doing this. One method is multistage continuous culture, and the system of Řičica, Nečinová, Stejskalová, and Fencl (Paper 18) promises to be a very useful tool for investigating the interrelationships of physiology with growth in continuous cultures. This versatile method has the facility for examining the hitherto largely unexplored area of "unrestricted" growth, both by itself and in relation to the "restricted" growth common to chemostat systems, apart from other systematic explorations of the physiological state (Papers 19 and 20).

In the progressive development of studies of microbial growth, mathematics has always played a part. This is demonstrated very well in the Monod formulation, a mathematical model that is necessarily restricted and excludes any consideration of physiological change. In a recent paper which considers the growth rate as a function of substrate concentration, Powell (Paper 21) moves significantly toward accommodating physiological change in the culture and attempts to modify the Monod formulation accordingly. However, as mathematical models are precise, no rigid formulation can be successfully advanced until the physiological parameters involved can be defined mathematically: unfortunately the enigma of the physiological state intrudes again.

Meanwhile, in Denmark and North America, in the later 1950s, further advances were taking place in growth studies that were to add new dimensions to microbial growth; techniques for synchronizing cell populations were being developed. In these cultures, the cells developed in unison in a population so that the culture served as an amplification of an individual cell's performance. At last the barrier that had separated the study of the cell and the cell population was removed; now they could be studied in relation to each other.

These early methods, largely empirical in their development and related to batch culture, were recognizably of two general types: synchronized and synchronous cultures, respectively. In synchronized cultures, developed initially, various constraints were applied to a microbial population to align the cells, so that upon subsequent release from the constraint, a simultaneous, coordinated development of the cells produced synchrony in the population. Scherbaum and Zeuthen (Paper 22) used temperature oscillations in the first example of this method, which unequivocally established synchrony for microorganisms in a culture.

However, objections to the use of constraints, because they produced "unnatural" or "convalescent" effects, led to the development of synchronous-type cultures. In these cultures, cells in the same stage of development were separated by mechanical or physical means from a growing population to form an inoculum for a subsequent batch culture, the first few generations of which maintained the synchrony. A filtration method initially developed by Maruyama and Yanagita, and improved by Abbo and Pardee, was the first example of this method (see Paper 14); but the more recent development by Helmstetter and Cummings (Paper 23) is included here because of its relative importance by widespread application in molecular biology. More recently centrifugal methods of separation have become popular [J. M. Mitchison and W. S. Vincent, *Nature*, **205**, 987–989 (1965)]. Unfortunately, the limitations of batch cultures are inherent in these methods: succeeding generations in the cultures are not exactly reproducible and a progressive randomization of the cells occurs.

To overcome the deficiencies, continuous methods for synchronization were soon developed; Padilla and James with protozoans, and Dawson with yeasts and bacteria, employed repeated dilutions of growing cultures to give continuously synchronized populations. Padilla and James (Paper 24), in a continuous development of Scherbaum and Zeuthen's method, used temperature and light changes to align the cells of *Astasia longa* and *Euglena gracilis*, respectively. Their paper rebutted the misgivings of some contemporaries concerning the feasibility of these studies for growth. Dawson's method of phased culture, described at this time, used nutrient regulation of the cell population by a modification of the chemostat. The method enabled continuous synchrony at any chosen growth rate to be attained so that cell cycles of different doubling times could be examined. The method was later adapted to permit postcycle developments to be followed, and the article reproduced here (Paper 25) outlines the compounded technique.

6

Another variation of the chemostat, which maintains a constant environment apart from the limiting nutrient in the culture, is that used by Hansche, and by Goodwin (Paper 26). This method can also be used to study synchrony at different growth rates, but not for postcycle studies.

On this new level of cellular growth studies, the perspectives of growth have changed again. Synchrony, by enabling the microbiologist to use the culture to study the cell, allows him to examine how the unit of growth performs, that is, how the microbe grows. This would appear to be the basic study of microbial growth, and at the frontier of contemporary studies of growth, the approach commands increasing attention and industry.

It is apparent now that microbial growth is not a simple study of primitive increments in microbial substance, but an involved conundrum of several dimensions and wide perspectives. The recent symposium on microbial growth sponsored by the Society for General Microbiology [*Microbial Growth* (19th Symp. Soc. Gen. Microbiol.), P. M. Meadow and S. J. Pirt, eds., Cambridge University Press, New York, 1969] was indicative of these many different prospects and reflects the contemporary scene. (The symposium is not represented here because it complements this collection of papers.)

Central to the different aspects of microbial growth is the present need to be able to bring them into focus with each other, a problem that might not be so difficult now that the cell can be used as the focal point. With the cell as the present focus for study in microbial growth, and with cellular metabolism intimately related to these developments, the boundary of microbiology becomes merged with those of molecular biology and cell physiology — so that the tenets of one become pertinent to the others. For example, it is important that the molecular biologist has properly defined materials to work with just as much as the microbiologist, cell physiologist, and biochemist.

As already noted, mathematics has always played a part in the progressive development of studies of microbial growth. Attempts to develop mathematical models of growth have appeared from time to time and have mostly related to the development of the population, but these must now embrace the cell and its physiology. There is a need to reconcile the development of the cell and its population with its environment — a problem that confronts the dogma of molecular biology with the variations of biology and the preciseness of mathematics. The paper by Dean and the late Sir Cyril Hinshelwood (Paper 27) is an attempt in this direction and, together with Professor Pirt's review of their book (Paper 28), admirably serves to portray the contemporary prospects here.

All the foregoing considerations of microbial growth, together with much of our knowledge of microbiology in general, were made on studies of pure cultures, which, as the late Professors Kluyver and Jerusalimsky and many others have pointed out, are laboratory artifacts; so it could be argued that these results do not necessarily reflect the relationships found in nature.

It is therefore of interest, and of importance, to acknowledge the pioneering studies conducted on organisms growing in nature; and the final landmark by Brock (Paper 29) outlines such beginnings. This paper, together with the others, underline the truth of Monod's perspicacious observation that the study of growth is ". . . the basic method of Microbiology" (Paper 6).

Editor's Comments on Papers 1 and 2

1 Ward: *On the Biology of* Bacillus ramosus *(Fraenkel), a Schizomycete of the River Thames*

2 Buchanan: *Life Phases in a Bacterial Culture*

At the outset of the study of microbial growth, the ability to observe and thereby study single organisms under a microscope confirmed growth to be a phenomenon related to the individual. It also showed, however, that if growth were considered to be an increase in size, it could not proceed indefinitely, because at a certain point the growth of individual microbes, such as bacteria, ceased when each divided into two, albeit the process thereupon repeated.

Ward was able to determine doubling times for his organisms, and by recognizing differences in these, to realize that growth was affected by temperature, by light from the violet end of the spectrum, by the food supply, by dilution effects, and by the addition of antiseptic substances to the culture. He also recognized the rudiments of a simple growth curve.

The study of microbial growth was also concerned with the multiplication of the individuals and of their increase in number in the population. Buchanan's paper reflects the emphasis placed upon numbers in the study of growth during this early period and of the attempt to define growth in mathematical terms. His paper also reflects the general tendency of workers to concentrate on the early (lag) phase of the growth curve, a trend that was to continue during the next 20 years.

1

Reprinted from *Proc. Roy. Soc. (London)*, **58**, 276–279, 288–290, 296–301, 395, 461–63 (1895)

On the Biology of *Bacillus ramosus* (Fraenkel), a Schizomycete of the River Thames

H. MARSHALL WARD

* * * * * * *

Methods.

As regards my own practice in making these cell-cultures, the following note may be of use to other students; though of course different workers may operate differently.

I first sterilise the plugged cells in the hot-air steriliser. When these are nearly cool enough to han lle, I heat the quartz or glass floor of the cell between two plates of talc (about 3 × 6 inches) held in a large bunsen flame, and allow it to cool slightly. The cell is then placed in position on its floor, and a small block of high-melting paraffin—cooled after sterilising—is placed just outside the cell. The temperature is still high. enough to melt this, and the liquid runs in by capillarity, and solidifies as the whole cools, cementing the cell to its floor.

When the culture is to be made I take such a prepared cell—several can be prepared and kept in sterile glass-covered dishes—and have ready the following: cover slips, a tube of infected gelatine (or other medium), platinum loops, sterile stiff gelatine, forceps, sterile water, and any convenient rest, such as a small ring of brass.

First, the necessary water is placed in the plugged cell, care being taken to wet the cotton plugs and that some water shall be retained on the inner side of each, whether a layer is spread over the floor or not.

Second, a cover-slip is sterilised between the talc-plates, and while it is cooling the platinum loops are held in the flame and set aside to cool.

Third, the cool sterile cover-slip is removed with the forceps, and laid on the brass ring support. Then the drop is quickly put on the centre of the slip, and the latter inverted on to the cell, so that the drop hangs over the centre.

Editor's Note: A row of asterisks indicates that material has been omitted from the original article.

Fourth, the slip with its hanging drop is now cemented as a roof to the cell by running melted sterile gelatine between it and the latter.

A little practice enables one to prepare such a culture in a few minutes, and very rarely need one go wrong if care is taken. The chief difficulties are with fluid drops, or very dilute gelatine, since they are apt to spread and run over the glass, especially when the air is moist and condenses quickly on the glass surface; in these cases, however, a little experience enables one to avoid letting the cover-slip get too cold before the drop is attached—though, of course the opposite danger has to be guarded against.

Having isolated a single spore, suspended in the drop of nutrient medium beneath the objective—the observations were made with Swift's 1/20th and Zeiss' 1/12th oil immersion, and with Zeiss' E, occ. 4—a drawing of the freshly sown spore was at once made. The culture was then left, with a bell-jar, darkened with black-paper over the whole, at a temperature of 15° to 20° C,* and further observations and drawings made at intervals. Naturally there were many failures, especially with the high-power immersion lens, and the following successful series were only obtained at intervals from cultures in which the thickness of the cover-slip, and of the hanging drop, the sufficient isolation of the spore, and the normal germination and further progress were suitable, and where no sudden changes of temperature interfered to check the growth, dry up the gelatine, or cause inconvenient condensations of moisture in the chamber, the relatively large size of which has again proved advantageous owing to the abundant supply of oxygen it ensures.

Under the conditions referred to, the spore without materially changing its ovoid shape begins to swell somewhat rapidly, and in from one to two hours has increased its dimensions from about $1\cdot5 \times 2\,\mu$ to $2 \times 2\cdot5\,\mu$ or more. As it does this the brilliant oil-drop-like contents become duller and more hyaline—like ground glass—and the sharply marked, almost black membrane, gradually loses some of the firmness of its contour, until it appears as a thinner limiting membrane. At the same time it becomes surrounded by an almost imperceptible pale halo-like investment which appears to be derived from the deliquescence of its most external layers, probably into a soft, transparent, swollen jelly. (*Cf.* figs. 9—12.)

In the course of the next one to two hours or so, the spore appears to be elongating. Close observation shows that this is due to a thinning out of the membrane at one of the ends, and soon afterwards the thinned out wall gives way, and the pale, hyaline, apparently homogeneous protoplasm, enveloped in an exceedingly

* These temperatures are somewhat low, and later results were got at 25—26° C. and higher.

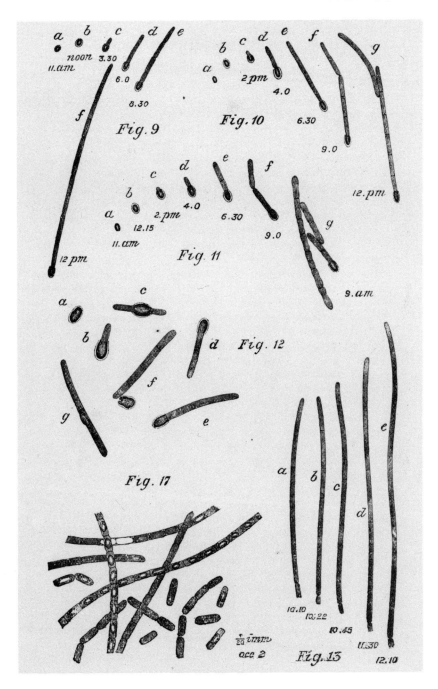

Fig. 9

Fig. 10

Fig. 11

Fig. 12

Fig. 17

Fig. 13

tenuous membrane, pushes its way out and grows as a blunt rod, about 1·75 μ broad, with rounded apex, in the direction of the longer axis of the spore.   (Figs. 9 *c*, 10 *c*, 11 *c*.)

In about four or five hours from the beginning of germination this straight rodlet has attained a length equal to twice that of the spore, and two or three hours later it has a length of approximately four times that of the spore (figs. 9 *e*, 10 *d*, 11 *e*), the membrane of which is still observable usually as a cap at the proximal rounded end of the rod.   (Same figures and fig. 12 *b, d, e*.)

The above is by far the commonest mode of germination, but in some cases this normal condition of affairs is so far modified that both ends of the spore are softened, and each gives rise to a germinal rodlet (fig. 12 *c* and *g*) in which case the remains of the spore-membrane may be found either encircling the germinal rodlet, much as a napkin ring does a rolled up serviette (fig. 12 *c*), or ruptured at one side and merely adhering to the rodlet as in fig. 12 *g*.   Occasionally, rodlets which have germinated out in the normal mode are found with the collapsed membrane lying loosely at one end, evidently having been thrown off, as in fig. 12 *f*; this seems to occur rather frequently in the later stages of germination in broth-cultures.   All these phenomena point to the elastic nature of the thin, but tough, spore membrane.

When the germinal rod or filament has attained a length equal to about four or five times that of the swollen spore, the first division wall is usually seen in the centre (figs. 11 *e*, 12 *d*, and *e*).   Whether the case illustrated in fig. 10 *d* is really an exception to this rule, or whether the apparent septum closer to the spore was really the rim of the burst spore (*cf.* fig 12 *e*), I cannot be certain; from the fact that I could not trace it in the next stage (fig. 10 *e*) it seems likely that the latter supposition is the correct one, and in any case the rule is that the first transverse septum divides the whole germinal filament into two cells approximately, but not necessarily exactly, equal in length, and measuring about 3 to 5 μ long by 1·75 μ broad.

The germination now rapidly proceeds by the growth in length of the stiff and nearly straight filament along its whole course, and in about six to eight hours from the commencement of the swelling of the spore, the filament is from 8 to 10 times as long as the spore (figs. 9 *e*, 10 *e*, 11 *f*), and each of the two cells into which it was segmented by the first transverse septum has been again bisected by a septum, thus cutting the filament into four segments (figs. as before).

*   *   *   *   *   *   *

Measurements of Growth of the Rods and Filaments.

As will be evident from what has been said above concerning the growth of the germinal filaments, they elongate in nearly a straight line so long as they are free to do so and meet with no mechanical obstruction. This fact, and the obviously rapid growth, led me to a method for measuring the rate of elongation; and not only did I succeed in doing this efficiently, but these growth measurements carried me on much further and to some unexpected and interesting results in another connection.

On placing the eye-piece micrometer so that its vertical division crossed the long axis of a filament shorter than the scale, it was easy to observe the gradual extension of the filament as its ends passed over the divisions. The value of each division was determined beforehand, by examining a stage-micrometer with the same combination as I used for the measurements.

Having selected a nearly straight filament which extended over twenty-seven of the fifty divisions on the scale, and having determined that each division was equal to 3μ for the power—Zeiss E occ. 2—employed, it was evident the filament was 81μ long.

It was growing in broth at 16° C. when put under observation at 10.10 A.M., and was watched for two hours, during which period the thermometer rose from 18° C. to 20° C.* The microscope stood under a shaded bell-jar at a south window, and the day was cloudy and dull.

At 10.22 A.M.—*i.e.*, twelve minutes after measuring the filament—it had elongated so as to cover thirty divisions instead of twenty-seven. In other words it had grown 9μ longer (fig. 13 *a, b*) and had slightly altered its slight curvature. At 10.45 it had grown another 12μ (fig. 13 *c*); at 11.30 it was longer by 18μ, and at 12.10 its elongation amounted to 24μ further.

That is to say, in the interval from 10.10 A.M. to 12.10 P.M. (two hours) the total growth in length of the filament amounted to 63μ. During the first twelve minutes the rate of growth was 0.75μ per minute; during the next twenty-three minutes the growth was at the rate of nearly 0.5μ per minute; during the next forty-five minutes it was at the rate of 0.4μ per minute; and during the last forty minutes at the rate of a little over 0.5μ per minute.

These facts may be conveniently tabulated as follows:—

* Air temperatures throughout, except where specially given as otherwise.

Time.	Length.	Interval.	Growth.	Approximate rate per minute.
	μ.	mins.	μ.	μ.
10.10 A.M.	81	. —	—	—
10.22 ,,	90	12	9	0·75
10.45 ,,	102	23	12	0·5
11.30 ,,	120	45	18	0·4
12.10 P.M.	144	40	24	0·6

It is obvious that the growth is moderately rapid (the length would be doubled in about $2\frac{1}{2}$ hours at same rate), but it seems to vary from time to time. An elongation of 63 μ in 120 minutes would give nearly 0·3 μ per minute at constant rate, whereas the rate varies considerably on either side of that.

These variations could not obviously be attributed to variations in temperature, for the thermometer was steadily rising the whole time, nor did I think they could be due to the measurements, for although the slight nutations do occasionally interfere with the strict accuracy of these, the disturbance can hardly be imagined to be so great as these variations imply, and repeated experience convinces me that this is not the explanation at all.

As already stated, the microscope was placed at a south window, under a bell-jar surrounded with black paper, and, except during the short periods necessary for drawing and recording, very little light could reach the object. The day was dull and rainy, but with somewhat brighter intervals. During the whole of the two hours the filament was describing the slight writhing movements which I regard as nutation curvatures, and the extent of which can be estimated by the drawings in fig. 13.

But another idea strikes one in connection with these measurements. If we take the germinating filaments, and draw them vertically to scale, on sectional paper, at any rate as regard the *lengths* attained during the various periods of growth, it is obvious that if their lengths are arranged as vertical lines (ordinates) on a base line divided into periods corresponding to the times (abscissæ) then the curve joining the tips of the filaments is *the curve of growth*, and clearly we may substitute mere vertical lines (ordinates) for the detailed drawing of the filaments themselves.

For instance, the curve of growth is got by straightening out the filaments of fig. 24, where all are drawn to scale at the indicated hours, and joining the tips—or, what amounts to the same thing, by joining the upper ends of ordinates of equal lengths erected on a base line.

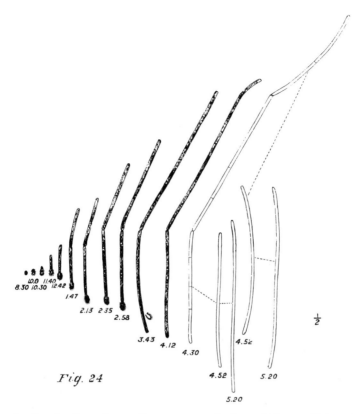

Fig. 24

In the further observations on growth, I availed myself largely of this idea and its consequences, with results of considerable interest and importance.

The following measurements of the rate of growth of the germinating filament (as contrasted with the longer older filament) were made. The spore was sown in 3 per cent. glucose solution (to which a little broth was added) shortly after 4 P.M.: the temperature was $20°-21°$ C. throughout.

From 5 P.M. to 6.5 P.M. the spore had swollen and elongated to 3 μ, and the measurements began at 8.10 P.M., when the germinal filament, still in the spore, measured 6 μ.

At 8.40 P.M. the young rodlet had elongated to 9 μ—*i.e.*, it had grown 3 μ in thirty minutes. At 9.45 it had grown another 9 μ. From 9.45 to 11.0 P.M. it grew 12 μ. From 11.0 to 11.45 it had grown 6 μ.

The measurements were then discontinued till 10.35 next morning, when 45 μ additional had been added.

* * * * * * *

15

It is clear from the foregoing results that the growth of the rods and filaments of this bacillus *can* be measured by the methods devised, and that the undertaking presents no particular difficulties so far, beyond those incident to all close and patient microscopic investigation.

But, beyond the fact that growth occurs at various rates, and can be thus measured, the results thus far give us far too little information to be of the value I anticipated from the application of the method. They simply raise a number of questions as to the action of various factors in influencing the course of growth and inducing the variations in the rate of growth which undoubtedly occur.

Such factors are (1) internal factors, such as the age of the filament, the process of cell-division, and possibly the vigour of the spore itself; and (2) external factors, such as temperature, light of various kinds, and the food-materials, &c.

Growth and Cell-division.

I made the following observations with a view to obtain more information concerning the question raised as regards the connection between growth and cell-division. The chief difficulty connected with them was, as might be expected, that very high powers have necessarily to be employed, and all the troubles of thin cover-slips, minute and not too deep drops, and so forth, arise.

I selected the tip of a filament which had been growing vigorously all the afternoon and evening, and which was in all about 600 μ long, and traced its behaviour under the 1/12th oil immersion. The piece I chose was the terminal portion from the tip to the first visible septum. Its length was measured as exactly as possible, and when the septum was under the division 0 of the scale the tip was exactly under the 14th division.

I had previously determined that six divisions on the scale used, each representing 4·5 μ with my ordinary measuring combination, are equal to fifteen divisions by this power, and therefore I was measuring nearly 2 μ per division (more exactly = 1·82 μ).

At 8.20 P.M. the segment in question measured (under the 1/12th immersion) fourteen divisions, *i.e.*, $14 \times 1·82 = 25·48$ μ, and I started

16

the observations. The scale consisted of fifty divisions, and I started with the septum at 0 and the tip at 14. The following table gives the positions of this septum and tip at successive intervals as (1) the growth of the terminal segment carried tip and septum further apart, and (2) the elongation of the rest of the filament as a whole pushed the entire segment forward over the scale.

It will be noticed that to do this I observed the successive periods during which the filament pushed the septum over five divisions, as I found that the easiest plan, and then recorded the position of the tip at the time (see fig. 26).

FIG. 26.

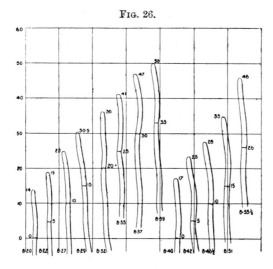

Time.	Division of scale over septum.	Division of scale over tip.	Growth of segment	Interval.	Rate.
			μ.	mins.	*μ.*
8.20 P.M.	0	14·0	—	—	—
8.22 ,,	5	19·0	—	—	—
8.27 ,,	10	25·0	1·82	7	0·26
8.29 ,,	15	30·3	—	—	—
8.32 ,,	20	36·0	1·82	5	0·36
8.35 ,,	25	41·0	—	—	—
8.37 ,,	30	47·0	1·82	5	0·36
8.40 ,,	0*	17·0	—	—	—
8.42 ,,	5	23·0	1·82	5	0·36
8.46½ ,,	10	28·0	1·82	4½	0·4
8.51 ,,	15	35·0	3·64	4½	0·8
8.55½ ,,	26	46·0	—	—	—

* *i.e.*, I brought the septum back again so as to lie under the division 0 on the scale.

Here we see the segment elongated from fourteen divisions to twenty divisions, *i.e.*, it grew six divisions ($= 6 \times 1\cdot82 = 10\cdot92\ \mu$) in the interval from 8.20 to 8.51—*i.e.*, in thirty-one minutes—giving an average rate of growth of about $0\cdot35\ \mu$ per minute, though the rate of growth varies from time to time.

Now the whole filament was 600 divisions long ($= 1,092\ \mu$), and consisted of at least forty such segments,* and if each of them was growing at anything like this rate, no wonder the filament pushed this segment forward so quickly, for it would be elongating as a whole at the rate of 14 μ per minute.

To gain further information in this connection, I exchanged the objective for the combination I usually employ for measuring, and measured the growth of about a third of the whole filament (including the part here concerned) during the seven minutes from 9 to 9.7 P.M., *i.e.*, beginning $4\frac{1}{2}$ minutes after the last measurement.

At 9 P.M. the piece observed was 210 μ long, and at 9.7 P.M. it had elongated to 250 μ, giving a growth of 40 μ, which is at the rate of $5\cdot7$ μ per minute. If this was going on through the rest of the length, the filament as a whole would be growing at a rate considerably in excess of my estimate.

Now if we look at the distance through which the measured segment was pushed during the thirty-one minutes' period given above, we find it amounts to forty-five divisions ($= 45 \times 1\cdot82 = 81\cdot90\ \mu$), so that even such numbers as I have proposed need not seem extravagant, and indeed I have reason to know they are much below the real ones in many cases.

One of the most interesting cases of rapid growth I have seen is the following.

Spores sown in normal gelatine at 22° C. had germinated out to filaments 80—100 μ long in $5\frac{1}{2}$ hours in the dark, and were then put into the dark Sachs' box† at 28° C. The temperature was then slowly raised, so that in two hours it had risen to 34° C., half an hour later to 38° C., and in the next quarter of an hour to 39° C. By chance I happened now to catch a broken-off segment which was growing at the maximum rate that precedes death at these high temperatures. The measurements were as follows:—

---

* Almost certainly more than 40, but I could not determine accurately because the last formed septa are not sufficiently distinct.

† See p. 394 for description of this box and the method of using it.

18

Time.	Length.	Interval.	Growth.	Rate.	Temperature.
	μ.	mins.	μ.	μ.	°C.
5.12 P.M.	139·5	—	—	—	39·0
5.14 ,,	153·0	2	13·5	6·7	39·1
5.15½ ,,	175·5	1½	22·5	15·0	39·25
5.17 ,,	198·0	1¼	22·5	15·0	39·2

Here we see growth going on at the enormous rate, hardly measurable, of 15 μ per minute. In such cases the growing tip is seen moving almost like an *Oscillatoria*.

Then the filament suddenly contracted and broke up, and in a short time presented the granular appearance of dead cells.

The following measurements under the 1/12th immersion were made on a germinal filament which had emerged from the spore during the night. The culture in broth was exposed to the daylight the whole time,* but the sun was obscured by haze and clouds. The temperature was rising slowly from 18° to 21° C., also during the whole period observed.

Each division of the micrometer scale was in this case again equal to 1·82 μ, and the measurements were made as 1, 2, 3, or 4 divisions were travelled over.

Time.	Length.	Interval.	Growth.	Rate.	Temperature of air.
	μ.	mins.	μ.	μ.	°C.
11.5 A.M.	27·30	—	—	—	18·0
11.17 ,,	29·12	12	1·82	0·15	18·0
11.30 ,,	32·56	13	3·64	0·28	18·0
11.40 ,,	34·38	10	1·82	0·18	18·5
11.49 ,,	36·20	9	1·82	0·20	18·5
12.1 P.M.	39·84	20	3·64	0·18	18·5
12.10 ,,	43·48	9	3·64	0·40	19·0
12.16 ,,	45·30	6	1·82	0·30	19·0
12.28 ,,	50·76	12	5·46	0·42	19·0
12.38 ,,	56·22	10	5·46	0·54	19·5
12.47 ,,	59·86	9	3·64	0·18	20·0
12.56 ,,	67·14	9	7·28	0·81	20·5
1.2 ,,	70·88	6	3·64	0·60	21·0
1.5 ,,	74·52	3	3·64	1·21	21·0

Here we have a total growth of the young filament amounting to 47·22 μ in 120 minutes, which would give an average rate of nearly

* It should be noted that in these high-power observations a bright illumination has of course to be employed.

Y 2

19

0·4 μ per minute if constant. But obviously the rate was not constant, as the table and curve show (see fig. 27).

Fig. 27.

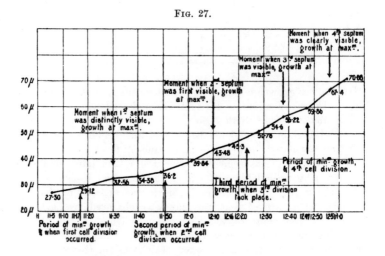

I made the following observations in addition to see if the variations were connected with cell-division as suspected. Up to 11.17 no septum was really visible* in the young filament, but a distinct median one was seen at 11.30 dividing the filament into a proximal and a distal half. By 12.10 a second septum was clearly visible bisecting the distal half, but none was as yet visible in the proximal half to which the spore membrane still clung. At 12.38 the proximal segment was also divided by a visible median septum. At 12.47 I measured both the primary segments and found the proximal, recently divided, one shorter than the distal one, in the ratio of 15 to 18; that is to say, the whole filament measured 59·86 μ, as seen, but that part of it to the proximal side of the first septum was only 27 μ, that to the distal 33 μ in length; so that already the symmetry of the filament was disturbed, and further measurements confirmed this.

At 12.56 the proximal segment had one septum, now very distinct, and measured 30·5 μ, whereas the distal one was by this time provided with two visible septa, and measured nearly 37 μ; at 1.5 P.M. the proximal one still had but one septum visible, and measured 32·5 μ, whereas the distal one, with its two visible septa, measured 42·5 μ.

Hence we see the two primary segments resulting from the first division of the germinal filament grow and (so far as *visible* segments

* Probably thin septa were present, but they were not *visible* in the living and rapidly growing filament.

20

show) divide at different rates from the first. I am disposed to regard the slower growth of the proximal segment, in part at any rate, to its being more especially concerned with the absorption of food-materials from the spore ; though the fact that it is still behind-hand, even after escaping from the spore-membrane, may indicate a deeper meaning—possibly that differences between basal and apical regions are more strongly defined in these organisms than we suppose.

But another point must be considered before the curve can be understood, and for this purpose it seems necessary to introduce a simple nomenclature for the divisions and segments.

We may term the first septum, which divides the whole germinal filament into its first two segments, the primary septum ; thus the primary septum was first *visible* at 11.30, dividing the filament into a proximal and a distal primary segment. At 12.10 the longer *distal* primary segment showed a further division by a *secondary* septum into two *secondary* segments ; but the corresponding secondary septum in the *proximal* primary segment was not visible until 12.38.

At 12.56 the distal secondary segment of the primary distal segment had a *tertiary* septum plainly visible, and the filament as a whole, therefore, consisted at this hour of five visible portions, two belonging to the primary proximal segment, and three belonging to the primary distal segment.

On turning to the curve of growth (fig. 27), it may now be possible to understand its principal features if we first accept as a fact that the period when a septum is first *distinctly visible* in these brilliant living cells is some time *after the moment of actual cell-division*. This, I think, must be accepted, because I find stained pre-parations of such filaments show many more septa to be actually present than can be seen in the living filaments, owing to the extremely high refrangibility of the protoplasm obscuring the view of the most recently formed and still tenuous walls.

In the curve referred to, as I understand it, its general form, with a higher and higher rate of ascent as time goes on, is due to the total increasing elongation of all the segments simultaneously—aided in this case by the slight continuous rise of temperature.

* * * * * * *

21

I adopted the following method of recording.

1. The table of growths was prepared, giving the time of observation, temperature, and length of filament, &c., as in preceding cases.

2. Then the curve of growth was plotted out on sectional paper, care being taken that the squares, &c., were all equal, and that the intervals between the observations were sufficiently short and numerous to give good curves. Since the measurements were all made with the same micrometer-scale and the same microscope, and taking into account the proofs of accurate measurement already given, no more need be said on that head.

3. From these curves and tables I then measured the period occupied by a rodlet or filament of any given length (to start with) in growing to *double its length*, and call this the *doubling period*.

4. The average *doubling period* for any temperature is then obtained in the usual way, by taking the sum of the times and dividing by the number of observations. Before saying more, however, it will be best to examine the actual results, which now follow.

I may add that I had already satisfied myself that the curves obtained at widely different temperatures, are markedly different, and less and less divergent as the temperatures of growth approach one another, facts which are in accordance with experience with other organisms, and which will be evident enough as we proceed.

In what follows I select a number of representative curves from larger series made to familiarise myself with the details; and for the sake of classification—and with reference to some conclusions later—I present them arranged according to the food-materials employed. The media chiefly used were four, viz.: (1) normal beef-broth; (2) the same with 1 per cent. of gelatine added to give a certain degree of stiffness; (3) a stiff 10 per cent. gelatine with mere traces of broth (referred to as weak gelatine); and (4) normal 10 per cent. broth-peptone gelatine.

* * * * * * *

An interesting result follows from the fact that the doubling period is simply the visible expression of the doubling in length and bipartition of all the cells composing the filament.

Suppose a filament to be 50 μ long, and composed of ten cells each 5 μ long, and that that filament doubles its length in thirty minutes at a given temperature: then the filament, now 100 μ long, consists of twenty cells, each of which has taken thirty minutes to divide and double itself; from this we can deduce the number of bacilli formed in a given time from the doubling periods, although the individual bacilli are themselves invisible, and when we find a curve like that of

August 4, where the filament grows from 10 μ to 652 μ in length, at 21—23·7°, with an average doubling period of about thirty-five minutes, it can be translated as meaning that the number of bacilli increased as follows :—

2 bacilli became
4 at the end of the first   35 minutes.

8	,,	,,	second	,,
16	,,	,,	third	,,
32	,,	,,	fourth	,,
64	,,	,,	fifth	,, .
128	,,	,,	sixth	,,

and so on, and we may assume that if the supply of food-material could be kept constant, and no disturbing conditions set in, this would go on. If it went on for only half a day—twelve hours—there would be nearly 4,000,000 of the bacilli produced from the pair started with above, and the filament would be nearly 40,000,000 μ in length—*i.e.*, nearly 40 metres—whence some idea may be obtained of the energy of the growth on the one hand, and of the limits imposed by the culture-drops on the other; for if we take the size of a drop as 1 cubic mm., which is approximately the volume of a hanging drop such as is used in the cultures, and remember that the bacilli in question are about 1·75 μ in diameter, it will be found that the above length of 40 metres, nevertheless, has plenty of room in the drop, for the filaments have a volume of only 96,250,000 cubic μ to pack away in the 1,000,000,000 cubic μ of the drop, so that we see the latter *could* hold ten times the quantity.

We are now in a position to resume the discussion of these growth-curves in detail, and the action of temperature, &c., on this schizo-mycete, with more hope of success.

It is evident that the normal growth-curve is one which begins to rise slowly, and gradually gets steeper and steeper, and then slowly rises less and less rapidly until the end. This gives a curve like a long drawn out $\int$.

At the optimum temperature the growth is very rapid, and lasts for a long time, and the organism uses the materials to maximum effect and produces from them the maximum amount of its own substance—in other words, the largest " crop."

At temperatures above the optimum, however, the growth, though at first as rapid as at the optimum temperature, lasts for a shorter and shorter time, according as the temperature is further and further removed from the optimum; consequently, the curve, though equally steep in its steepest parts, begins to fall sooner, and growth ceases sooner, and the crop obtained from the same amount of original food-material is smaller and smaller according as the temperature is higher.

At length a temperature is reached where the curve is infinitely short, *i.e.*, no growth occurs at all. This temperature is, however, above 39° C., and indicates the death-point.

Taking temperatures below the optimum. There is a point, somewhere below 8° C., where the curve is indefinitely postponed, *i.e.*, no growth can occur at all. Then comes a temperature, also below 8—10° C., where the curve ascends slowly and never attains the steepness of the curve at optimum temperature. This is the minimum temperature.

At temperatures above the minimum the curve attains more and more nearly, and in shorter and shorter times, to the steepness of the normal curve, the nearer the temperature in question is to the optimum temperature.

This optimum temperature is either 25° C. or some point very near it.

The above case of the normal curve is the hypothetical one where *all* the conditions are constant, a state of affairs never realised.* During the growth, between the period when the germination is completed and the organism no longer obtains any supplies from the spore, but is totally dependent on the food-materials given it, and the period when the curve begins to ascend less rapidly, there is a period of maximum growth, during which the filament doubles its length in equal minimum times. This is the critical period of the curve. The more closely the curve approximates to the normal curve the longer this phase of equal minimum doubling periods lasts; the more external conditions affect the curve the shorter this phase is, and the longer the doubling periods become.

The factors affecting the curve may be regarded as of two kinds, internal and external, though they probably never vary entirely independently.

The internal factors are such as (1) irregularities of cell-divisions: if a single cell fails to divide in due order, the curve is at once affected, because the regularity of the intercalary growth of the filament is destroyed, and this occasionally happens. (2) The separation of the segments: several observations suggest that the growth is slowed at once when the new surfaces of the broken ends come in contact with the food-medium. (3) Nutations and oscillatory movements, though possibly these affect the *measurements* rather than actual growth. (4) Unknown internal factors which affect the rapidity of germination, the ability to assimilate the food-materials, and so forth. In some cases these may be due to pathological conditions, as in the case given on p. 392.

* Theoretically, with absolute uniformity of conditions, including food supply, the curve would go on to infinity, and the doubling periods be equal throughout; the fall above would then be indefinitely postponed.

Editor's Note: Ward then goes on to discuss the external factors of temperature, light, nutrients, oxygen, and toxic substances.

2

Reprinted from *J. Infect. Diseases*, **23**, 109–125 (1918)

LIFE PHASES IN A BACTERIAL CULTURE

R. E. BUCHANAN

From the Bacteriological Laboratories of Iowa State College, Ames, Ia.

Several important contributions to our knowledge of the numbers of bacteria present in a culture medium at certain stages in the development of the culture have been made recently, but with the exception of the work of Slator (1917), apparently there has been no effort to coordinate these results and develop a complete mathematical theory of such changes in numbers. The present paper is an attempt to analyze the results of these authors and to present certain phases which have apparently been neglected heretofore.

When bacteria, particularly cells from an old culture, are inoculated into a suitable culture medium, as broth, the bacteria will at first remain unchanged in numbers; then multiplication begins, the numbers increase at first slowly, then more rapidly until a certain minimum average generation time is reached; this after a time begins to increase, and there is a negative acceleration in growth, which finally ceases; the numbers remain constant for a time, then the bacteria begin to die off.

Lane-Claypon[1] recognized four periods or phases in the life of a bacterial culture as follows:
1. Initial period of slow growth or even no growth.
2. Period of regular growth.
3. Period during which numbers remain more or less stationary.
4. Period during which the numbers of living bacteria are decreasing.

It would seem, however, that the life phases are somewhat more complex than indicated by the preceding statement. A study of the results secured by various authors indicates that seven relatively distinct periods may be differentiated. These may be recognized easily by plotting the logarithms of the numbers of bacteria against time. Chart 1 is such a plot with the seven phases indicated. It will be noted that points designating the beginning and end of each phase are points where a curve changes to a straight line and vice versa.

Received for publication Feb. 22, 1918.

[1] Jour. Hyg., 1909, 9, p. 239.

These various growth phases may be designated as follows:

1. *Initial Stationary Phase.*—During this phase the number of bacteria remains constant, and the plot is a straight line parallel to the x axis indicated by 1 — a.

2. *Lag Phase, or Positive Growth Acceleration Phase.*—During this phase the average rate of increase in numbers per organism increases with the time, giving rise to the curve a — b. This increase in rate of growth per organism does not continue indefinitely but only to a certain point determined by the average minimal generation time per organism under the conditions of the test.

3. *Logarithmic Growth Phase.*—During this phase the rate of increase per organism remains constant, in other words, the minimal average generation time is maintained throughout the period. This gives rise to the straight line b — c.

4. *Phase of Negative Growth Acceleration.*—During this phase the rate of growth per organism decreases, that is, the average generation time is increased.

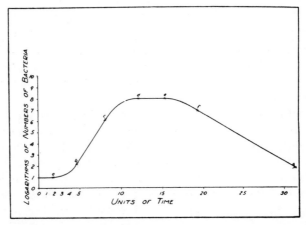

Chart 1.—Diagrammatic plot of logarithms of numbers of bacteria present in a culture.

The bacteria continue to increase in numbers, but less rapidly than during the logarithmic growth phase. This is the curve c — d.

5. *Maximum Stationary Phase.*—During this period there is practically no increase in the numbers of bacteria. The plot gives a straight line d — e parallel to the x axis. The rate of increase per organism is zero and the average generation time infinity.

6. *Phase of Accelerated Death.*—During this phase the numbers of bacteria are decreasing, slowly at first and with increasing rapidity, until the establishment of a logarithmic death rate. In the terminology used in the growth phases, the average "rate of death per organism" is increasing to a certain maximum. This gives the curve e — f.

7. *Logarithmic Death Phase.*—During this phase the "rate of death per organism" remains constant, the plot of the logarithms gives a straight line with a negative slope. This is represented by f — g.

If death and life in the bacterial cell could be regarded as reversible processes, we might expect the appearance of an eighth phase, a negative acceleration of the death rate.

It may be noted that the seven phases previously defined are in a sense arbitrary. The curve, if it could be plotted with data absolutely accurate, would probably be smooth; in other words, the portions designated as straight lines are probably curves, but with curvature so slight that they may be treated mathematically as straight lines without the introduction of any error commensurate in value with the inevitable experimental errors.

These various life phases of the bacterial culture will be discussed in some detail.

I. Initial Stationary Phase

Spore-producing bacteria exhibit this growth phase particularly well. If a suspension of bacterial spores be placed in a suitable culture medium microscopic observation will show that growth does not apparently begin immediately. There can be no increase in numbers until the spores have germinated and begun to multiply. Samples of equal volume taken during this period show no increase in numbers. While this stage is most prominent with sporulating organisms it is by no means always absent in nonspore formers, as is shown by results of Lane-Claypon,[1] Penfold,[2] and others. In other words, there is evidence that in old cultures of many bacteria the cells are in a relatively dormant stage, the physiologic equivalent of sporulation though without the spore morphology. When such cells are planted in a suitable medium there will be an appreciable interval before a single cell will have resumed growth sufficiently to divide. During this phase the rate of increase per cell would be zero, and the average generation time infinity. The equation of the curve which represents this phase would be

$$b = B$$

where $b = $ Number of bacteria after time t,
          $B = $ Initial number of bacteria.

Little work has been done on this phase. The conditions which determine its length are probably those influencing the length of other phases.

It should be noted that some cultures will not show this phase at all. If there are any actively dividing bacteria in the inoculum at the time of inoculation it will be absent or very transitory. It is therefore probable that the phase of the culture from which the transfer is made will affect the length of this phase.

[2] Jour. Hyg., 1914, 14, p. 215.

II. Lag Phase, or Phase of Positive Growth Acceleration

This phase apparently has not been differentiated from the preceding by previous writers. This is illustrated by the definition of latent period given by Chesney:[3] "By latent period or lag is meant the interval which elapses between the time of seeding and the time at which maximum rate of growth begins." The necessity for differentiation of the two phases is not urgent except when the first is long continued.

The lag phase may be defined as that period elapsing between the beginning of multiplication and the beginning of the maximum rate of increase per organism.

The phenomenon of bacterial lag was apparently first noted by Müller,[4] and was later studied by Rahn[5] and by Coplans.[6] Penfold[2] gave the first adequate discussion of the various theories which might be suggested to explain the phenomenon. Chesney[3] later made a careful study of the lag phase with special reference to the growth of the pneumococcus. A mathematical analysis of the lag phase was given by Penfold and Ledingham[7] and elaborated by Slator.[8]

Theories of Bacterial Lag.—Penfold[2] has enumerated some nine different theories as to the cause of bacterial lag, all of which he discards as inadequate. Inasmuch as certain of these have been maintained by other writers, and perhaps some discarded hastily, they will be briefly summarized and reasons for discarding given under the following seven heads.

1. The organism must excrete some essential substance into the medium before maximal growth can occur. Experiments show, however, that subcultures taken from cultures showing maximal growth do not show any lag period.

2. Adaptation to a new medium requires time. This must be discarded, inasmuch as transfers to the same medium may show lag.

3. Some of the bacteria transferred are not viable, and die off early. Inasmuch as enumeration is by plating and not by direct counting, the organisms not viable would never be enumerated.

4. Bacteria may agglutinate and plating would then be an enumeration of clumps and not of individual bacteria. While this may be a factor in some cases, it cannot explain the lag which still persists when adequate precautions against confusion from this source have been taken.

[3] Jour. Exper. Med., 1916, 24, p. 387.
[4] Ztschr. f. Hyg. u. Infektionskr., 1895, 20, p. 245.
[5] Centralbl. f. Bakteriol., 1906, Abt. 2, 16.
[6] Jour. of Path. and Bacteriol., 1909, 14, pp. 1-27.
[7] Jour. Hyg., 1914, 14, p. 242.
[8] Ibid., 1917, 16, p. 100.

5. Accumulated products of metabolism may injure the bacterial cell, the length of the lag phase is the time required to recover from the injury. Penfold rejects this explanation as inadequate. Chesney, however, insists that lag is "an expression of injury which the bacterial cell has sustained from its previous environment." This conception may well be an approximation of the truth, though probably not entirely accurate.

6. "The inoculum consists of organisms having individually different powers of growth, and during the lag the selection of the quick growing strain occurs in response to some selecting agent in the peptone." It is possible that this might occur in cultures which were not "pure lines," or which contained several strains, but there is no proof of its occurrence in pure strains. While there are undoubtedly some differences in the rates of multiplication of individual bacteria in the same culture, they are insufficient to account for the great differences characteristic of the lag period.

7. Bacteria must overcome an "inertia" before reaching maximal growth rate. Penfold dismisses this on the basis of certain experiments in which he chilled rapidly growing bacteria, and thus stopped multiplication, which was resumed at its former rate when the optimum temperature was restored. While it is probable that Penfold is correct in discarding inertia in this sense as a factor, nevertheless a modification of this theory is in the opinion of the writer the only adequate one suggested.

The explanation favored by Penfold is in fact a variant of the last. He believes that certain essential constituents of the bacterial protoplasm, probably synthesized in steps, must be present in the bacterial cell in optimum concentration or at least the intermediate bodies of the steps of the synthesis. When the bacteria cease growing these intermediate bodies diffuse from the cell and disappear, and before maximal growth can begin in a new medium these bodies must again be synthesized. This theory in effect holds that the loss of these substances gives rise to inertia. During the lag phase the bacteria are gradually recovering from injury.

It is probable that none of the preceding are wholly satisfactory explanations of the lag phase. An explanation more in accord with observed facts may be found in the assumption by the bacterial cells of a "rest period" comparable to the resting stages so often assumed by higher forms. It is a well known fact that at certain stages in the life history of many plants certain cells or tissues are developed which pass into a resting stage. When these are morphologically well differentiated they are termed spores, sclerotia, etc., in the lower forms, and seeds, bulbs, tubers, etc., among the higher types. In many other cases cells or tissues pass into a similar resting stage as a result of certain environmental influences, without showing marked morphologic differentiation. These resting cells are usually aroused to renewed growth and activity only as the result of certain stimuli. The cold of winter followed by the warmth of spring may be the stimulus which causes buds to develop. Some seeds will germinate only after the seed coat has decayed or has been scratched or corroded by acid. Bacterial spores form at certain stages in the life history of the bacteria, but do not usually germinate in the parent culture in spite of abundant moisture, food and optimum temperature. Germination takes place under the stimulus of change to some new medium. It is altogether probable that most bacteria, whether spore producers or not, enter into such a resting stage. When not morphologically differentiated as a spore this resting period is probably more transitory than in a spore, but it is nevertheless just as real.

What happens, then, when a considerable number of bacteria in the resting stage are transferred to a medium suitable for development? If we were to examine the culture microscopically we would find that the bacteria would not all begin development at once, probably for the same reason that seeds placed under uniform favorable conditions for growth do not all germinate at the same instant. Cell division will occur in a few cells first, followed by larger and larger numbers at succeeding intervals of time until a maximum has been reached and passed, and at last all the cells have "germinated." As soon as a cell has actually germinated, there would seem to be no a priori reason why the cells should not thereafter multiply rapidly, showing practically at once a minimum generation time. There is no more reason to suppose that the length of time it takes a bacterial cell to germinate will affect its subsequent rate of growth than to assume that plants derived from seeds slow in sprouting grow more slowly than those from seeds soonest sprouted. After any cell had once "germinated" then, it would proceed to increase in numbers in geometrical progression. Theoretically the lag period would continue until the last viable cell had started to multiply; practically, however, it ceases before this as the rapid increase in the bacteria which have germinated soon makes the ungerminated cells such a small fraction of the whole number that their inclusion is within the limits of error of measurement of the numbers present.

MATHEMATICAL ANALYSIS OF THE LAG PERIOD

The lag period has been previously defined as that period during which there is an increase in the average rate of multiplication of the bacteria, an increase from zero to some constant which is maintained during the succeeding period. Another statement is that it includes the period during which there is a decrease in the average generation time. It should be noted that when used in this sense, the term generation time means the time required for the bacteria to double in numbers, if they continued growing at the same rate. At any given instant during this period there will be some cells not multiplying at all, these at that particular instant would have an infinite generation time, and the term average generation time would cease to have any meaning. In other words, during this period there is an acceleration in the rate of growth.

Let us first examine the equation of growth if the rate of increase per organism should remain constant, that is, the average generation time should not vary. Let

$$b = \text{number of bacteria after time t,}$$
$$B = \text{initial number of bacteria.}$$

It is evident that the rate of increase in number of bacteria at any instant will vary directly as the number of bacteria, or expressed in terms of the calculus:

$$db/dt = kb \text{ where k is a constant. Now}$$
$$\frac{db/dt}{b} = k$$

Therefore k is the rate of growth per organism, or the velocity coefficient of growth. On integration this becomes,

$$\ln b = kt + \text{constant of integration.}$$

The constant of integration is found to be $\ln B$ by taking $t = 0$. The equation then becomes

$$\ln b/B = kt$$

This may be interpreted as the equation of a straight line, hence when $\ln b$ is plotted against t, a straight line with slope k will be secured.

The curve showing the number of bacteria after any time may be derived from the above equation

$$\ln b/B = kt$$
$$b/B = e^{kt}$$
$$\text{and} \qquad b = Be^{kt}$$

This equation may be derived without resort to the calculus as follows:

$$\text{Let } n = \text{number of generations in time t}$$
$$g = \text{generation time.}$$

At the end of time t one organism will have produced 2^n bacteria, then

$$b = B2^n$$
$$\text{Now} \quad n = t/g$$
$$\therefore \quad b = B \, 2^{t/g}$$
$$\text{Let} \quad 2^{1/g} = e^k$$
$$\text{then} \quad b = Be^{kt}$$
$$\text{and} \quad k = \ln 2/g$$

Since the rate of growth varies inversely as the generation time, k may be regarded as this rate of growth per organism, or the velocity coefficient.

Inasmuch as rate of growth per organism is a function of time, it is a matter of interest to determine just what the relationship may be existing between them. Penfold and Slator have suggested relationships empirically determined from experimental data. Apparently there has been no attempt to derive the relationship from theoretical considerations.

Assume that the lag phase represents the time required for all the viable bacteria planted to "germinate." Take as the time of "germination" the instant that the cell first divides to form two individuals. It is assumed that as soon as an organism begins dividing its rate of increase is at once constant. Let this be k'.

Let $w = $ number of bacteria that are dividing after time t, $=$ progeny of all bacteria that have germinated within time t

$z = $ number of bacteria that have not germinated.

31

The rate of increase per organism [f(t)] at any instant is given by the following equation:

$$f(t) = \frac{k'w}{w+z} = k'. \quad 1/(1+z/w) \qquad (1)$$

It is apparent that if the numbers of bacteria "germinating" during each unit of time are plotted against time, a curve may be secured resting on the x axis at both ends, one of the forms of a probability curve. The general equation for such a curve has been shown by Karl Pearson to be

$$y = c(1 + x/a_1)^{m_1}(1 - x/a_2)^{m_2}$$

in which m_1 and m_2 are constants, c equals the maximum ordinate, and $-a_1$ and a_2 are intercepts of curve with x axis.

Let y be the number of bacteria germinating at time t, then

$$y = c(1 + t/a_1)^{m_1}(1 - t/a_2)^{m_2}$$

and $a_1 + a_2$ is the total length of the lag period.

The total number of bacteria which will germinate in time dt is ydt. Since the number of bacteria developing after time t from one organism is $2^{t/g}$, those which will develop after time t from those beginning growth during time dt is $y2^{t/g}dt$, and the total number of bacteria developed after time t from those starting growth within that time is

$$\int_{-a_1}^{t} y2^{t/g}dt$$

Therefore $$w = \int_{-a_1}^{t} c(1 + t/a_1)^{m_1}(1 - t/a_2)^{m_2}\, 2^{t/g}dt$$

The total number of bacteria which "germinate" within time t is

$$\int_{-a_1}^{t} ydt$$

The total number not germinated is

$$B - \int_{-a_1}^{t} ydt$$

$$\therefore \quad z = B - \int_{-a_1}^{t} ydt$$

The relationship probably existing between rate of growth per organism and t may be shown by substituting the values secured for w and z in the equation (1).

$$f(t) = k'\cfrac{1}{1 + \cfrac{B - \int_{-a_1}^{t} c(1 + t/a_1)^{m_1}(1 - t/a_2)^{m_2}dt}{\int_{-a_1}^{t} c(1 + t/a_1)^{m_1}(1 - t/a_2)^{m_2}\, 2^{t/g}dt}}$$

All efforts to simplify this expression or put it into usable form have thus far failed. The only points where the exact relationship is known are when t = 0, f (t) = 0, and when $t = a_1 + a_2$, $f(t) = k'$. It is evident that the relationship existing between rate of growth per organism and t during the lag period is quite complex.

The problem may also be attacked by the empirical derivation of a formula for a plotted curve by a critical examination of the data of the lag phase. This has been done by Ledingham and Penfold. These authors first reduced all figures to a seeding of 1, that is, the numbers of bacteria found at successive stages of the lag phase were divided by the initial number of bacteria. The logarithms of these numbers were plotted against the logarithms of times. This gave a curve which appeared to be logarithmic. The logarithms of the logarithms of the numbers of bacteria were then plotted against the logarithms of the times. These points were found to lie approximately on a straight line. If n is the slope of this line, and c the intercept with the x axis, the equation of the line is

$$n = \frac{\log (\log b)}{(\log t) - c} \qquad (2)$$

From this they derive the equation

$$t^n = k \log b$$

$$\text{Since} \quad \ln b = \ln 10. \log b$$
$$\ln b = \ln 10.t^n/k = k't^n$$

$$\text{where} \quad \frac{\ln 10}{k} = k'$$

Therefore $b = e^{k't^n}$ and for a seeding of B bacteria $b = Be^{k't^n}$
It is evident that this equation and the equation for regular growth

$$b = Be^{kt}$$

are special forms of the equation

$$b = Be^{\mu kt}$$

in which $\mu = f$ (t). In the equation for constant rate of growth per organism, $\mu = 1$, and in the Ledingham-Penfold equation $\mu = t^{n-1}$, and in the equation of initial stationary phase $\mu = 0$ and $b = B$. The equation developed,

$$b = Be^{k't^n}$$

has two constants which must be evaluated for each particular experiment. An equation of this general form was tested out by Ledingham and Penfold (1914) on data from eight series of experiments, and was found to give remarkably consistent results. The value of n in these experiments varied from 1.56 to 2.7 six being below 2.0. The value of k in the equation

$$t^n = k \log b/B,$$

varied from 2329 to 1,045,000.

The tables given by Chesney for increase of bacteria during the lag period afford an opportunity for testing independently the validity of the Ledingham-Penfold equation, or its generalized form.

Slator after a study of the data of Penfold (1914) concluded that in every experiment recorded there existed a relationship between the two constants n and k such that an equation could be derived in which there would appear but one undetermined constant n. He found by examination that the following relationship always held:

$$\frac{\log k/n}{n} = \text{constant} = 2.024$$

$$k = 105.7^n$$

Substituting the value for k in the Ledingham-Penfold equation

$$t^n = k \log b/B$$
$$t^n = n\ 105.7^n \log b/B = 105.7^n \log b^n/B^n$$

Slator uses the general form of equation

$$kt^n = \log b/B$$

This becomes $(.00945)^n t^n = \log b^n/B^n$.

While the equation as developed holds for the work of Penfold, Slator generalizes into the form

$$C^n t^n = \log b^n/B^n$$

in which C might have some value other than .00945. This can be put into the form of the equations

$$b^n = B^n 10^{C^n t^n}$$
$$\text{or}\quad b^n = B^n e^{k^n t^n}$$

The advantage of Slator's generalized equation over that of Ledingham and Penfold, at least for the lag period, is not apparent.

MEASUREMENT OF LAG

A numerical expression indicating the amount of lag may be secured in either of two principal ways: (*a*) An expression may be secured which will involve directly the length of the lag period, this may be termed "period of lag measurement;" (*b*) an expression may be secured which will give a numerical value to the degree of depression of rate of multiplication at any time during the progress of the lag period. This may be termed the "time index of lag."

(*a*) Period of Lag Measurement: Three suggestions have been made as to methods of measuring lag in terms involving the length of the lag period. These have been defined by Penfold.

1. The actual length of the lag period may be measured.

2. Coplans (1909) measured the restraint of growth in terms of minimum generation time. It may be expressed by the formula

$$\frac{t - ng}{g}$$

where t = length of lag period

n = number of generations during lag period.

g = minimum generation time.

3. The average generation time for the first part of the period may be compared with that of any succeeding period.

(b) Time Index of Lag: The degree or amount of lag at any instant during the lag period may have a numerical value assigned to it in either of two ways; the ratio of the generation time at any instant to the minimal generation time characteristic of the logarithmic period of increase may be determined, or, the rate of change or increase per organism at any given instant during the lag period may be compared with the similar rate of increase per organism during the logarithmic period. Inasmuch as the rate of growth must vary inversely as the generation time, it is evident that these two methods of expressing results will have a constant ratio.

1. *Measurement of Lag by Comparison of Generation Times.*—The problem is to secure the ratio of the generation time of the bacteria at any time during the lag phase to the minimal generation time. It should be recalled that the term generation time as used here is not a time average, but that length of time required for the bacteria present to double in number if the average rate increase per individual remained constant.

It was earlier developed that the expression

$$b = B \; 2^{t/g}$$

represents the equation of growth if the rate of increase per individual remains constant. Differentiating and solving for g,

$$g = b \ln 2 \; dt/db \qquad (1)$$

The value of dt/db may be determined for the lag phase by differentiation of either of the equations

$$b = B_0 e^{kt^n} \qquad (2)$$
$$\text{or} \quad b^n = B^n e^{k^n t^n} \qquad (3)$$

Differentiating (2)

$$db/b = knt^{n-1}dt$$
$$dt/db = 1/bknt^{n-1} \qquad (4)$$

Substituting the value of dt/db in (1)

$$g = \frac{b \ln 2}{bknt^{n-1}} = \ln 2/knt^{n-1}$$

The ratio between the value of generation time as determined by this formula during the period of lag and the minimum value of g as determined during the logarithmic period gives a numerical index to the degree of lag at any instant.

If the equation

$$b^n = B^n e^{k^n t^n}$$

be chosen as the more general for the lag period (as developed from the work of Slator), the expression for generation time becomes:

$$g = \ln 2/k^n t^{n-1}$$

2. *Measurement of Lag by Comparisons of Rates of Increase Per Cell.*—The work of Slator suggests the possibility of measuring lag at any instant during the lag phase by a comparison of the rates of increase per cell with similar rates for the logarithmic period.

This may be determined from either lag phase equation

$$b = Bekt^n$$
$$\text{or} \quad b^n = Bne^{k^n t^n}$$

Differentiate

$$db/dt = bknt^{n-1}$$

The rate of increase per organism at any instant is therefore

$$\frac{db/dt}{b} = knt^{n-1}$$

The corresponding rate of increase per organism during the logarithmic period is

$$\frac{db/dt}{b} = k'$$

The ratio knt^{n-1}/k' gives the numerical index desired.

If the second equation of the lag phase be employed the ratio becomes

$$k^n t^{n-1}/k'$$

It may be noted that the so-called "constant of growth" during the lag period, the expression $\dfrac{db/dt}{b}$, used by Slator and termed z is directly proportional to the μ of the equation

$$b = Be^{\mu kt}$$

III. The Logarithmic Phase

The logarithmic phase of bacterial growth in a culture is that time during which there is a maximum rate of growth per organism, that is, the time during which a certain minimum generation time is maintained. The various relationships which define this period have for the most part been developed in the discussion of the lag phase. They are as follows:

If B = number of bacteria at beginning of logarithmic period,
b = number of bacteria after time t,
n = number of generations in time t,
g = generation time,
k = velocity coefficient of growth,
$$b = B\ 2^n = B\ 2^{t/g} = Be^{kt}$$
$$g = \frac{t \ln 2}{\ln b - \ln B}$$
$$n = \frac{\ln b - \ln B}{\ln 2}$$
$$k = 1/t.\ \ln.\ b/B$$

This phase of bacterial growth has perhaps been more investigated than any other. The mathematical relationships during this period are comparatively simple. It is evident that any effect of change of environmental conditions on the rate of increase of bacteria will be manifested through a change in the generation time. For every variable in the environment there is of course an optimum for each kind of organism, that is, a condition or concentration such that the generation time is minimal.

There is need for careful mathematical study of the effect of temperature changes, changes in concentration of nutrients, of hydrogen ions, of inhibiting substances, etc., on the rate of growth. It will be noted that the equation

$$k = 1/t \cdot \ln \cdot b/B$$

is one form of the expression for the value of the velocity coefficient of a monomolecular reaction. It has been shown that a similar (not identical) expression holds for the logarithmic death period of bacteria. Will the following expression

$$k = 1/tC^n \cdot \ln b/B$$

hold where C is the concentration of some nutrient or inhibiting substance, and n a constant?.

The temperature coefficient per degree or per 10 degree rise in temperature is in need of study, particularly near the minimum and maximum growth temperatures. This temperature coefficient over certain ranges has been determined for some bacteria. Lane-Claypon gives the value per 10° and 35° as 2 to 3 with B. coli. Similar results were secured from 20° to 30° by Hehewerth (1901) and Barber (1908).

IV. PHASE OF NEGATIVE ACCELERATION OF GROWTH

It is a matter of common laboratory observation that bacteria do not long maintain their maximum rate of growth, the logarithmic phase does not usually persist more than a few hours in quick growing types of bacteria. The average generation time apparently lengthens until at the close of the period the bacteria are no longer dividing.

The general equation of this portion can be written, as for the preceding phases

$$b = Be^{\mu_k t}$$

During this phase the μ varies as some function of the time, from the l of the logarithmic period to 0. Apparently the exact relationship between μ and t during this phase has not been studied. It is apparent that as t increases μ must decrease, but a mathematical characterization has not been successful. The reasons for the decreased rate of growth per organism are complex. Among them may be enumerated the following:

1. The average rate of growth per cell will decrease with the increase in concentration of the injurious products of metabolism.

2. The average rate of growth per organism will decrease with decrease in the available food supply, or with some single limiting factor of this food supply.

3. As the period progresses a larger and larger proportion of the cells go into the "resting stage" and are withdrawn from those dividing or multiplying.

4. It is probable that before this period is completed some cells die.

Slator has suggested that the curve might be described by

$$b^n = B^n e a k^n t^n$$

where n, k and a are constants suitably adjusted. Until further data are accumulated an attempt to evaluate these constants will prove difficult. From analogy with preceding and succeeding equations, it is possible that the growth equation of this phase might be

$$b = Be^{kt^{-n}} \text{ and } \mu = t^{-n-1}$$

V. The Maximum Stationary Phase

During this period there is theoretically no change in the total number of bacteria present. If we still employ the useful general expression

$$b = Bc^{\mu kt}$$

μ during this time remains zero, and the number of bacteria is constant.

Persistence of this phase must mean the balancing of increase and death. The rate of increase of bacteria must be such as to quite exactly make good the loss from death.

Investigations as to the length of this phase, and the influence of environment upon it are needed. With some organisms the phase is very transitory if it can be said truly to occur at all, with other forms apparently it persists for some time before there is marked any tendency to decrease in numbers.

VI. Phase of Accelerated Death Rate

Sooner or later the number of bacteria which die in a unit time will exceed the increase. In other words, as soon as bacteria reach the "resting stage" we may assume that they begin to die off, but they do not all reach this stage at the same instant. For some time there is an acceleration in the rate of death. The μk of the equation

$$b = Be^{\mu kt}$$

varies from zero to the velocity coefficient (constant) of the logarithmic death period. It also becomes negative in sign. It increases numerically in value during this period as time increases. During this stage the curve apparently is just the reverse of that of the lag period.

It is not improbable that the equation of the curve during this period will be found to be

$$b = Be^{-k't^n}$$

When $t = 0$, $b = B$. As t increases, b will be found to decrease more and more rapidly. Data are not at hand to prove the reliability of this equation. This stage probably does not persist long in most cultures, the velocity coefficient of death soon reaching a certain maximum.

VII. LOGARITHMIC DEATH PHASE

It was first shown by Madsen and Nyman and later by Chick that when bacteria are subjected to the action of unfavorable environment such as the presence of disinfectants they die off in accordance with the law which governs monomolecular reactions. If the logarithms of the numbers of surviving bacteria after various lengths of time are plotted against time, they will be found to lie on a straight line. The slope of this line is negative. This slope is the velocity coefficient of the reaction.

$$-k = 1/t \cdot \ln b/B$$
$$\text{or} \quad k = 1/t \cdot \ln B/b$$

The equation of the curve of the surviving bacteria is

$$b = Be^{-kt}$$
$$\text{or} \quad B = be^{kt}$$

This behavior of the bacteria has been abundantly verified by experimentation. It has been found to be of great service in the evaluation of disinfectants.

The effect of concentration of disinfectants has been developed principally by the work of Paul, Bierstein and Reuss, and of Chick[9] and the results generalized by Phelps. It is found that a change in the concentration of a particular disinfectant will change the velocity coefficient of the death rate in accordance with the following relationship:

$$k = k'C^n$$

where k' is the velocity coefficient of the original, and k the velocity coefficient with new concentration C, and n is a constant. For a different concentration the equation then becomes,

$$k' = \frac{1}{C^n t} \ln \frac{B}{b}$$

and the equation of the curve of surviving bacteria becomes

$$b = Be^{-kC^n t}$$

[9] Jour. Hyg., 1912, 12, p. 414.

Determination of the values of k and n for a disinfectant and a comparison of these values with those determined for some standard, as phenol, constitute efficient characterization of the disinfectant.

The Rideal-Walker and the Hygienic Laboratory phenol coefficients of disinfectants are determined by the use of facts inherent in these formulae. If the same concentration of two disinfectants are to be compared, we may place the same number, B of bacteria per unit of solution in each, and determine the length of time it takes to reduce the number of living bacteria to less than one per loop. Under these conditions the time required to change b to a certain number b' is determined. The only undefined quantities left are t and k in the equations

$$b' = Be^{-k't'}$$
$$b' = Be^{-k''t''}$$
$$\therefore \quad k't' = k''t''$$
$$\text{or} \quad k'/k'' = t''/t'$$

that is, the velocity coefficients are inversely proportional to the times required for "disinfection." By determining variations in the values of these ratios with different concentrations one may approximate the values of n in the equation

$$b = Be^{-kC^{n}t}$$

If, in addition, the effect of heat be determined in accelerating the death rate of the bacteria, a relatively complete diagnosis of the characteristics of the disinfectant is at hand.

Summary

1. There are at least seven life phases during the development of a culture of bacteria.

2. The general equation which represents the curve of the plot of numbers of bacteria against time is $b = Be^{\mu kt}$.

3. During the first or initial stationary stage μ is equal to zero, b is equal to B and there is no change in the numbers of bacteria.

4. During the second or lag phase μ is a function of time, increasing with time from 0 to 1. The relationship between μ and time is complex, but it is approximated by the equation

$$\mu = t^{n-1}$$

and the growth curve equation becomes

$$b = Be^{kt^{n}}$$

5. During the third or logarithmic growth phase $\mu = 1$ and the equation becomes

$$b = Be^{kt}$$

where k is related to the minimum generation time as follows:

$$k = \frac{\ln 2}{g}$$

and the equation of the growth curve is

$$b = Be^{kt} = Be^{(t \ln 2)/g}$$

6. During the fourth period or phase of negative growth acceleration μ decreases from the 1 of the logarithmic period of growth to 0. It is a function of time, decreasing with time. The relationship is complex, and no satisfactory evaluation of μ in terms of constants and time has been secured. It is possible the equation of growth may assume the form

$$b = Be^{kt^{-n}} \text{ and } \mu = t^{-n-1}$$

7. During the fifth period or maximum stationary phase μ remains equal to 0 and b equals B.

8. During the sixth period or phase of accelerated death rate μ varies from 0 to -1. From analogy with the lag phase, the equation of growth during this phase may be

$$b = Be^{-kt^{n}}, \text{ and } \mu = -t^{n-1}$$

9. During the seventh period or phase of logarithmic decrease μ remains constant at -1, the growth curve having the equation

$$b = Be^{-kt}$$

10. The lag phase is interpreted as the time during which bacteria are gradually emerging from a resting stage. It is not improbable that the numbers of bacteria which emerge at various successive periods of time are distributed in accordance with some probability curve.

Editor's Comments on Papers 3, 4, and 5

3   **Winslow and Walker:** *The Earlier Phases of the Bacterial Culture Cycle*

4   **Stephenson:** *Growth and Nutrition*

5   **Van Niel:** *The Kinetics of Growth of Microorganisms*

Papers 3 through 5 summarize the trends and developments of the period 1915–1945. The introductory section of Winslow and Walker's review, reproduced here as Paper 3, outlines the developments treated in greater detail later in their paper. The predominance of studies of the early stages of the growth curve is reflected in some 20 pages of the paper devoted to this period, compared with less than 4 pages for that of the logarithmic and subsequent stages. In the section reproduced here it is noteworthy that Winslow recognized binary fission to be the only valid life cycle in microbial growth and considered morphological and physiological change to be governed by environmental rather than inherent factors.

Stephenson's chapter (Paper 4) documents the experimental results of most of the important developments of the period.

Van Niel's chapter (Paper 5) contributed a synthesis, rather than a review, of the knowledge at the end of the period, at the time when it was becoming evident that growth was not just a simple study of empirical increase in mass and numbers. The predominance given to the earlier growth stages had declined, and now a balanced consideration of the whole sequence of the growth curve attempted to relate the facts emerging from the increasing biochemical and physiological explorations of batch cultures.

3

Reprinted from *Bacteriol. Rev.*, **3**, 147–153 (1939)

THE EARLIER PHASES OF THE BACTERIAL CULTURE CYCLE

C.-E. A. WINSLOW

Yale School of Medicine, New Haven, Conn.

AND

HAROLD H. WALKER

University of Tennessee, Nashville

Received for publication March 15, 1939

THE BACTERIAL CULTURE CYCLE AND ITS SIGNIFICANCE

Our knowledge of this subject may be considered to have begun with recognition of the fact that bacteria transferred to a culture medium suitable for their growth exhibit a period of delayed multiplication or "lag." Specific study of the problem dates at least as far back as the observations of Müller (1895). He recognized three phases of the early bacterial culture cycle, lag, logarithmic increase and slackened growth and followed Buchner, Longard and Riedlin (1887) in computing generation times during the logarithmic phase by the formula

$$G = \frac{T \log 2}{\log b - \log a};$$

where T represents the time interval, and b and a the final and initial numbers of cells. In the same year, Ward (1895) in a brilliant and exhaustive study of *Bacillus ramosus* also determined generation times and demonstrated the three fundamental phases of slow acceleration, maximum acceleration and reduced acceleration. Furthermore, he indicated clearly the effect upon the culture cycle of temperature, of light rays from the violet end of the spectrum, of food, of oxygen, of dilution and of antiseptic substances.

The phases of the culture cycle were analyzed in a broader sense

147

43

by Rahn (1906) and Lane-Claypon (1909), who recognized four phases, lag, logarithmic increase, stationary population and decrease. McKendrick and Pai (1911) attempted to explain assumed changes in growth rate during the culture cycle as manifestations of autocatalytic reactions, the governing factors at any moment being bacterial numbers and amounts of available nutriments. Chesney (1916) made valuable contributions to our knowledge of the culture cycle, as did Buchanan (1918 and 1925) who recognized seven phases, instead of four,—initial stationary phase, lag phase, phase of logarithmic increase, phase of negative acceleration, phase of maximum stationary population, phase of accelerating mortality and phase of logarithmic mortality.

The lag period was specifically studied by Hehewerth (1901) and Whipple (1901), although the phenomenon had been noted by many earlier bacteriologists in connection with multiplication of bacteria in water samples and in relation to the "bactericidal properties of milk," (for further references, see Winslow, 1928). It was carefully analyzed by Rahn (1906), Coplans (1910) Penfold (1914), Ledingham and Penfold (1914) and Chesney (1916).

That the phenomenon of lag had a biological basis was indicated by Müller (1895) who showed that when cultures of differing ages were used for inoculation into a new medium, the generation times in the new medium differed widely. When the source culture of typhoid bacilli was $2\frac{1}{2}$ to 3 hours old, the generation time was 40 minutes in the new medium; when the same culture was $6\frac{1}{4}$ hours old the generation time was 80 to 85 minutes; when the source culture was 14 to 16 hours old, the generation time was 160 minutes. Similarly, Barber (1908) and others showed that when transfer was made from a culture in the logarithmic phase, to the same medium and under the same conditions, the new culture multiplied at once at a logarithmic rate.

It was Müller again, in 1903, who first studied the chemical activity of bacteria at various stages of the culture cycle. He did not draw any conclusions as to the ratio of end-products formed to bacterial numbers but comparison of the various tables in his paper makes it clear that the amounts of carbon dioxide and hydrogen sulfide per cell must have been much greater in the earlier phases of the culture cycle. This phenomenon was clearly recognized at a much later date by Bayne-Jones and Rhees (1929) for heat production, by Cutler and Crump (1929) for liberation of carbon dioxide and by Stark and Stark (1929a) for acid production.

A third differential characteristic of specific phases of the culture cycle is resistance to various harmful environmental influences. Many early observations showed that very old cultures were characterized by low resistance; but this might be due merely to degenerative changes. Much more significant was the discovery by Reichenbach (1911) that young cultures in the lag and early logarithmic phase were more sensitive to heat treatment than those at the peak of their population curve and that resistance increased again in a late stage of the phase of stable maximum population. Even more striking results were reported by Schultz and Ritz (1910). These authors found that a given heat treatment (53°C. for 25 minutes) killed about 95 per cent of colon bacilli from a 20-minute culture. In slightly older cultures, resistance decreased, so that at 4 hours 100 per cent destruction occurred. Then, resistance increased again so strikingly that 7-hour to 13-hour cultures showed no reduction whatever under the same heat treatment. A markedly low resistance to harmful chemical agents was demonstrated by Sherman and Albus (1923) to be characteristic of the lag and early logarithmic phase and these authors, on the basis of their experiments, developed in a highly fruitful manner the concept of "physiological youth" as applied to the bacterial culture cycle.

A fourth characteristic of the phase of "physiological youth" had meanwhile been described by Clark and Ruehl (1919) and by Henrici (1921 to 1928). These investigators demonstrated that certain early phases of the culture cycle are characterized not only by rapid multiplication, high metabolic activity and low resistance to certain harmful environmental conditions but also by highly characteristic types of morphology, the individual cells being in general much larger than in the maturing culture. Wilson (1926) demonstrated the same phenomena by a comparison of plate counts with measurements of the opacity of bacterial cultures.

Finally, a fifth differential characteristic was described by Mac-Gregor (1910) and by Sherman and Albus (1923), who recorded that young cultures were more resistant than old cultures to acid agglutination. Shibley (1924) demonstrated that in the early phases, the electrophoretic charge on the cells was much less than at a later period.

The fundamental biological significance of the bacterial culture cycle was perhaps first clearly recognized by Henrici. In one of his earlier papers (Henrici, 1925a) he says:

The cells of bacteria undergo a regular metamorphosis during the growth of a culture similar to the metamorphosis exhibited by the cells of a multicellular organism during its development, each species presenting three types of cells, a young form, an adult form and a senescent form; that these variations are dependent on the metabolic rate, as Child has found them to be in multicellular organisms, the change from one type to another occurring at the points of inflection in the growth curve. The young or embryonic type is maintained during the period of accelerating growth, the adult form appears with the phase of negative acceleration, and the senescent cells develop at the beginning of the death phase.

The same theme was developed in his later monograph (Henrici, 1928) as follows:

The acceptance of this theory demands the acceptance of certain corollaries. If it be granted that the cells of bacteria undergo a metamorphosis of the same kind as that exhibited by multicellular organisms, then it must be granted that to this degree a population of free one-celled organisms, even though those cells have no connection other than the common nutrient fluid which bathes them, behaves like an individual. There has already been accumulated a great deal of evidence of other kinds to support the idea that there is no essential difference, that there can be drawn no hard and fast line, between populations of one-celled organisms and multicellular individuals; that a higher plant or animal is but a population of more highly differentiated cells. But there has been, in the past at least, a tendency to look upon cell differentiation in multicellular organisms as being the result of some organizing agency peculiar to such individuals. If, however, we find in cultures of micro-organisms where no such governing agency can be supposed to exist, a differentiation of cells, even though very primitive, we are forced to conclude that such is not the case; that the high degree of organization of higher organisms is a result and not a cause of the high degree of cell differentiation.

Acceptance of the validity of this analogy between a bacterial culture cycle and a multicellular organism clarifies very greatly the long conflict between pleomorphists and monomorphists in the field of bacteriology. One group has assumed a bacterial "life cycle" governed by some inherent biological tendencies; the other group denies that a cycle of any kind exists. Both perhaps are wrong and both right. One of us (Winslow, 1935) has pointed out that variations in bacterial morphology and physiology certainly do exist but that their succession is governed by environmental and not automatic inherent factors. Furthermore, this is precisely what occurs with higher forms of life where the organism as a whole forms the environment for its individual cells.

May we not assume then, that with all living cells, the "life cycle"— so far as the individual cell is concerned—is a cycle of simple binary fission. Other phenomena involving change in cell morphology and physiology of a cyclical nature are responses to changing environmental conditions and not the result of any inherent time mechanism. If a unicellular organism shows a definite series of morphological and physiological alterations in response to certain changes in environment which are likely to occur with reasonable frequency in its natural life we may call it a "life cycle" if we wish or we may call it something else. In any case, this is the only kind of life cycle (other than binary fission) which can occur in unicellular and relatively simple multicellular forms. In this sense, the bacteria have life cycles. When we find a more complex and more regular life cycle in the higher plants and animals (relatively independent of external environment), it is because the interrelationships of the complex organism produce a cyclical change in the internal environment which is comparable with the change which takes place in a bacterial culture and which affects the individual body cell very much as the cultural environment affects the unicellular organism. (Winslow, 1935.)

The study of the bacterial life cycle is, then, the bacteriological equivalent of the study of embryology, adolescence, maturity and senescence in the higher forms.   It is the purpose of the present article to review in orderly fashion some of the things we know about the earlier parts of this cycle—those included under the Buchanan phases of initial stationary, lag and logarithmic growth.   No attempt will be made to cover all the literature, which would be impossible in so vast a field; but certain significant and typical data will be cited in regard to each essential point.

References

Barber, M. A.   1908   The rate of multiplication of *Bacillus coli* at different temperatures.   J. Infectious Diseases, **5**, 379–400.

Bayne-Jones, S., and Rhees, H. S.   1929   Bacterial calorimetry. II. Relationship of heat production to phases of growth of bacteria.   J. Bact., **17**, 123–140.

Buchanan, R. E.   1918   Life phases in a bacterial culture.   J. Infectious Diseases, **23**, 109–125.

Buchanan, R. E.   1928   Growth curves of bacteria. In Jordan, E. O., and Falk, I.S., The Newer Knowledge of Bacteriology and Immunology, University of Chicago Press, pp. 46–57.

Buchner, H., Longard, K., and Riedlin, G.   1887   Ueber die Vermehrungsgeschwindigkeit der Bacterien. Zentr. Bakt. Parasitenk., **2**, 1–7.

Chesney, A. M.   1916   The latent period in the growth of bacteria.   J. Exptl. Med., **24**, 387–418.

Clark, P. F., and Ruehl, W. H.   1919   Morphological changes during the growth of bacteria.   J. Bact., **4**, 615–629.

Coplans, M.   1910   Influences affecting the growth of micro-organisms. — Latency: Inhibition: Mass action.   J. Path. Bact., **14**, 1–27.

Cutler, D. W., and Crump, L. M.   1929   Carbon dioxide production in sands and soils in the presence and absence of amoebae.   Ann. Applied Biol., **16**, 472–482.

Hehewerth, F. H.   1901   Die mikroskopische Zählungsmethode der Bacterien von Alex. Klein und einige Anwendungen derselben.   Arch. Hyg., **39**, 321–389.

Henrici, A. T.   1921–22   A statistical study of the form and growth of a spore-bearing bacillus.   Proc. Soc. Exptl. Biol. Med., **19**, 132–133.

Henrici, A. T.   1922–23   A statistical study of the form and growth of a diphtheroid bacillus.   Proc. Soc. Exptl. Biol. Med., **20**, 179–180.

Henrici, A. T.   1923–24a   A statistical study of the form and growth of Bacterium coli. Proc. Soc. Exptl. Biol. Med., **21**, 215–217.

Henrici, A. T.   1923–24b   Influence of age of parent culture on size of cells of *Bacillus megatherium*. Proc. Soc. Exptl. Biol. Med., **21**, 343–345.

Henrici, A. T.   1923–24c   Influence of concentration of nutrients on size of cells of *Bacillus megatherium*. Proc. Soc. Exptl. Biol. Med., **21**, 345–346.

Henrici, A. T.   1925a   On cytomorphosis in bacteria.   Science, **61**, 644–647.

Henrici, A. T.   1925b   A statistical study of the form and growth of the cholera vibrio.   J. Infectious Diseases, **37**, 75–81.

Henrici, A. T.   1928   Morphologic variation and the rate of growth of bacteria. C. C. Thomas, Springfield, Ill.

Hershey, A. D.   1938   Factors limiting bacterial growth.   II. Growth without lag in *Bacterium coli* cultures.   Proc. Soc. Exptl. Biol. Med., **38**, 127–128.

Lane-Claypon, J. E. 1909 Multiplication of bacteria and the influence of temperature and some other conditions thereon. J. Hyg., **9**, 239 – 248.

Ledingham, J. C. G., and Penfold, W. J. 1914 Mathematical analysis of the lag-phase in bacterial growth. J. Hyg., **14**, 242 – 260.

MacGregor, A. S. M. 1910 Immunity-phenomena in cerebro-spinal meningitis: opsonins and agglutinins in their relation to clinical features, prognosis, and therapy. J. Path. Bact., **14**, 503 – 521.

McKendrick, A. G., and Pai, M. K. 1911 The rate of multiplication of micro-organisms. A mathematical study. Proc. Roy. Soc. Edinburgh, **31**, 649 – 655.

Müller, M. 1895 Ueber den Einfluss von Fiebertemperaturen auf die Wachsthumsgeschwindigkeit und die Virulenz des Typhus-bacillus. Z. Hyg. Infektionskrankh., **20**, 245 – 280.

Müller, M. 1903 Über das Wachstum und die Lebenstätigkeit von Bakterien, sowie den Ablauf fermentativer Prozesse bei niederer Temperatur unter spezieller Berücksichtigung des Fleisches als Nahrungsmittel. Arch. Hyg., **47**, 127 – 193.

Penfold, W. J. 1914 On the nature of bacterial lag. J. Hyg., **14**, 215 – 241.

Rahn, O. 1906 Ueber den Einfluss der Stoffwechselprodukte auf das Wachstum der Bakterien. Zentr. Bakt. Parasitenk., II, **16**, 417 – 429 and 609 – 617.

Reichenbach, H. 1911 Die Absterbeordnung der Bakterien und ihre Bedeutung für Theorie und Praxis der Desinfektion. Z. Hyg. Infektionskrankh., **69**, 171 – 222.

Schultz, J. H., and Ritz, H. 1910 Die Thermoresistenz junger und alter Coli-Bacillen. Zentr. Bakt. Parasitenk., I, Orig., **54**, 283 – 288.

Sherman, J. M., and Albus, W. R. 1923 Physiological youth in bacteria. J. Bact., **8**, 127 – 139.

Shibley, G. S. 1924 Studies in agglutination. II. The relationship of reduction of electrical charge to specific bacterial agglutination. J. Exptl. Med., **40**, 453 – 466.

Stark, C. N., and Stark, P. 1929a Physiological difference between young and old bacterial cells. J. Bact., **17**, 2 – 3. (Abstract)

Ward, H. M. 1895 On the biology of *Bacillus ramosus* (Fraenkel), a schizomycete of the River Thames. Proc. Roy. Soc. London, **58**, 265 – 468.

Whipple, G. C. 1901 Changes that take place in the bacterial contents of water during transportation. Tech. Quart., **14**, 21.

Wilson, G. S. 1926 The proportion of viable bacilli in agar cultures of B. aertrycke (mutton), with special reference to the change in size of the organisms during growth, and in the opacity to which they give rise. J. Hyg., **25**, 150 – 159.

Winslow, C.-E. A. 1928 The rise and fall of bacterial populations. In E. O. Jordan, and I. S. Falk. The Newer Knowledge of Bacteriology and Immunology. University of Chicago Press, pp. 58 – 83.

Winslow, C.-E. A. 1935 What do we mean by a bacterial life cycle? Science, **81**, 314 – 315.

49

Reprinted from *Bacterial Metabolism*, 3rd ed., Longmans, Green & Co., Ltd. London, 1949, pp. 159–178.

GROWTH AND NUTRITION

MARJORY STEPHENSON

PHASES OF BACTERIAL GROWTH IN CULTURE MEDIA

BACTERIAL growth is usually estimated by a "viable count." This consists in transferring a small volume from the culture to a known volume of sterile Ringer's solution or other liquid which is non-toxic and non-nutrient ; after successive dilutions in the same diluent, a known volume is finally transferred to sterile nutrient agar which is either poured on to a plate or rolled in a thin layer inside a test-tube (roll tube method) ; after incubation each viable cell originally present is represented by a colony ; these are counted and the number of viable organisms in the original culture calculated. In order to obtain a total count a drop, after suitable dilution, is mounted in a special chamber of known volume and counted under the microscope, either after staining or by the use of dark ground illumination ; full critical description of both these methods is given by Wilson.[1] A measure of growth may also be obtained by dry weight or by a determination of total nitrogen by a micro-Kjeldahl method ; in the latter case it is necessary to be sure that the nitrogen content of the organism is constant. With organisms which do not clump the growth may be measured in a turbidimeter, preferably by means of a photoelectric cell ; the method must be checked against total count or dry weight for each organism used.

When culture medium is seeded with a small inoculum and the total count plotted against time, a curve of the form shown in Fig. 1 is obtained.

Fig. 1 shows eight phases of growth :

1. An initial stationary phase during which no multiplication occurs.

2. The lag phase or period of positive growth acceleration during which the rate of multiplication increases with time.

3. The logarithmic growth phase during which the rate of increase remains constant.

4. The phase of negative growth acceleration during which the rate of multiplication decreases.

[1] Wilson, 1922.

5. The maximum stationary phase during which no change in population occurs.

6. The phase of accelerated death during which the numbers are falling off with increasing rapidity.

7. The logarithmic death phase during which the rate of death is constant.

8. The phase of decreasing death-rate.

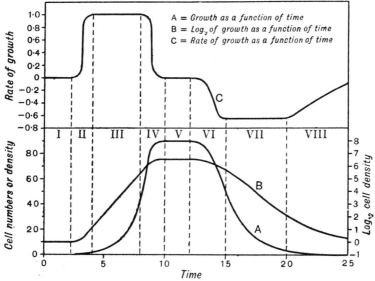

FIG. 1.—Typical growth curve of bacteria (Roman numerals and dotted vertical lines mark the growth phases)[1]

The lag phase

This comprises all that period (1 and 2) between the moment of inoculation and the establishment of a constant and maximum rate of cell division. A large amount of work has been done on this phase in order to explain why the viable cell when removed to fresh media does not immediately begin to divide at the maximum rate. In the vast majority of cases growth has been estimated by cell numbers, either total or viable, and it has been assumed that growth, i.e. increase in cell material, is proportional to increase in cell numbers. Actually, during the logarithmic and later phases this is true, but in the lag phase this relation does not hold. This was first discovered by Henrici, who showed that in the cases of *B. megatherium* and *Bact. coli* the average cell size increased very markedly during the lag phase, especially along the major axis,

[1] Monod (1942), *La Croissance des cultures bactériennes* (Hermann et Cie, Paris).

the average length of *B. megatherium* at the end of lag being six times that of the inoculated cells. This period of increased size is also marked by great fluctuations in form, as shown by the area-length index $\left(\dfrac{\text{area}}{\text{length}}\right)$.[1] On passing into the logarithmic phase, the cells decrease in size, approximating to that of the inoculation ; simultaneously, the great fluctuations in form cease. In some instances the largest cells occur in the logarithmic phase. Fig. 2, from Henrici's studies (on *Bact. coli*), illustrates these points.

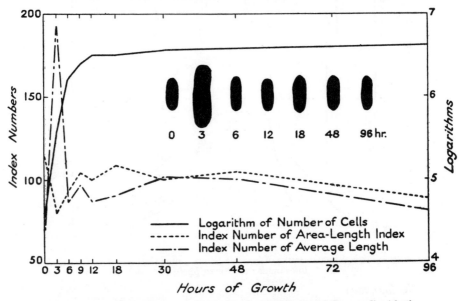

FIG. 2.—Graph illustrating variations in form and size of *Bact. coli* with the age of the culture grown on beef extract agar[1]

These results suggest that some cause operates in fresh media favourable to increase of cell material but inhibitory to cell division, leading the majority of cells to attain an abnormal size before splitting occurs. Large inocula tend to decrease this effect, the average maximum size being smaller and attained earlier than with small inoculations.[2] Moreover, if in the latter case the culture is reinoculated into fresh medium before the critical point of maximum size is reached, the cells attain a larger size, and the critical point occurs later than in the case of the parent culture.[3] Hence it appears that the size of the cell is partly conditioned by the density of the cell population. Another controlling factor is con-

[1] Henrici, *Proc. Soc. exp. Biol. and Med.*, 1923, **21**, 216.
[2] Ibid., 1921, 1923 (1), (2).   [3] Ibid., 1923 (2).

M

centration of nutrient material ; in broth agar of one-quarter the normal strength, the maximum size of the organisms is smaller and the critical point reached earlier (see Fig. 3). Thus crowding of cells and decreased concentration of nutriment produce the same effect.

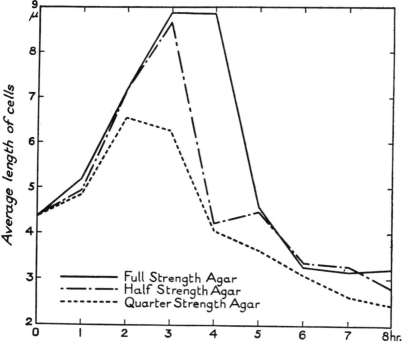

Fig. 3.—Influence of concentration of nutrients on average size of cells of *B. megatherium*[1]

Hershey and Bronfenbrenner applied Henrici's initial observation to the study of the lag phase.[2] These workers followed the bacterial development through a 22-hour period by the viable count and by the estimation of total nitrogen (Table 1).

TABLE 1[3]

Age of culture, hours	Viable count, V.C.	Nephelometric count, N.C.	Nitrogen per ml., N	Apparent size, NC/VC	Ratio, N/NC
	× 10⁶/ml.	× 10⁶/ml.	mg.		
24·0	940	835	0·25	0·9	0·30
3·5	310	739	0·25	2·4	0·34

[1] Henrici, *Proc. Soc. exp. Biol. and Med.*, 1923, **21**, 346.
[2] Hershey & Bronfenbrenner, 1937, 1938.   [3] Hershey, 1939.

Fig. 4 shows the usual phenomenon of lag when cell numbers are plotted against time, but no lag when cell material as measured by total nitrogen is in question. This discrepancy is explained when the volume of the cell is considered, the period of lag corresponding to the presence of large cells and high generation time, and the logarithmic phase to the return to normal cell size and low generation time.

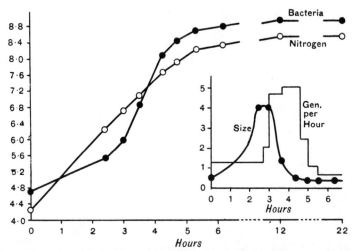

FIG. 4.—Increase of population and of bacterial nitrogen in *Bact. coli* culture. The logarithms of numbers and of nitrogen in mg. $\times$ 10^{10} per ml. are plotted against time. The insert shows the changes in average size expressed as nitrogen in mg. $\times$ 10^{10} per cell, and in the average rate of growth as reciprocal of generation time during the intervals observed[1]

From the above observations it appears that a new definition of lag is needed, viz. that phase of growth during which cell multiplication is retarded though increase in cell material may be occurring at the maximum rate. This new view of lag makes the discussion of much work on that phase very difficult since it is probable that much of it is based on a misconception, whilst it is not certain that some instances of lag may not involve retarded growth as well as cell division. With this uncertainty in mind some characteristics of the phase may now be considered.

Early observations showed that the older the culture from which the inoculant is taken the longer the lag, and this has received recent corroboration. It must be noted, however, that Hershey[2] has shown that this form of lag relates to cell division and not to growth. Table 2 shows that in the early period of a culture sown from a 24-hour inoculation there are 0·7 generations per hour,

[1] Hershey, *Proc. Soc. exp. Biol. and Med.*, 1938, **38**, 128.   [2] Ibid., 1939.

whilst in the corresponding period of a culture sown from a 3-hour inoculant there are 3·4 generations per hour, the rates of growth as measured by increase of total nitrogen being unaffected.

TABLE 2[1]

Parent culture		Subculture									
Age, hours	Viable count, cells/ml.	Period of incubation, hours	Viable count, ×10⁻⁸/ml.		Nitrogen, ×10⁻⁴mg/ml.		Nitrogen, ×10⁻¹⁰mg/cell		Rate of multiplication, gen/hr.	Rate of growth, twofold increase N/hr.
			Initial	Final	Initial	Final	Initial	Final		
3	7·4 × 10⁶	1·50	0·37	51·0	0·49	14·0	1·32	0·28	3·4	3·2
24	2·2 × 10⁹	1·58	2·20	58·0	0·57	21·0	0·26	0·36	0·7	3·3

TABLE 3[2]

INFLUENCE OF AGE OF INOCULANT ON RATES OF MULTIPLICATION AND GROWTH

Age of culture, hours	Rate of multiplication, gen/hr.	Rate of growth, twofold increase N/hr.
3·0	3·4	3·2
3·3	3·3	3·0
3·0	3·5	3·3
3·0	3·4	2·8
3·3	3·3	3·0
24·0	0·7	3·3
24·0	0·8	2·5
36·0	1·2	2·9
36·0	1·2	2·8
36·0	2·5	2·4
42·0	1·7	2·4

The type of lag due to age of inoculant is illustrated in an experiment of Salter[3] in which subcultures were made at the end of 8 hours and the generation time compared with that of the parent culture (Table 4). If inoculations were made from older cultures the generation time in the initial phase was much prolonged (Table 5).

TABLE 4[4]

Time, hours	Average number of bacteria per c.c.	Generation time in minutes
Culture, B. coli communis		
0	305	—
4	9,450	51·5
8	2,570,000	31·5
Sub-culture		
4	945	—
8	238,500	31·9

[1] Hershey, *J. Bact.* (1939), **37**, 290.   [2] Ibid., 1939.   [3] Salter, 1919.
[4] Ibid., *J. Inf. Dis.* (1919), **24**, 260.

TABLE 5[1]

Time, hours	Age of culture from which inoculations were made	Average number of bacteria per c.c.	Average generation time in minutes
	Days		
0	1	388	
2		1,070	72·1
4		13,000	33·3
6		177,000	31·8
8		4,400,000	25·9
0	4	349	
2		520	179·7
4		7,500	26·9
6		167,000	23·1
8		1,750,000	30·5

Highly instructive in this connection are the experiments of Sherman and Albus.[2] These workers showed that " old " cultures of *Bact. coli*, i.e. those over a week old, which had ceased for some days from appreciable multiplication, were markedly less sensitive to unfavourable conditions, such as exposure to cold (2° C.), heat (53° C.), 2% sodium chloride or 5% phenol, than were those taken from rapidly growing cultures a few hours after reinoculation. This was shown by removing samples from the culture and making a viable count (1) immediately, and (2) after exposure for 1 hour to the unfavourable condition. An example of their results is shown in Table 6.

TABLE 6[3]

EFFECTS OF EXPOSURE TO 2% NaCl UPON MATURE AND YOUNG CELLS OF *Bact. coli*

Exp. No.	Age of culture	Temperature incubated	No. of Bacteria per c.c.	
			At beginning	After 1 hour
1	21 hours	37° C.	890	770
2	2 days	Laboratory	378	356
3	7 ,,	,,	1470	1410
4	7 ,,	,,	880	960
1	3½ hours	37° C.	2530	27
2	3½ ,,	,,	1110	72
3	3½ ,,	,,	2190	55

The direct bearing of these results on the problem of the lag phase is displayed clearly in an experiment in which an old peptone culture of *Bact. coli* was sown into a new peptone medium and samples withdrawn at intervals and counted (1) direct, (2)

[1] Salter, *J. Inf. Dis.* (1919), **24**, 260.
[2] Sherman & Albus, 1923 and 1924. [3] Ibid., *J. Bact.* (1923), **8**, 127.

after exposure for 1 hour to 5% sodium chloride. Here the bacteria exhibit a decreasing tolerance for the salt solution as the lag phase advances, showing that they are undergoing a slow physiological change either of permeability or other surface condition evidenced by their reaction to the salt solution.

It has frequently been asserted that cells in the later stages of lag are in a high state of metabolic activity in respect of O_2 uptake, CO_2 elimination, deamination, etc.[1, 2, 3] It is now clear that this is true only of chemical or metabolic activity related to cell numbers and disappears if related to quantity of cell material.[4] This

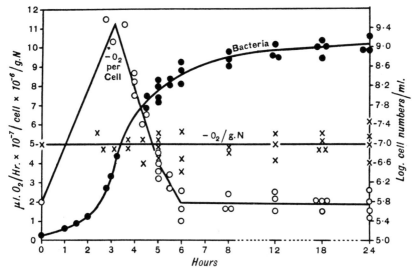

FIG. 5.—Rates of oxygen uptake per viable cell and per g. bacterial N in relation to phase of growth of *Bact. coli* cultures[5]

is shown in Fig. 5 in respect of respiratory activity and in Fig. 6 where rate of disappearance of glucose is shown to be strictly proportional to increase of cell material throughout the growth period.

Prolonged lag, actually a prolonged stationary phase, has been shown to occur when growth takes place in a synthetic as opposed to a broth medium.[6] This effect is not attributable to change of medium and it is worthy of remark that the subsequent growth rate is not affected. It has been suggested that the effect is due to traces of poisonous metals which are in time removed by adsorption on to the colloidal particles introduced with the in-

[1] Martin, 1932.   [2] Walker & Winslow, 1932.   [3] Walker *et al.*, 1934.
[4] Hershey & Bronfenbrenner, 1937.
[5] Ibid., *J. Gen. Physiol.*, **21**, 726 (1937.)   [6] Monod, 1942.

oculum ; in broth media, well furnished with colloids, this removal occurs more quickly. This explanation is supported by the observation that the length of lag in brain broth was increased from 30 minutes to 2 hours by the presence of a strip of silver foil, the subsequent rate of growth and total crop being unaffected.[1]

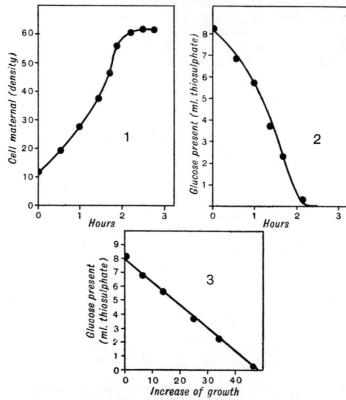

FIG. 6.—*Bact. coli* in synthetic media and glucose[2]

1. Rate of growth. 2. Consumption of glucose as function of time. 3. Consumption of glucose as a function of increase in growth.

In a synthetic medium the length of lag is decreased if the number of cells in the inoculant is increased ; this may be due to adsorption of metal poisons as suggested above.

The relation of the viable to the total count

Another aspect of growth rate has been disclosed by the careful studies of Wilson.[3] Working with *B. suipestifer* in a tryptic

[1] Monod, 1942.
[2] Ibid., *La Croissance des cultures bactériennes* (Hermann & Cie, Paris).
[3] Wilson, 1926.

broth, he made both total and viable counts, and found that even during the logarithmic phase the total count exceeded the viable ; an example of his results is given in Fig. 7 and Table 7. These

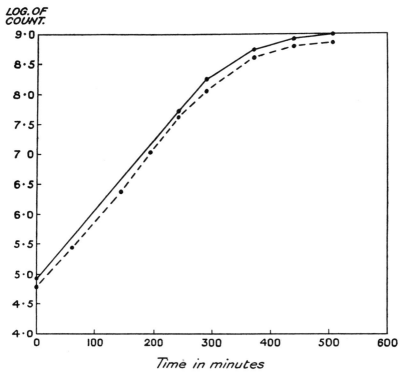

FIG. 7.—Showing the total and viable counts in a broth culture of *Bact. suipestifer*[1]

Total count = continuous line.   Viable count = interrupted line

TABLE 7[2]

Time after inoculation,	Viable count per c.c.	Total count per c.c.	Relation of viable to total
minutes			per cent
0	62,150	81,470	76·28
60	249,700	—	—
140	2,417,000	—	—
190	10,570,000	—	—
240	43,290,000	50,780,000	85·23
290	116,900,000	176,000,000	66·37
370	416,100,000	535,100,000	77·76
440	646,800,000	860,400,000	75·16
510	781,500,000	1,045,000,000	71·90

[1] Wilson, *J. Bact.*, 1922, 7, 434.   [2] Ibid.

results agree with the theory that in every generation the majority of the organisms continue to divide, whilst a constant percentage fail to do so. To make this clear, Wilson supposes 1000 organisms per c.c. alive at the beginning of the logarithmic phase ; at the end of the first generation there would be 2000 organisms, of which, say, 80% or 1600 would live and divide, whilst 20% or 400 would die. The continuation of the process is seen clearly in Table 8.

TABLE 8

	Viable	Organisms per c.c.	
		Non-viable	Total
Start . . .	1,000	0	1,000
End of 1st generation	1,600	400	2,000
„ 2nd „	2,560	640+ 400= 1,040	3,200+ 400= 3,600
„ 3rd „	4,096	1,024+1,040= 2,064	5,120+1,040= 6,160
„ 4th „	6,555	2,064+1,637= 3,701	8,192+2,064=10,256
„ 5th „	10,488	3,701+2,622= 6,323	13,110+3,701=16,811
„ 6th „	16,781	6,323+4,195=10,518	20,976+6,323=27,299

If the logarithms of these counts are plotted against the time, the curve for the viable organisms lies along an oblique straight line, whereas that for the total rises at first slightly more rapidly and then continues along an almost straight line very slightly divergent from that of the viable count (Fig. 8).

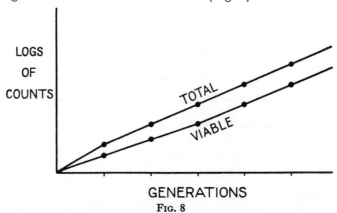

GENERATIONS

FIG. 8

It was pointed out by Wilson that this view of the discrepancy between the total and viable counts involves a modification in our method of calculating the generation time of an organism. According to the assumption that during the logarithmic phase all the organisms produced in each generation are viable, then the number of generations in a given period is calculated from the formula $n = \dfrac{\log b - \log a}{\log 2}$, where $n =$ the number of generations, *b* the number of organisms at the end, and *a* the number at the beginning of the given period. But if (say) 80% only are dividing, the number of organisms in each generation rises by 1·6 instead of 2, hence the formula required is $n = \dfrac{\log b - \log a}{\log 1·6}$, which decreases the figure for the average generation time. Thus Wilson calculates from one of his experiments a generation time of 25·5 minutes, according to the old formula, and of 19·7 minutes according to the new.

Causes affecting reproductive rate

The rate of reproduction is largely influenced by temperature. Thus measurable reproduction in *Bact. coli*, for example, begins at

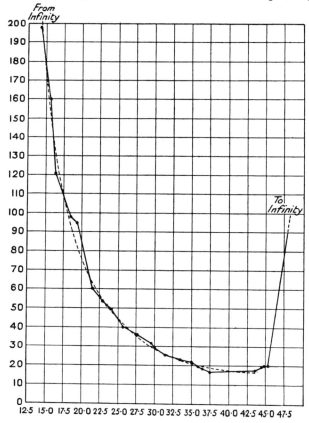

Fig. 9.—The growth rate of *Bact. coli* at different temperatures[1]

[1] Barber, *J. inf. Dis.*, **5**, 396 (1908).

61

about 10° and increases rapidly to 37°, where it reaches a maximum at a generation time of 17 minutes ; this remains nearly constant to 45°, when it falls rapidly till growth practically ceases at 49° (Fig. 9). The rates of growth of *Bact. coli* on synthetic media with glucose, mannitol, sorbitol and maltose are very close ; the Q_{10} is given in Table 9. The log of the growth rate plotted against the reciprocal of the absolute temperature is shown in Fig. 10.

TABLE 9

			Q_{10}	
			23°–33°	27°–37°
Source of carbon .	.			
Glucose	.	.	2·1	1·90
Mannitol	.	.	2·0	1·85
Sorbitol	.	.	2·1	1·90
Maltose	.	.	2·1	1·80

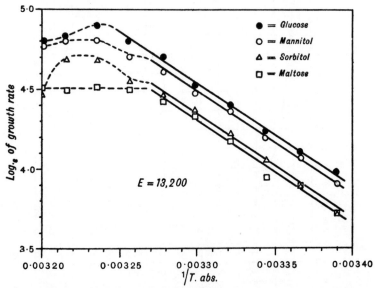

FIG. 10.—*Bact. coli* on synthetic medium with different sources of carbon. Log. of the growth rate as a function of the reciprocal of the absolute temperature[1]

The effect of pH

The effect of pH on multiplication rate is very marked but varies with different organisms. Much published work on this subject is vitiated because it takes account only of the initial pH. In some

[1] Monod (1942), *La Croissance des cultures bactériennes* (Hermann & Cie, Paris).

media, e.g. those containing fermentable carbohydrate, this changes so rapidly that the pH at which the reproduction rate is measured is far removed from that at the start. Fig. 8, Chapter XI, shows that *E. coli* grows between pH (initial) 4·5 and 9·0. Table 10 shows how far these values vary during the course of the experiment.[1]

TABLE 10

VARIATION OF pH OF MEDIA (TRYPTIC DIGEST OF CASEIN) DURING GROWTH OF *E. coli*

Buffer	pH		
	Initial	Final	Mean
M/60 phthalate . .	4·55	4·71	4·63
M/60 phthalate . .	4·95	5·58	5·26
M/45 phosphate . .	6·16	6·58	6·37
M/45 phosphate . .	7·01	7·10	7·05
M/45 phosphate . .	8·21	7·82	8·01
M/60 borate . .	8·82	8·70	8·75
M/60 borate . .	9·10	9·10	9·10

Effect of food concentration on growth rate

The rate of growth in the logarithmic phase is largely independent of concentration of carbon nutrients until this reaches a low level. This has been shown in brain broth and also in synthetic medium for both *coli* and *subtilis*. In the former the minimum generation time for *subtilis* is 36 minutes (see Table 11), in synthetic medium with sucrose 45 minutes, whilst *coli* on glucose attains a generation time of 45 minutes. Fig. 11 shows the influence of glucose concentration on generation time for *coli*. Maximum growth rate is reached at about 25 mg./litre, half-rate

TABLE 11

Bac. subtilis ON BRAIN BROTH. RATE OF GROWTH, ETC., IN THE LAG PHASE AS A FUNCTION OF THE INITIAL CONCENTRATION OF THE MEDIUM

Initial conc. of medium × 0·005 g./ml.	Rate of growth No. divs./H	Gen. time, minutes
15·0	1·67	36
12·0	1·65	36
8·0	1·65	36
5·0	1·64	37
4·0	1·43	42
3·0	1·18	52
1·5	0·87	69

[1] Gale & Epps, 1942.

at 4 mg./litre. The resemblance between this curve and one in which enzyme activity is plotted against concentration of substrate is obvious and indicates that growth rate may be controlled by the rate of action of one enzyme in the chain of glycolytic reactions ; this idea has been critically discussed by Monod.

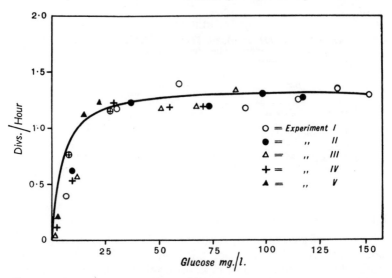

FIG. 11.—Rate of growth of *Bact. coli* in synthetic medium as a function of the concentration of glucose[1]

Factors influencing total crop

In studying the factors influencing the total bacterial substance produced on any given media the problem of aeration becomes very important[2] where aerobes or facultative anaerobes are in question. There is little doubt that insufficient aeration has, in the past, often been the limiting factor when cessation of growth was attributed to other causes. Mechanical agitation—as in a Warburg tank—suffices if the depth of the medium does not exceed 8 mm. ; when this is exceeded bubbling sterile air from a pressure cylinder through a porous glass filter or similar device must be employed. The necessity for adding CO_2 in the early stages must also be emphasised ; CO_2 as a growth factor has been discussed elsewhere (pp. 86 and 213). The accompanying graph[3] (Fig. 12) shows the relation of CO_2 to growth rate. The amount of bacterial growth is usually measured by a photo-electric turbidimeter. The relation of this to cell material and also to cell numbers once

[1] Monod, 1942, p. 70.   [2] Ibid. 1942.   [3] Dagley & Hinshelwood, 1938.

the logarithmic phase is reached has been tested by a number of workers and may now be taken for granted.[1]

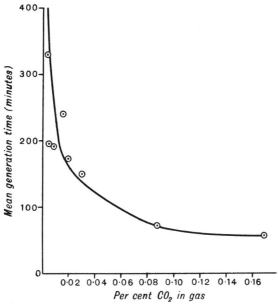

Fig. 12.—Effect of concentration of CO_2 on the growth rate of *Bact. lactis aerogenes*[2]

The relation of total crop to concentration of medium has been tested out by Monod.[3] It is important to select for this purpose an organism which can grow on a simple medium in the absence of growth factors and amino-acids, otherwise absence or short supply of any of these may become limiting. Using *coli* growing on brain broth, the total crop was shown to be strictly proportional to the concentration of the medium between 0·005 and 0·1 g./ml. (wet weight). (The units in which the crop was measured are not supplied.) A similar result was obtained with *Bac. subtilis*. It was shown in both cases that the limiting factor was carbon supply, glucose added after growth had ceased causing its immediate resumption, whilst added NH_4Cl or phosphate had no effect.

Studies of both *coli* and *subtilis* on synthetic media showed that the crop is strictly proportional to the concentration of glucose or sucrose between 0·2 and 0·02 mg./ml., the mean value for the ratio mg. bacteria (dry) per mg. carbohydrate being 0·233 in the case of *coli* and glucose and 0·218 for *subtilis* and sucrose. No

[1] Monod, 1942, p. 38.   [2] Dagley & Hinshelwood, *J. Chem. Soc.*, 1938.
[3] Monod, 1942, p. 38.

tendency was found for this ratio to alter as the concentration of the carbohydrate fell off, and the author concludes that the sole factor limiting crop is the amount of carbon material, and that the ratio of crop to concentration of medium is independent of the dilution of the food material. This implies that there exists no demand for maintenance apart from growth. Thus if M = total crop, C the concentration of food and Co the concentration required for maintenance apart from growth, then M = K (C–Co) and Co is the point where the straight line cuts the abscissa. But since in all cases tested the line passes through the origin it may be deduced that Co is O and hence that the sole use of the carbon constituent of this medium is for growth, the conception of energy of maintenance becoming unnecessary.

TABLE 12[1]

TOTAL GROWTH OF *Bact. coli* ON SYNTHETIC MEDIUM + GLUCOSE

Conc. glucose, mg./litre A	Total growth, units of density	Total growth, mg. dry wt. per litre B	Ratio $\frac{B}{A}$	Error % digression from mean
200	58·7	47·0	0·235	+ 0·9
180	53·0	42·4	0·236	+ 1·3
160	46·8	37·4	0·234	+ 0·43
140	41·3	33·0	0·236	+ 1·3
120	35·0	28·0	0·234	+ 0·43
90	26·0	20·8	0·230	− 1·3
70	20·0	16·0	0·228	− 2·1
50	14·4	11·5	0·230	− 1·3
25	6·7	5·4	0·258	+ 10·7
			Mean 0·233	

TABLE 13[2]

TOTAL GROWTH OF *Bac. subtilis* ON SYNTHETIC MEDIUM + SUCROSE

Conc. sucrose, mg./litre A	Total growth, units of density	Total growth, mg. dry wt. per litre B	Ratio $\frac{B}{A}$	Error % digression from mean
300	82·5	61·0	0·203	− 6·9
250	68·0	54·4	0·218	0·0
200	56·5	45·2	0·225	+ 3·2
150	38·3	30·6	0·204	− 6·4
100	26·2	21·0	0·210	− 3·7
50	15·5	12·4	0·248	+ 13·7
25	8·0	6·4	0·256	+ 17·4

[1] Monod, 1942, p. 44.   [2] Ibid., p. 43.

66

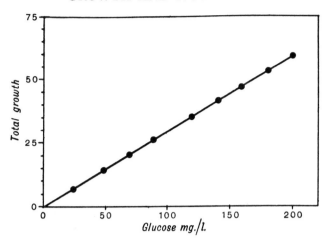

FIG. 13.—Total growth of culture of *Bact. coli* in synthetic medium as a function of the concentration of glucose[1]

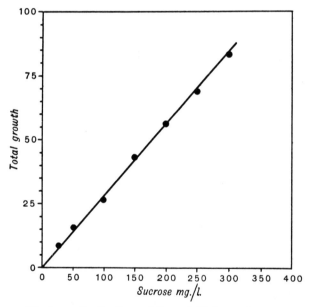

FIG. 14.—Total growth of culture of *Bac. subtilis* in synthetic medium as a function of the concentration of sucrose[2]

[1] Monod (1942), *La Croissance des cultures bactériennes* (Hermann & Cie, Paris).
[2] Ibid.

The effect of growth rate on crop

It has been shown that in synthetic media in optimal conditions bacterial crop is proportional to the concentration of the carbon compound and that the rate is independent of concentration within wide limits. If now the growth rate is limited by some factor such as suboptimal oxygen supply or change of pH, it is found that the total crop is unaffected, though of course taking longer to attain. This is true both for synthetic and complex media[1] (see Fig. 15). These results provide additional evidence against the idea that the cell requires energy for maintenance apart from

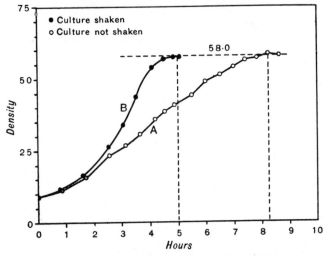

FIG. 15.—Growth of two cultures of *Bact. coli* on synthetic medium + ammonium lactate. A at rest, B shaken[2]

growth. Such energy is proportional to t, the time of the growth period, and D, the amount or density of living material produced in time t. Since D is the same whether t = 5 or 8 hours, the evidence that the cell requires some part of the energy derived from the food material for purposes other than growth falls to the ground. For further discussion of this subject see Monod.[3]

Influence of temperature on crop

The influence of temperature on crop is shown in Fig. 16. For the four carbon sources given, between 29° and 33° the crop is unaffected by temperature, between 33° and 41° it falls off and

[1] Monod, 1942.
[2] Ibid. (1942), *La Croissance des cultures bactériennes* (Hermann & Cie, Paris).
[3] Ibid., 1942, p. 92.

N

between 29° and 23° it rises. Closely parallel results are obtained for the four compounds tried.

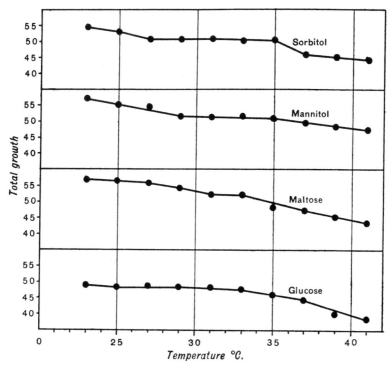

FIG. 16.—Total growth as a function of temperature. *Bact. coli* on synthetic medium in presence of different carbon sources[1]

NUTRITION

General considerations

The food requirements of bacteria differ profoundly from species to species. At one end of the scale are the strict autotrophants whose needs are met by inorganic materials, carbon being supplied in the form of carbonate or carbon dioxide and nitrogen as ammonium, salts, nitrates or nitrites. At the other end of the scale are organisms which can be grown at present only in complex protein digests to which are added blood or other tissues, culminating in such organisms as *M. lepræ*, which has so far defied all efforts at cultivation *in vitro*.

[1] Monod (1942), *La Croissance des cultures bactériennes* (Hermann & Cie, Paris).

References

Page

Barber, M. A. (1908), *The rate of multiplication of Bacillus coli at different temperatures*   170
J. inf. Dis., **5**, 379.

Dagley, S. and Hinshelwood, C. N. (1938), *Physico-chemical aspects of growth. II. Quantitative dependence of the growth rate of Bact. lactis aerogenes on the carbon dioxide content of the gas atmosphere*   174
J. Chem. Soc., 1936.

Gale, E. F. and Epps, H. M. R. (1942), *The effect of the pH of the medium during growth on the enzymic activities of bacteria (Escherichia coli and Micrococcus lysodeikticus) and the biological significance of the changes produced*   172, 304, 305, 306, 30
Biochem. J., 36, 600.

Henrici, A. T. (1921, 1), *A statistical study of the form and growth of a spore-bearing bacillus*   161
Proc. Soc. exp. Biol. and Med., **19**, 132.

Henrici, A. T. (1921, 2), *Influence of the age of the parent culture on the average size of cells of Bacillus megatherium*   161, 162
Proc. Soc. exp. Biol. Med., **21**, 343.

Henrici, A. T. (1923), *A statistical study of the form and growth of Bacterium coli*   161
Proc. Soc. exp. Biol. Med., **21**, 215.

Henrici, A. T. (1923, 2), *Influence of age of the parent culture on the size of cells of Bacillus megatherium*   161
Proc. Soc. exp. Biol. Med., **21**, 343.

Henrici, A. T. (1928), *Microbiology Monographs, London. Morphologic variation and the rate of growth of bacteria*   218
Ballière, Tindall and Cox.

Hershey, A. D. and Bronfenbrenner, J. (1937), *On factors limiting bacterial growth*   162, 166, 287
Proc. Soc. exp. Biol. Med., **36**, 556.

Hershey, A. D. and Bronfenbrenner, J. (1938), *Factors limiting bacterial growth. III. Cell size and "physiologic youth" in Bacterium coli cultures*   162, 287
J. gen. Physiol., **21**, 721.

Hershey, A. D. (1939), *Factors limiting bacterial growth*   162, 163, 164
J. Bact., **37**, 290.

Martin, D. S. (1932), *The oxygen consur.ption of Escherichia coli during the lag and logarithmic phases of growth*   166
J. gen. Physiol., **15**, 691.

Monod, J. (1942), *La Croissance des cultures bactériennes*   160, 166, 142, 167, 171, 173, 176, 177, 178
Hermann et Cie, Paris.

Salter, R. C. (1919), *Observations on the rate of growth of B. coli*   164, 165
J. inf. Dis., **24**, 260.

Sherman, J. M. and Albus, W. R. (1923), *Physiological youth in bacteria*   165
J. Bact., **8**, 127.

Sherman, J. M. and Albus, W. R. (1924), *The function of lag in bacterial cultures*   165
J. Bact., **9**, 304.

Walker, H. H. and Winslow, C.-E. A. (1932), *Metabolic activity of the bacterial cell at various phases of the population cycle*   166
    J. Bact., **24**, 209.

Walker, H. H., Winslow, C.-E. A., Huntington, E. and Mooney, M. G. (1934). *The physiological youth of bacteria as evidenced by cell metabolism*   166
    J. Bact., **27**, 303.

Wilson, G. S. (1922), *The proportion of viable bacteria in young cultures with especial reference to the technique employed in counting*   159, 168
    J. Bact., **7**, 405.

Wilson, G. S. (1926), *The proportion of viable bacilli in agar cultures of B. aertrycke (mutton) with special reference to the change in size of the organisms during growth and in the opacity to which they give rise*   167
    J. Hyg., **25**, 150.

Reprinted by permission of Princeton University Press from *The Chemistry and Physiology of Growth*, Princeton University Press, Princeton, N.J., 1949, pp. 91–105

V. THE KINETICS OF GROWTH OF MICROORGANISMS

BY C. B. VAN NIEL[1]

THE consensus of opinion, expressed in the preceding chapters, seems to have been that "growth is a very complex phenomenon." I see no reason for amending this verdict when considering the growth of microorganisms. It is true that these creatures appear, even when studied with the best optical or electronic equipment, as rather simple structures; and, compared with higher plants and animals, they certainly do not seem to be as highly differentiated. Nevertheless, they display an astonishing diversity of functions which in many respects they share with the trees, oysters, butterflies, and elephants.

Generally they are found swimming or floating about in an environment containing a more or less abundant variety of chemical compounds. From these they pick up or absorb special ones which as a result of metabolic activities are subsequently converted into cellular constituents. When a large enough quantity and variety of the latter have been produced, and when the individual organism has sufficiently increased in size, it divides into halves, each one of which then repeats the process. This continues until the food supply, or specific ingredients thereof, gives out, or until inhibitory metabolic products begin to interfere with normal development.

The amazing feature of growth, in a microbial culture as in living beings in general, lies in the fact that it involves the manufacture and orderly arrangement of the manifold structural and functional components which characterize the organism in such a way as to duplicate accurately an existing pattern. How this is accomplished is one of the major biological problems, fundamentally as complex in the lower as in the higher forms of life. On the other hand, the comparative ease of handling microorganisms, their rapid growth, and the opportunity for rigorously controlling various environmental factors all combine to render them favorable material for experimental studies of the basic mechanism of the growth process.

In several instances the successful resolution of a complex phenomenon into its constituent components, the latter more readily amenable to scientific analysis than the overall event, has resulted from kinetic

[1] Hopkins Marine Station of Stanford University, Pacific Grove, Calif.

studies. By relating the rate of the process to a number of variables, preferably one at a time, regularities and irregularities are often disclosed which serve as useful indicators of component steps. It therefore seems appropriate to examine what kinetic studies on the growth of microorganisms have contributed and may contribute to a deeper penetration into the problem of growth in general.

Nearly all that is known about the kinetics of growth of microorganisms has been learned from studies of so-called growth curves. By inoculating a specified culture medium with a suitable organism and determining, at the start and after various time intervals, the number of individuals present—frequently restricted to a determination of the number of viable cells—the requisite data are collected for constructing a graph which represents the relationship between the microbial population (or density) and the time elapsed since the initiation of the culture. The most convenient curves are those in which the number of individuals per unit volume is plotted on a logarithmic scale against time. Figure 1 demonstrates the commonly encountered shape of such growth curves.

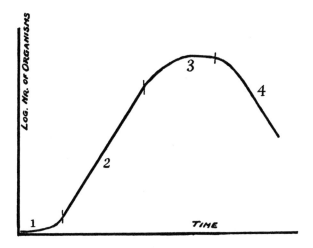

Figure 1. A representative growth curve.

Much attention has in the past been paid to an interpretation of the various phases of these curves. It will be clear that the first phase [1] denotes a period in which a gradual transition takes place from a condition where an increase in numbers does not occur to one where the increase is exponential [phase 2]. This second phase is easily accounted for; it means that during this period all organisms divide regularly, with the daughter cells behaving in an identical fashion. Towards the end of this phase the slope changes back to zero, evidence that when this point is attained [phase 3] either no further divisions take place, or that the increase in cell numbers is exactly balanced by the disappearance or death of individuals. The transition period, often brief and rather sharp, would imply that:

(*a*) either fewer and fewer cells are still in a position to divide, or
(*b*) that the division rate of the individuals becomes progressively less.

During the final phase [4] the death of the cells is predominant; from the point of view of studies on growth this phase is of no interest.

A considerable amount of time has been devoted to the first and originally quite enigmatic phase, referred to as the latency and lag period. As a result of various investigations, mostly before 1930, the attitude towards the problems it presents can be summed up as follows.

In many cases the cause for the existence of a latency and lag period resides in the cells used as inoculum. It can be completely eliminated by inoculating from a culture which itself is in the exponential phase. This has been demonstrated by Chesney (1) for bacterial, and by Phelps (2) for protozoan cultures. The careful observations of Henrici (3) have shown that the morphology of a bacterial species differs markedly according to the phase of the culture. It is therefore clear that the initial lag must be ascribed to a transformation of aged cells into "embryonic forms" (3, p. 140). Depending upon the age of the culture from which the inoculum is prepared, this change may require anywhere from minutes to many hours.

Similarly, a period of preliminary adjustment generally follows upon a change of environmental conditions, such as composition of the culture medium, change in temperature of incubation, etc. In that event a lag phase can be observed even if the new cultures are inoculated from a culture in the exponential phase, and the length of the lag depends upon the extent and nature of the change in conditions.

Growth curves of microorganisms are as a rule accurately reproducible. Chiefly responsible for this feature are (1) the rigorous control which the experimenter can exercise over the environmental conditions, and (2) the small size of the organisms. The small size eliminates many complications connected with the accessibility of foodstuffs. In higher plants and animals transport of nutrients, often over considerable distances, may be a controlling factor for growth; in microorganisms this is not the case. Furthermore, the small size insures that large numbers of individuals are always involved. Hence a statistical treatment of results becomes superfluous; every experiment automatically yields data in which individual variations are represented. This will be obvious if it be realized that with bacterial cultures populations of about 10^9 organisms per ml. are the rule, and that even for pure cultures of protozoa numbers up to 10^6 per ml. are far from rare.

That not all individuals in a culture behave in an identical manner, and that the results obtained really express statistical averages, follow from the fact that growth curves of microorganisms are always smooth even if cultures are started with a single individual, where one might expect a discontinuity such as was first observed by Ellis and Delbrück (4) in their studies on bacteriophage multiplication. More direct evidence for the existence of variation among individuals in bacterial cultures can be found in the investigations of Vaas (5) and Doudoroff (6).

In spite of these advantages, the kinetic approach has not yet been used very effectively or extensively. It is probable that this must chiefly be ascribed to the unwieldy and laborious methodology that used to be necessary for the determination of a growth curve, requiring either many sets of culture plates, or the actual counting under the microscope of thousands of cells for each point on the curve. The newer methodology of determining population densities by turbidity measurements with the aid of photoelectric apparatus, though not always entirely free from objections, may well bring about a renewed interest in this line of work, as is foreshadowed by the important studies of Monod (7) during the last few years.

Kinetic investigations on cultures of microorganisms are eminently suited for establishing relations between growth and environmental factors, especially the nature and amount of nutrients. Based upon the concept developed by Liebig about a century ago, it could be anticipated that in microbial cultures growth would be limited by those ingredients which are present in "minimum" quantity. Experiments such as those of Meyer (8) on the growth of *Aspergillus niger* in media with graded concentrations of phosphate and ammonium sulfate have gone far towards establishing the validity of Liebig's "law of the minimum" for microorganisms. But these were not, strictly speaking, kinetic studies; the measurements dealt with total crops as dependent upon nutrient concentrations rather than with rates of growth. Furthermore, these researches were carried out at a time when little was known concerning the exact mineral requirements of the organism used, which makes the interpretation of the results somewhat difficult.

Decidedly more conclusive in this respect have been the recent studies in connection with growth factor requirements of microorganisms. The mere fact that microbiological assay methods have been developed for many of the known vitamins and amino acids provides convincing evidence for the thesis that a single nutrient factor may determine the extent of growth. However, with a very few exceptions, these contributions too are based upon measuring final yields rather than rates. And where the latter are employed, as for example in the thiamin assay by means of yeast fermentation (9), it is the metabolic, not the growth rates, that are the basis for the determination.

Before 1942 only a few investigations on the kinetics of microbial growth in relation to nutrient concentration were reported. The results have been well summarized by Rahn (10, p. 251) in the following statement:

"With increasing amounts of food, the crop, i.e. the final number of cells in a culture, usually increases.

"With a liberal supply of building material, increase of energy food will cause no acceleration in growth, but a longer growth period, and, therefore, a larger crop. This is limited by accumulation of fermentation products or of inhibiting cell secretions.

"With all food components increasing, the crop will still be limited by fermentation products or cell excretions. The law of diminishing returns becomes quite evident."

While these conclusions were based upon experiments with bacteria and yeasts, they apply equally to the growth of protozoa in pure culture, as was first established by Phelps (11). His studies also revealed a direct proportionality between cell yield and food concentration within very wide limits, a situation which, wrote Phelps, "has not yet been demonstrated in the bacteria nor in most of the work on yeasts." (11, p. 494).

Two years later, however, Dagley and Hinshelwood (12) supplied the experimental evidence that in bacterial cultures also the rate of growth is independent of the concentration of nutrient materials over a wide range, while the final crop of organisms is directly proportional to this concentration. The latter part of these conclusions was corroborated by studies on the growth of *Escherichia coli* in beef broth (13).

All of the earlier studies on the kinetics of growth in relation to nutrient concentration were carried out with complex media in which not one but all constituents were present in varying amounts. It was only recently that Monod (7) started his investigations in a manner more likely to furnish basic information. By using bacteria which can develop in a strictly defined medium, it was possible to determine the effect of variations in the concentration of a single nutrient constituent upon both the rate of growth and the final crop. So far, this approach has been restricted to experiments with sugars and related compounds, applied over a wide range of concentrations. With three different bacterial species it was established that, for a given carbon source, the rate of growth is independent of the substrate concentration, and the total yield is strictly proportional to it, down to such low levels as 20 mg. substrate per liter.

There are on record a number of observations which suggest a different response of microorganisms to the concentration of nutrients. Thorne (14), for example, claims to have shown that the rate of yeast growth in sugar media falls uniformly with increasing concentration of the yeast, a behavior entirely at variance with the above-mentioned results. And Porter (15) has recently resuscitated the concept of "biological space" as earlier developed by Bail (16). The argument here is that, because the total crop of bacteria in liquid cultures reaches a certain maximum regardless of the concentration of foodstuffs, growth must be limited by a space factor rather than by nutrients. The following passage from Porter's book (15, p. 136) is worth quoting in this connection:

"There is considerable evidence in favor of some of Bail's claims concerning space theory and M(maximum)-concentration, but many of his points are not accepted at this time. Further work will have to be done on this subject—why bacteria stop dividing at a maximum speed when a certain population is reached—before any definite conclusions can be reached."

Most, if not all, of these aberrant results can be readily ascribed to the experimental conditions imposing a limiting factor which, in spite of repeated exhortations, is still insufficiently taken into account. This limiting factor is oxygen. The supply of this gas is governed by its very low solubility coefficient and unless special measures are adopted to insure a greatly increased rate of diffusion (e.g. by forced aeration, continuous shaking, etc.), its availability in liquid media is apt to fall below that required for an optimum growth rate of the organisms. The studies of Phelps (11), Rottier (17), Rahn and Richardson (18), and Anderson (19) attest to the correctness of this statement. It seems more than likely that the "biological space" referred to above is simply a measure of the oxygen supply. Particularly the close packing of bacteria in colonies provides a convincing argument for the contention that space *per se* cannot be considered seriously as an important factor in growth.

Thus the best accessible evidence indicates that the growth rate is not affected by the concentration of nutrients between very wide limits, and that the total yield, over an equally wide range, is proportional to the amount of food. This situation carries important implications which will next be considered.

Let me reiterate that growth consists in part of the transformation of foodstuffs into cellular constituents. This, in turn, depends upon metabolism. In Chapter III Thimann has presented a number of cases in which growth responses are strictly paralleled by metabolic responses following upon the exposure of plants to auxin solutions. It is therefore of the greatest significance that the rate of metabolism, as has been repeatedly ascertained, is generally unaffected by concentration differences of the substrate down to extremely low levels.[2] And it is not surprising that growth would follow the same pattern.

In many respects the relation between metabolism and growth is quite obvious. "No growth without metabolism" was a dictum as familiar since the latter part of the nineteenth century as the phrase "No thought without phosphorus" had been earlier. The reason for this has become apparent as the concepts of thermodynamics were developed. The growing cell must synthesize cellular constituents, and for these syntheses energy is required.

As a first approximation the relation between growth and metabolism could thus be understood as one of energy linkage; the metabolic activities of the cell provide the energy necessary for the performance of synthetic reactions. Based upon this concept, various investigations have been conducted aimed at measuring the "energetic yield" of growth processes. By determinations of the amount of cell material that could be synthesized from a known quantity of substrate, of the caloric value of the cell material, and of the energy liberated by oxidations accompanying the observed growth, data were assembled which permitted the computation of such "energetic yields." In subsequent studies of the same general nature more precise relations were calculated by distinguishing between a "metabolism for upkeep" and a "metabolism for growth" (20, 21).

However useful and important this sort of information may have been, recent trends have brought about a considerable change in the

[2] Some interesting cases have been found in which this is not true. These, however, present problems which are outside the scope of this discussion.

primary approach to the problems of syntheses and growth. Biochemists in particular have come to recognize that the thermodynamic formulations beg the question which has become paramount in today's biochemistry, viz. that of the mechanism whereby the result is achieved. Applied to the problem under discussion, this means that the concept of an energetic coupling between synthesis and breakdown, no matter how sound fundamentally, is too vague to be permanently satisfactory, and that ultimately our task is to discover the mechanisms whereby such coupling is accomplished.

The most fruitful hypothesis that has been advanced in this connection assumes the existence of material links between breakdown and synthesis. It may, somewhat primitively perhaps, be paraphrased in this manner: instead of regarding biological syntheses as very special processes which, because they require energy, are basically different from the breakdown reactions in which energy is liberated, it would be more appropriate to describe syntheses as series of enzyme-controlled step reactions, each step comparable in every way but one with those which collectively compose the breakdown. The differentiating feature is that, whereas the breakdown reactions can always start with any one of the utilizable substrates in the food, the synthetic processes require particular substances which are not usually provided but which arise from the initial substrates as intermediate products in the normal course of their metabolic degradation. Such products are characterized by being more unstable, i.e. more reactive, than their ultimate precursors, the foodstuffs proper, thus implying that their energy level has been raised with respect to the initial materials.

This concept, first enunciated by Kluyver (22, 23) some fifteen years ago, does not conflict with the earlier one of energetic coupling. Its great value lies in the penetrating insistence upon the functioning of an intelligible chemical mechanism which is capable of accounting for the transfer of energy.

During the past few years tremendous strides have been made in this direction. Ranking high among these is the discovery of the phosphorylated carbohydrate now known as "Cori-ester," a compound which can serve as the immediate raw material for the spontaneous enzymatic synthesis of a number of di- and polysaccharides (24-30). No less significant are the studies of Lipmann (31, 32) on the labile intermediary metabolic products characterized by a high-energy phosphate bond. It has now been shown that these substances, too, can be used for performing a number of fundamental biochemical syntheses with isolated enzyme systems.

A special corollary of Kluyver's concept is that the molecular structure of the substrate and of the intermediate products to which this can lead during its breakdown, rather than the amount of energy liberated during this process, becomes the all important determinant of its potentialities as the starting point for the synthesis of cell materials. The investigations by Clifton and Logan, Doudoroff, and Bernstein on the assimilation of chemically related compounds at different energy levels (33-36) provided suggestive results in this connection. Better yet can this situation be illustrated by recent studies on the synthesis of some characteristic cell constituents.

Having recognized a number of vitamins as parts of enzyme systems, it has become increasingly clear that the biosynthesis of such entities proceeds in a chemically intelligible manner, even though many of the steps involved in their manufacture may still be obscure. The mere fact that certain microorganisms are unable to develop in simple media without being supplied with preformed parts of enzymes shows conclusively that the supply of energy is not the governing factor, and that the availability of appropriate building blocks must be reckoned among the requirements for growth. Such phenomena are now generally interpreted to mean that the cells lack the machinery for the synthesis of essential compounds, which must consequently be supplied. It has frequently been shown that the same organisms can also develop in an environment which contains, instead of the enzyme part itself, substances that are chemically more or less closely related to it, and from which the cellular constituent in question is elaborated.

Corresponding observations have been made pertaining to the synthesis of specific amino acids, as the epoch-making investigations of Beadle and Tatum and their collaborators with "biochemical mutants" have so amply demonstrated. Through studies of this kind it has been possible to elucidate various steps in the biosynthesis of a number of cell materials and to develop the thesis that each step is controlled by a specific gene. It is impossible to do this field justice here; reference is made to some recent reviews (37, 38).

No student of the chemistry and physiology of growth is likely to deny the importance of these advances in our knowledge of biosynthetic mechanisms. Nevertheless the inclusion of this phase may appear as a digression from my topic. What have such problems to do with the kinetics of growth of microorganisms?

It will be evident that even the growth of a microbe is subjected to the "principle of limiting factors." If the synthesis of all cell constituents can proceed faster in one medium than in another, growth will obviously be faster in the former. Similarly, the rate at which a single component is being synthesized may, under special circumstances, be the determining factor for the rate of growth of the organism. In that case our understanding of biosynthetic mechanisms might be advanced by kinetic studies on growth, as well as by the more strictly chemical ones which hitherto have been the sole method applied. And through a kinetic approach it may be possible to learn something about the synthesis of these cell constituents by organisms whose external environment does not need to contain a supply of closely related substances because the biosyntheses can proceed from a variety of simple ingredients of the medium.

While no investigations have yet been conducted with this purpose in mind, some data are available which support the idea here advanced. For example, Monod includes in his monograph (7) figures on the rate of growth of *Bacillus subtilis* in a complex and in a synthetic medium, the latter composed of inorganic substances and sucrose. Growth proceeds faster in the former than in the latter medium; the generation times are 36 and 50 minutes respectively. Such differences could well be due to differences in the rate at which particular cell constituents become available, especially in view of the fact that in the synthetic solution they must all be built up from inorganic ingredients and sucrose. If it were possible to increase the growth rate in the simple medium through the addition of known chemical entities, it might also become clear what special syntheses are responsible for the lower growth rate.

Another instance is provided by Whelton and Doudoroff's (39) study on the assimilation of sugars and related compounds by *Pseudomonas saccharophila*. It includes a determination of the generation times of this bacterium in a standard mineral medium with any one of a number of carbon compounds. These were found to vary from 105-110 minutes in the presence of trehalose and sucrose, to 196 minutes in the presence of acetate; cultures in glucose and lactate exhibited an intermediate behavior with generation times of 178 and 136 minutes respectively. Studies of this kind, if sufficiently extended, could yield information concerning the nature of intermediate products which serve as building blocks and the rate at which these are generated. Correlation of such data with overall growth rates, supplemented with experiments on the influence of special additions selected for the purpose of furnishing similar or different building blocks, might then reveal the nature of limiting synthetic reactions.

I do not mean to convey the impression that the above-suggested approach is an easy and certain way of solving the problem of growth. They are intended to emphasize that kinetic studies of the growth of microorganisms, hitherto rather neglected, should be seriously considered as potentially capable of supplying information which can contribute toward a more detailed analysis of the manifold and interlinking events which, in their totality, constitute the growth process.

The foregoing reflections on the effects of external factors on the growth rate of microorganisms, and on the interpretations and deductions that can be derived from the observed results, may have fostered the impression that internal factors can be regarded as more or less constant. But this is not the case; there exists an impressive body of indisputable evidence to the contrary. Changes in the morphological, physiological, and biochemical behavior of microbial cultures, attributed to adaptations, modifications, mutations, etc., have long been recognized as manifestations of the intrinsic variability of the organisms. Monod's recent studies (7, 40-42) on the growth rate of bacteria in media containing two different carbon sources instead of a single one offer an important advance in this field.

Depending upon the nature of the organism and of the substrates, either of two types of growth curves describes the observed response of the bacteria. One represents a single period of exponential growth extending over the entire time interval during which substrate is available, with the sharp break at the end marking the complete utilization of the two carbon sources. The other kind consists of two distinct parts, each in itself a characteristic growth curve, with its phases of accelerated, maximal, and retarded growth rates, the two curves being conspicuously separated by a horizontal stretch which indicates the cessation of growth during the corresponding period. Monod has designated this latter behavior by the term "phénomène de diauxie" ("diphasic growth") and has shown that it involves the occurrence of an adaptation process.

By chemical analyses of the medium at different times it was ascertained that the first part of the growth curve corresponds to growth of the organism at the expense of only one of the two substrates, to the complete exclusion of the other. The second, separate growth curve coincides with the decomposition of the second substrate. And the intervening period during which the population remains constant represents the interval needed by the bacteria to adapt themselves to the utilization of this second substance.

Adaptations of this sort have been known for several years and have been ascribed to an adaptive enzyme formation (43, 44). They are strictly dependent upon the presence of a specific substrate, an observation which has led Yudkin (45) to develop an ingenious hypothesis concerning the mechanism by which adaptive enzymes are generated, based upon differences in stability of the free enzyme molecules and of the enzyme-substrate complex. Monod's experiments (40) have shown that the rate of such adaptations in bacterial cultures depends not only upon the presence but also upon the concentration of the substrate. In view of the well-established fact that metabolic rates are generally independent of this last-named factor, and also that the growth rate of the adapted bacteria is constant over a wide concentration range, Yudkin's hypothesis may have to be modified.

Later experiments by Monod (42) have made it probable that adaptive enzyme formation must be interpreted as a competitive phenomenon. This follows from observations on diphasic growth in relation to variations in the concentration ratio of the two substrates. The higher this ratio in favor of the substrate inducing the adaptation, the shorter is the period of suspended growth: the measure of the adaptation rate. The lag can even be eliminated completely by sufficiently magnifying the difference in concentrations of the substrates.

From these results it has been inferred that the two different substrates compete for a common precursor of the specific enzymes. Differences in affinity of the substrates for the precursor would then be responsible for the quantitative variations in the enzyme systems of organisms exposed to solutions which contain both substrates simultaneously but in different concentration ratios.

Growth is the expression *par excellence* of the dynamic nature of living organisms. Among the general methods available for the scientific investigation of dynamic phenomena, the most useful ones are those which deal with the kinetic aspects. The relative paucity of such data on the growth of microorganisms at the present time cannot be denied.

Nevertheless a few contributions of considerable importance have been made in this respect. It can also be asserted with some confidence that the recent advances in methodology for the determination of growth curves will help in overcoming the justifiable hesitation on the part of investigators, previously confronted with cumbersome techniques, to adopt this approach for future studies on microbial growth. Only by further experimentation can the potentialities of kinetic studies in this field be duly appreciated and more clearly defined.

REFERENCES

1. Chesney, A. The latent period in the growth of bacteria. *Jour. Exper. Med., 24,* 387-418, 1916.
2. Phelps, A. Growth of protozoa in pure culture. I. Effect upon the growth curve of the age of the inoculum and of the amount of the inoculum. *J. Exp. Zool., 70,* 109-130, 1935.
3. Henrici, A. T. *Morphologic Variation and the Rate of Growth of Bacteria.* Charles C. Thomas, Springfield, Ill., 1928.
4. Ellis, E. L., and M. Delbrück. The growth of bacteriophage. *J. Gen. Physiol., 22,* 365-384, 1939.
5. Vaas, K. F. *Studies on the Growth of Bacillus megatherium de Bary.* Diss. Leiden, 1938.
6. Doudoroff, M. Experiments on the adaptation of Escherichia coli to sodium chloride. *J. Gen. Physiol., 23,* 585-611, 1940.
7. Monod, J. *Recherches sur la croissance des cultures bactériennes.* Hermann & Cie., Paris, 1942.
8. Meyer, R. Zum Ertragsgesetz bei Aspergillus niger. *Arch. f. Mikrobiol., 1,* 277-303, 1930.
9. Schultz, A. S., L. Atkin, and C. N. Frey. Determination of vitamin B_1 by yeast fermentation method. *Ind. Eng. Chem.,* Anal. Ed., *14,* 35-39, 1942.
10. Rahn, O. *Physiology of Bacteria.* Blakiston, Philadelphia, 1932.
11. Phelps. A. Growth of protozoa in pure culture. II. Effect upon the growth curve of different concentrations of nutrient materials. *J. Exp. Zool., 72,* 479-496, 1936.
12. Dagley, S., and C. N. Hinshelwood. Physicochemical aspects of bacterial growth. I. Dependence of growth of Bacterium lactis aerogenes on concentration of medium. *J. Chem. Soc., 1938,* 1930-1936, 1938.
13. Korinek, J. Zum Problem der Bakterienpopulation. *Centr. f. Bakt.,* II. Abt., *100,* 16-25, 1939.
14. Thorne, R. S. W. The nitrogen nutrition of yeast. *Wallerstein Laboratories Communic., 9,* 97-114, 1946.
15. Porter, J. R. *Bacterial Chemistry and Physiology.* Wiley, New York, 1946.
16. Bail, O. O. možnosti a základech pokusné nauky populační. Čas. Čes. lék. 1927; *Immunitäts f., 60,* 1-22, 1929.
17. Rottier, P. B. Recherches sus les courbes de croissance de *Polytoma uvella.* L'influence de l'oxygénation. *C. rend. Soc. Biol., 122,* 65-67, 1936.
18. Rahn, O., and G. L. Richardson. Oxygen demand and oxygen supply. *Jour. Bact., 41,* 225-249, 1940; *ibid., 44,* 321-332, 1942.

19. Anderson, E. H. Nature of the growth factor for the colorless alga Prototheca zopfii. *J. Gen. Physiol., 28*, 287-296, 1945.

20. Algera, L. Energiemessungen bei Aspergillus niger mit Hilfe eines automatischen Mikro-Kompensations-Calorimeters. *Diss. Gronigen; Rec. Trav. botan. néerland., 29*, 47-163, 1932.

21. Tamiya, H. *Le Bilan matériel et l'énergétique des synthèses biologiques.* Paris, Hermann & Cie., 1935.

22. Kluyver, A. J. Atmung, Gärung und Synthese in ihrer gegenseitigen Abhängigkeit. *Arch. f. Mikrobiol., 1*, 181-196, 1930.

23. Kluyver, A. J. *The Chemical Activities of Microorganisms.* Univ. of London Press, 1931.

24. Cori, G. T., and C. F. Cori. The kinetics of the enzymatic synthesis of glycogen from glucose-1-phosphate. *J. Biol. Chem., 135*, 733-756, 1940.

25. Cori, C. F. Phosphorylation of glycogen and glucose. In *Biological Symposia, 5*, 131-140. Lancaster, Jacques Cattell Press, 1941.

26. Hanes, C. S. The breakdown and synthesis of starch by an enzyme system from pea seeds. *Proc. Roy. Soc., B, 128*, 421-450, 1940.

27. Hassid, W. Z. The molecular constitution of starch and the mechanism of its formation. *Quart. Rev. Biol., 18*, 311-330, 1943.

28. Doudoroff, M., N. Kaplan, and W. Z. Hassid. Phosphorolysis and synthesis of sucrose with a bacterial preparation. *J. Biol. Chem., 148*, 67-75, 1943.

29. Doudoroff, M., W. Z. Hassid, and H. A. Barker. Synthesis of two new carbohydrates with bacterial phosphorylase. *Science, 100*, 315-316, 1944.

30. Hassid, W. Z., M. Doudoroff, and H. A. Barker. Enzymatically synthesized crystalline sucrose. *J. Amer. Chem. Soc., 66*, 1416-1419, 1944.

31. Lipmann, F. Metabolic generation and utilization of phosphate bond energy. *Advances in Enzymology, 1*, 99-162, 1941.

32. Lipmann, F. Acetyl phosphate. *Advances in Enzymology, 6*, 216-268, 1946.

33. Clifton, C. E., and W. A. Logan. On the relation between assimilation and respiration in suspensions and in cultures of Escherichia coli. *Jour. Bact., 37*, 523-540, 1939.

34. Doudoroff, M. The oxidative assimilation of sugars and related substances by Pseudomonas saccharophila. *Enzymologia, 9*, 59-72, 1940.

35. Bernstein, D. E. Studies on the assimilation of dicarboxylic acids by Pseudomonas saccharophila. *Arch. Biochem., 3*, 445-458, 1944.

36. Clifton, C. E. Microbial assimilations. *Advances in Enzymology, 6*, 269-308, 1946.

37. Knight, B. C. J. G. Growth factors in microbiology. *Vitamins and Hormones, 3*, 105-228, 1945.

38. Beadle, G. W. Biochemical genetics. *Chem. Rev., 37*, 15-96, 1945.

39. Whelton, R., and M. Doudoroff. Assimilation of glucose and related compounds by growing cultures of Pseudomonas saccharophila. *Jour. Bact.*, *49*, 177-186, 1945.

40. Monod, J. Influence de la concentration des substrats sur la rapidité d'adaptation chez le B. coli. *Ann. Inst. Past.*, *69*, 179-181, 1943.

41. Monod, J. Remarques sur le problème de la spécificité des enzymes bactériens. *Ibid.*, *70*, 60-61, 1944.

42. Monod, J. Sur la nature du phenomène de diauxie. *Ibid.*, *71*, 37-40, 1945.

43. Karström, H. Über die Enzymbildung in Bakterien und über einige physiologische Eigenschaften der untersuchten Bakterienarten. Diss., Helsingfors, 1930.

44. Karström, H. Enzymatische Adaptation bei Mikroorganismen. *Ergebn. Enzymforsch.*, *7*, 350-376, 1938.

45. Yudkin, J. Enzyme variation in microorganisms. *Biol. Rev.*, *13*, 93-106, 1939.

Editor's Comments on Paper 6

6 Monod: *The Growth of Bacterial Cultures*

This paper by Monod, being more widely distributed than his "Recherches," probably had a much greater impact and influence: it was outstanding as a lucid, concise, and incisive study of bacterial growth. Monod's work was a significant advance in many ways. In his "Recherches," simple, but precise experiments used carefully controlled environmental conditions for examining growth. The organisms were grown on chemically defined nutrient substrates and the growth obtained was determined by nephelometry. Measurements, made well within the range of his instrument, gave results that could be repeated. From the quantitative data obtained, Monod was able to recognize the importance of the nutrient in limiting growth, and of the control of growth by the growth-limiting nutrient. His review conveniently summarizes these results.

By recognizing the supreme importance of the exponential phase, its control by the nutrient limiting to growth, and the specific growth constants by which it could be defined, Monod rationalized the simple growth curve and put it in perspective as an analytical tool. He had resolved three important conditions [(1) the start, (2) the rate, and (3) the end point] of growth in a culture. The exponential phase, as the optimal and constant growth phase, was flanked on either side by two transient conditions: an initial "winding up," where cells adjusted to growth at a maximal rate for the prevailing nutrient condition; and a "running down," when the nutrient was being exhausted. The many phases of the classical growth curve had suddenly become largely irrelevant and redundant.

THE GROWTH OF BACTERIAL CULTURES

By Jacques Monod

Pasteur Institute, Paris, France

INTRODUCTION

The study of the growth of bacterial cultures does not constitute a specialized subject or branch of research: it is the basic method of Microbiology. It would be a foolish enterprise, and doomed to failure, to attempt reviewing briefly a "subject" which covers actually our whole discipline. Unless, of course, we considered the formal laws of growth for their own sake, an approach which has repeatedly proved sterile. In the present review we shall consider bacterial growth as a method for the study of bacterial physiology and biochemistry. More precisely, we shall concern ourselves with the quantitative aspects of the method, with the interpretation of quantitative data referring to bacterial growth. Furthermore, we shall consider exclusively the positive phases of growth, since the study of bacterial "death," i.e., of the negative phases of growth, involves distinct problems and methods. The discussion will be limited to populations considered genetically homogeneous. The problems of mutation and selection in growing cultures have been excellently dealt with in recent review articles by Delbrück (1) and Luria (2).

No attempt is made at reviewing the literature on a subject which, as we have just seen, is not really a subject at all. The papers and results quoted have been selected as illustrations of the points discussed.

DEFINITION OF GROWTH PHASES AND GROWTH CONSTANTS

Division Rate and Growth Rate

In all that follows, we shall define "cell concentration" as the number of individual cells per unit volume of a culture and "bacterial density" as the dry weight of cells per unit volume of a culture.

Consider a unit volume of a growing culture containing at time t_1 a certain number x_1 of cells. After a certain time has elapsed,

371

all the cells have divided once. The number of cells per unit volume (cell concentration) is then

$$x = x_1 \cdot 2;$$

after n divisions it will be

$$x = x_1 \cdot 2^n.$$

If r is the number of divisions per unit time, we have at time t_2:

$$x_2 = x_1 \cdot 2^{r(t_2 - t_1)}$$

or using logarithms to the base 2.

$$r = \frac{\log_2 x_2 - \log_2 x_1}{t_2 - t_1} \quad \cdots\cdots\cdots\cdots\cdots [1]$$

where r is the mean division rate in the time interval $t_2 - t_1$. In defining it we have considered the increase in cell concentration. When the average size of the cells does not change in the time interval considered, the increase in "bacterial density" is proportional to the increase in cell concentration. Whether growth is estimated in terms of one or the other variable, the growth rate is the same[1].

However, as established in particular by the classical studies of Henrici (3), the average size of the cells may vary considerably from one phase to another of a growth cycle. It follows that the two variables, cell concentration and bacterial density, are not equivalent. Much confusion has been created because this important distinction has been frequently overlooked. Actually, one or the other variable may be more significant, depending on the type of problem considered. In most of the experimental problems of bacterial chemistry, metabolism, and nutrition, the significant variable is bacterial density. Cell concentration is essential only in problems where division is actually concerned, or where a knowledge of the elementary composition of the populations is important (mutation, selection, etc.).

---

[1] The use of log base 2 in place of log base 10 simplifies all the calculations connected with growth rates. It is especially convenient for the graphical representation of growth curves. If $\log_2$ of the bacterial density ($\log_2 = 3.322 \log_{10}$) is plotted against time, an increase of one unit in ordinates corresponds to one division (or doubling). The number of divisions that have occurred during any time interval is given by the difference of the ordinates of the corresponding points. It is desirable that this practice should become generalized.

Although the two variables are not equivalent, it is convenient to express growth rates in the same units (i.e., number of doublings per hour) in both cases. When cell concentrations have been estimated, it is equivalent to the true division rate. When bacterial density is considered, it expresses the number of doublings of bacterial density per unit time, or the division rate of cells postulated to be of constant average size. In all that follows, unless specified, we shall consider growth and growth rates in terms of bacterial density.

These definitions involve the implicit assumption that in a growing culture all the bacteria are viable, i.e., capable of division or at least that only an insignificant fraction of the cells are not capable of giving rise to a clone. This appears to be a fairly good assumption, provided homogeneous populations only are considered. It has been challenged however [Wilson (4)] on the basis of comparisons of total and viable counts. But the cultures examined by Wilson were probably not homogeneous (see p. 378), and the value of the viable count in determining the "absolute" number of cells which should be considered viable under the conditions of the culture is necessarily doubtful (see p. 378). Direct observations by Kelly & Rahn (5) contradict these findings and justify the assumption. [See also Lemon (42) and Topley & Wilson (43).]

GROWTH PHASES

In the growth of a bacterial culture, a succession of phases, characterized by variations of the growth rate, may be conveniently distinguished. This is a classical conception, but the different phases have not always been defined in the same way. The following definitions illustrated in Fig. 1 will be adopted here:

1. lag phase: growth rate null;
2. acceleration phase: growth rate increases;
3. exponential phase: growth rate constant;
4. retardation phase: growth rate decreases;
5. stationary phase: growth rate null;
6. phase of decline: growth rate negative.

This is a generalized and rather composite picture of the growth of a bacterial culture. Actually, any one or several of these phases may be absent. Under suitable conditions, the lag and acceleration phases may often be suppressed (see p. 388). The retardation phase is frequently so short as to be imperceptible. The same is sometimes true of the stationary phase. Conversely, more complex growth cycles are not infrequently observed (see p. 389).

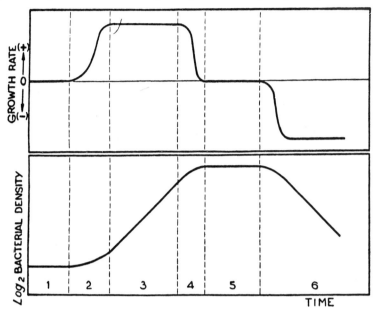

Fɪɢ. 1.—Phases of growth. Lower curve: log bacterial density. Upper curve: variations of growth rate. Vertical dotted lines mark the limits of phases. Figures refer to phases as defined in text (see p. 373).

GROWTH CONSTANTS

The growth of a bacterial culture can be largely, if not completely, characterized by three fundamental growth constants which we shall define as follows:

Total growth:[2] difference between initial (x_0) and maximum (x_{max}) bacterial density:

$$G = x_{max} - x_0.$$

Exponential growth rate: growth rate during the exponential phase (R). It is given by the expression

$$R = \frac{\log_2 x_2 - \log_2 x_1}{t_2 - t_1}$$

[2] "Croissance totale," Monod 1941.

91

when $t_2 - t_1$ is any time interval within the exponential phase.

Lag time and growth lag.—The lag is often defined as the duration of the lag phase proper. This definition is unsatisfactory for two reasons: (a) it does not take into account the duration of the acceleration phase; (b) due to the shape of the growth curve, it is difficult to determine the end of the lag phase with any precision.

As proposed by Lodge & Hinshelwood (6), a convenient lag constant which we shall call lag time (T_l) may be defined as the difference between the observed time (t_r) when the culture reaches a certain density (x_r) chosen within the exponential phase, and the "ideal" time at which the same density would have been reached (t_i) had the exponential growth rate prevailed from the start, i.e., had the culture grown without any lag $T_l = t_r - t_i$, or

$$T_l = t_r - \frac{\log_2 x_r - \log_2 x_0}{R}.$$

The constant thus defined is significant only when cultures having the same exponential growth rate are compared. A more general definition of a lag constant should be based on physiological rather than on absolute times. For this purpose, another constant which may be called growth lag (L) can be defined as

$$L = T_l \cdot R.$$

L is the difference in number of divisions between observed and ideal growth during the exponential phase. T_l and L values are conveniently determined graphically (Fig. 2).

ON TECHNIQUES

ESTIMATION OF GROWTH

Bearing these definitions in mind, a few general remarks may be made about the techniques employed for the estimation of bacterial density and cell concentrations.

Bacterial density.—For the estimation of bacterial density, the basic method is, by definition, the determination of the dry weights. However, as it is much too cumbersome (and accurate only if relatively large amounts of cells can be used) it is employed mainly as a check of other indirect methods.

Various indirect chemical methods have been used. Nitrogen determinations are generally found to check satisfactorily with dry weights. When cultures are grown on media containing an ammonium salt as sole source of nitrogen, estimations of the decrease of free ammonia in the medium appear to give adequate results (7). Estimations of metabolic activity (oxygen consumption, acid production) may be convenient (8), but their use is obviously

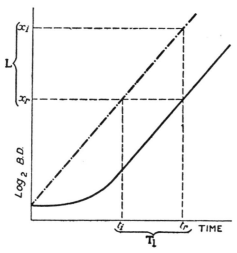

Fig. 2.—Lag time and growth lag. Solid line = observed growth. Dotted line = "ideal growth" (without lag). T_l = lag time. L = growth lag. (See text p. 375.)

very limited. Centrifugal techniques have been found of value (9).

The most widely used methods, by far, are based on determinations of transmitted or scattered light. (Actually, the introduction around 1935 of instruments fitted with photoelectric cells has contributed to a very large degree to the development of quantitative studies of bacterial growth.) We cannot go here into the physical aspects of this problem [for a discussion of these see (10)]. What should be noted is that in spite of the widespread use of the optical techniques, not enough efforts have been made to check them against direct estimations of cell concentrations or bacterial densities. Furthermore a variety of instruments, based on different principles, are in use. The readings of these instruments are often quoted without reference to direct estimations as arbitrary units of turbidity, the word being used in an undefined sense, or as "galvanometer deflections" which is worse. This practice introduces no little confusion and indeterminacy in the interpretation and comparison of data. It should be avoided.

Whatever instruments are used, the readings should be checked against bacterial density or cell concentration determinations, and the checks should be performed not only on different dilutions of a bacterial suspension, but at various times during the growth of a control culture. Only thus will the effects of variations of size of the cells be controlled. Without such controls it is impossible to decide whether the readings can be interpreted in terms of bacterial density or cell concentration, or both, or neither.

Actually, the instruments best fitted for the purpose appear to be those which give readings in terms of optical density (log I_0/I). With cultures well dispersed, it is generally found that optical density remains proportional to bacterial density throughout the positive phases of growth of the cultures (11). When this requirement is fulfilled, optical density determinations provide an adequate and extremely convenient method of estimating bacterial density.

It is often convenient to express optical density measurements in terms of cell concentrations. For this purpose, the two estimations should be compared during the exponential phase. The data, expressed as cell concentrations, may then be considered as referring to "standard cells," equal in size to the real bacteria observed during the exponential phase, larger than bacteria in the stationary phase and probably smaller than those in the acceleration phase.

Cell concentration.—Cell concentration determinations are performed either by direct counts (total counts) or by indirect (viable) counts. The value of the first method depends very much on technical details which cannot be discussed here. Its interpretation depends on the properties of the strains (and media) and is unequivocal only with organisms which do not tend to remain associated in chains or clumps. Total counts are evidently meaningless when there is even a slight tendency to clumping.

The same remarks apply to the indirect, so called viable, counts made by plating out suitable dilutions of the culture on solid media. The method has an additional difficulty, as it gives only the number of cells capable of giving rise to a colony on agar under conditions widely different from those prevailing in the culture. Many organisms, such as pneumococci (12), are extremely sensitive to sudden changes in the composition of the medium. The mere absence of a carbon source will induce "flash lysis" of *Bacillus subtilis* (13). Such effects may be, in part at least, responsible for the discrepancies often found between total and viable counts.

In spite of these difficulties viable counts retain the undisputed privilege of being by far the most sensitive method and of alone permitting differential counting in the analysis of complex populations. In the latter case, relative numbers are generally the significant variable, and whether or not the counts give a reasonably accurate estimation of absolute cell concentrations is unimportant.

Methods of Culture

Although the methods of culture will vary according to the problems investigated, certain general requirements must be met in any case. The most important one is that the cultures should be constantly mixed, homogenized, and in equilibrium with the gas phase. This is achieved either by shaking or by bubbling air (or other gas mixtures) or both. Bubbling is often found inefficient unless very vigorous, when it may provoke foaming which should be avoided. Slow rocking of a thin layer of liquid is the simplest and probably the best procedure. [For detailed descriptions of techniques, see (14).]

Various techniques for the continuous renewal of the medium have been described (15) and should be found useful for certain types of experiments (see p. 385).

The composition of the medium is largely determined by the nature of the experiment, and the properties of the strains. One general rule should however, so far as possible, be followed in the planning of quantitative growth experiments. As a culture grows, the conditions in the medium alter in a largely uncontrollable and unknown way. Therefore, the observations should be performed while the departure from initial conditions may still be considered insignificant. The more dilute the cultures, the closer will this requirement be met. The sensitivity of optical density measurements makes it practicable to restrict most experiments to a range of bacterial densities not exceeding 0.25 mg. dry weight per ml.

THE PHYSIOLOGICAL SIGNIFICANCE OF THE GROWTH CONSTANT

Total Growth

Limiting factors.—The metabolic activity of bacterial cells modifies the composition of the medium in which they grow. Depending on the initial conditions, and on the properties of the strains, one or another, or several, of these changes will eventually result in a decrease of the growth rate, bringing the exponential phase to an end, and leading more or less rapidly to the complete cessation of growth.

The factors most commonly found to be limiting can, as a rule, be classified in one of the following groups: (a) exhaustion of nutrients; (b) accumulation of toxic metabolic products; and (c) changes in ion equilibrium, especially pH.

The physiological significance of the constant G (total growth) depends on the nature of the limiting factor. It is uninterpretable when the limiting factor is unknown, or when several factors cooperate in limiting growth. The conditions of an experiment where G is to be estimated must therefore be such that a single limiting factor is at work. This may be considered to be the case only where it can be shown that no change, other than the one considered, plays a significant part both in breaking the exponential phase and in stopping the growth. Provided these requirements are met, the utilization of G as a measure of the effect of a limiting factor is warranted.

Actually, the estimation of G is especially useful when the limiting factor is a single, known, essential nutrient. Under such conditions, it can be a most convenient tool for the study of many aspects of nutritional problems. The principles of this technique and some general results will be considered in the next paragraphs.

Nutrients as limiting factors.—The bacteria most commonly studied are chemoorganotrophs[3] requiring an organic compound as carbon and energy source, a hydrogen acceptor, inorganic ions, and carbon dioxide. Most of the parasitic (and many saprophytic) bacteria are chemoorganoheterotrophs requiring, in addition to the above diet, certain specific organic molecules (growth factors).

Any one of the essential nutritional requirements of an organism is, by definition, a potential limiting factor. With organisms able to grow on simple defined media (whether they are organoautotrophs or organoheterotrophs), the composition of a medium is easily adjusted so that the concentrations of all essential nutrients are in large excess compared to one of them, which then becomes the sole limiting factor, provided its concentration is kept

[3] The Cold Spring Harbor Nomenclature (16) is adopted here.

low enough to eliminate interference from other potential limiting factors (pH changes, accumulation of metabolic products, etc.). Within the limits thus defined, the relation between G and the initial concentration (C) of the nutrient is, as a very general rule, found to be the simplest possible, namely, linear and to conform to the equation:

$$G = KC.$$

This relation implies that the amount of limiting nutrient used up in the formation of a unit quantity of cell substance is independent of the concentration of the nutrient. It implies also that growth stops only when the limiting nutrient is completely exhausted, or, in other words, that there is no threshold concentration below which growth is impossible (11).

Neither of these conclusions can be considered strictly true of course, and the linear relation cannot be taken for granted a priori. But it does seem to be a general approximation, and even a remarkably accurate one in many cases (Fig. 3 and Table I). Where

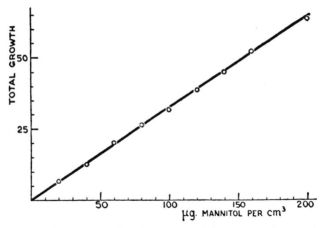

Fig. 3.—Total growth of *E. coli* in synthetic medium with organic source (mannitol) as limiting factor. Ordinates: arbitrary units. One unit is equivalent to 0.8 µg. dry weight per ml. (11).

it holds, the estimation of G affords a simple and direct measure of the growth yield (K) on the limiting nutrient, or

$$\frac{G}{C} = K = \frac{\text{amount of bacterial substance formed}}{\text{amount of limiting nutrient utilized}}.$$

When the proportion of the dry weight representing substance derived from the limiting nutrient is known, it is a measure of the fraction assimilated. If G is expressed as "standard" cell concentration, $1/K$ represents the amount of limiting nutrient used up in the formation of a "standard" cell. Thus, when determined under proper conditions, G is a constant of perfectly clear and fundamental significance; it is a measure of the efficiency of assimilatory processes.

TABLE I

Total growth of purple bacteria with acetate as
limiting factor [after Van Niel (9)]

Acetate (mg/ml.)	0.5	1.0	2.0	3.0
Total growth (mg/ml.)	0.18	0.36	0.70	1.12
K	0.36	0.36	0.35	0.37

Extensive data on G and K values are available only with respect to the organic source (9, 11). Little is known of K values in the case of inorganic sources. Owing to the development of microbiological assay methods, abundant data are available on the quantitative relations between growth of many bacteria and concentration of a variety of growth factors. But the major part of these data do not bear any known relation to G or any other definable growth constant, which is most unfortunate. It does seem at least probable that in many instances, the measurement of total growth, under conditions insuring homogeneity and limitation of growth by a single factor, could with advantage replace estimations of "turbidity at 16 hours," or "galvanometer deflections at 24 hours." It can be predicted with confidence that in most cases linear relations would be found [see e.g. (44)], permitting the estimation of K, and on which simpler and more reproducible methods of assay could be based. Furthermore, an intelligible and very valuable body of quantitative data on nutritional requirements of bacteria would thus become accumulated.

The remarkable degree of stability and reproducibility of K values, for a given strain and a given compound under similar conditions, should be emphasized. Contrary to the other growth constants, it seems to be very little affected by hereditary variability (45).

In general, of the three main growth constants, total growth is the easiest to measure with accuracy and the most stable. Its interpretation is simple and straightforward, provided certain experimental requirements are met. These are remarkable properties, which could, it seems, be put to much wider use than has hitherto been done, especially with the focussing of attention on problems of assimilatory and synthetic metabolism.

Exponential Growth Rate

The exponential phase as a steady state: rate determining steps.— The rate of growth of a bacterial culture represents the over-all velocity of the series of reactions by virtue of which cell substance is synthesized. Most, if not all, of these reactions are enzymatic, the majority probably are reversible, at least potentially. The rate of each, considered alone, depends on the concentrations of the reactants (metabolites) and on the amount of the catalyst (enzyme).

During the exponential phase, the growth rate is constant. It is reasonable to consider that a steady state is established, where the relative concentrations of all the metabolites and all the enzymes are constant. It is in fact the only phase of the growth cycle when the properties of the cells may be considered constant and can be described by a numeric value, the exponential growth rate, corresponding to the over-all velocity of the steady state system.

It has often been assumed that the over-all rate of a system of linked reactions may be governed by the slowest, or master, reaction. That this conception should be used, if at all, with extreme caution, has also been emphasized (17, 18). On theoretical grounds, it can be shown that the over-all rate of a system of several consecutive reversible enzymatic reactions depends on the rate and equilibrium constant of each. The reasons for this are obvious, and we need not go into the mathematics of the problem. A master reaction could take control only if its rate were very much slower than that of all the other reactions. Where hundreds, perhaps thousands, of reactions linked in a network rather than as a chain are concerned, as in the growth of bacterial cells, such a situation is very improbable and, in general, the maximum growth rate should be expected to be controlled by a large number of different rate-determining steps. This makes it clear why exponential growth rate measurements constitute a general and sensitive physiologic test which can be used for the study of a wide variety of effects, while, on the other hand, quantitative interpretations are subject to severe limitations. Even where the condition or agent studied may reasonably be assumed to act primarily on a single rate determining step, the over-all effect (i.e., the growth rate) will generally remain an unknown function of the primary effect.

Although very improbable, it is of course not impossible that the exponential growth rate could in certain specific cases actually

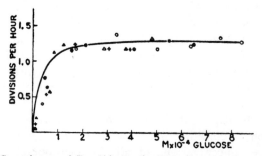

FIG. 4.—Growth rate of *E. coli* in synthetic medium as a function of glucose concentration. Solid line is drawn to equation (2) with $R_K = 1.35$ divisions per hour, and $C_1 = 0.22 \, \text{M} \times 10^{-4}$ (11). Temperature 37° C.

be determined by a single master reaction. But such a situation could hardly be assumed to prevail, in any one case, without direct experimental evidence. Some recent attempts at making use of the master reaction concept in the interpretation of bacterial growth rates are quite unconvincing in that respect (19).

Rate-concentration relations.—Notwithstanding these difficulties, relatively simple empirical laws are found to express conveniently the relation between exponential growth rate and concentration of an essential nutrient. Examples are provided in Figs. 4 and 5. Several mathematically different formulations could be made to fit the data. But it is both convenient and logical to adopt a hyperbolic equation:

$$R = R_K \frac{C}{C_1 + C} \quad \text{...................}[2]$$

similar to an adsorption isotherm or to the Michaelis equation. In the above equation C stands for the concentration of the nutrient. R_K is the rate limit for increasing concentrations of C. C_1 is the concentration of nutrient at which the rate is half the maximum.

The constant R_K is useful in comparing efficiency in a series of related compounds as the source of an essential nutrient. So far extensive data are available only with respect to the organic source (11). The value of R_K may vary widely when different

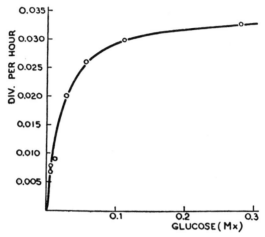

FIG. 5.—Growth rate of *M. tuberculosis* in Dubos' medium, as a function of glucose concentration. Solid line drawn to equation (2) with $R_K = 0.037$ and $C_1 = M/45$ (20).

organic sources are compared under otherwise identical conditions. There is no doubt that it is related to the activity of the specific enzyme systems involved in the breakdown of the different compounds, and it can be used with advantage for the detection of specific changes (e.g., hereditary variation) affecting one or another of these systems (30).

The value of C_1 should similarly be expected to bear some more or less distant relation to the apparent dissociation constant of the enzyme involved in the first step of the breakdown of a given compound. Furthermore, since a change of conditions affecting primarily the velocity of only one rate-determining step will, in general (but not necessarily), be only partially reflected in the over-all rate, one might expect C_1 values to be lower than the corresponding values of the Michaelis constant of the enzyme catalysing the reaction. This may explain why C_1 is often so small, compared to the concentrations required for visible growth, that its value may be difficult to determine, and the exponential growth rate appears practically independent of C (19).

It may be of interest to note that in a few instances exceptionally large values of C_1 have been obtained. For instance for *Mycobacterium tuberculosis*, on Dubos' medium, the value of C_1 for glucose is $M/45$, i.e., some 1,000 times its value for *Escherichia coli*. Whether this is due to a very low affinity of an enzyme or whether it reflects a peculiar permeability property of the membrane of this organism is not known (20).

Growth rate determinations as a null point method.—Although the growth rate is an unknown function of a large number of variables, quantitative comparisons of the effects of conditions or agents affecting it through the same rate-determining reaction (or system of reactions) are possible (at least in principle) by using growth rate measurements as a null point method, that is to say by determining the equivalent conditions at which a certain, conveniently chosen, value of R obtains. This general method is susceptible of many applications, especially in the study of antagonistic effects. Here reliable and sensitive methods for distinguishing between various types of antagonistic effects, and determining the relative activities of different antagonists, are needed. Theoretically the most sensitive comparisons should be afforded by determining, at various absolute concentrations, the ratios of inhibitor and antagonist at which a given per cent decrease of R (over uninhibited controls) occurs.

Although this may not always prove practicable, there is little doubt that growth rate measurements do yield data, not only more accurate, but essentially more informative, than "turbidity at 16 hours" or "galvanometer deflections at 72 hours." The studies of McIllwain on the pantoyl taurine-pantothenate antagonism (8) adequately illustrate this point. They clearly show, in particular, the importance of distinguishing between effects on growth rate and on total growth [see also (21 to 24)].

Linear growth.—Since we are discussing the interpretation of exponential growth rates, it may be worthwhile to consider the case when growth is linear with time, although, to the reviewer's knowledge, this has been clearly observed only once (25), actually during the residual growth of a streptomycin-requiring *B. subtilis* in a medium containing no streptomycin (Fig. 6). The interpretation is obvious, albeit surprising. Growth must be limited by one enzyme or system of enzymes, the activity of which is constant. In other words, in the absence of streptomycin, one rate-determining enzyme ceases to be formed, so that by being outgrown by the

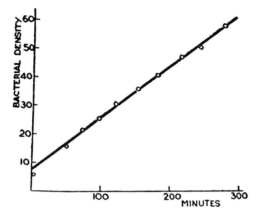

FIG. 6.—Residual growth of a streptomycin requiring strain of *Bacillus subtilis* in the absence of streptomycin. Growth is linear for over 4 hr. (25).

other enzymes, it eventually achieves true mastery and sets the system at its own constant pace, disregarding the most fundamental law of growth.

Similar systems could be artificially set up by establishing a constant, limited supply of an essential metabolite (using an organism incapable of synthesizing it), while all other nutrients would be in excess. Such a technique should prove useful for certain studies of metabolism (see p. 378).

LAG TIME

Types of lag.—The lag and acceleration phases correspond to the gradual building up of a steady state. The growth lag (*L*) may be considered a measure of the physiological distance between the initial and the steady state. Depending on the specific conditions and properties of the organism, one or several or a large number of reactions may determine the rate of this building up process. Furthermore each rate-determining reaction may be affected in either or both of two ways: (*a*) change in the amount and activity of the catalyst; (*b*) change in the concentration of the reactants (metabolites):

When the phenomenon is associated with the previous ageing of the cells of the inoculum, the chances are that it involves at once a large number of reactions, and specific interpretations are impossible. Furthermore an apparent lag may be caused if a large fraction of the incoulated cells are not viable (18). When, however, the lag can be shown to be controlled primarily by only one reaction, or system of reactions, the measurement of lag times becomes a useful tool for the study of this reaction. This may often be achieved by a careful preconditioning of the inoculated cells, and appropriate choice of media [see e.g. (26)]. In point of fact this technique amounts to artificially creating conditions where one or a few rate limiting steps become true master reactions, at least during the early stages of the lag.

Theoretically, the lagging of a reaction may be due either to insufficient supply of a metabolite or to the state of inactivity of the enzyme. In the first case, the technique may be used for the study of certain essential metabolites synthesized by the cell itself during growth, and consequently difficult to detect and identify otherwise. Few examples of this sort are available besides the glutamine effects studied by McIlwain *et al.* (27) and the detection of metabolites able to replace carbon dioxide (26), but it is probable that the method could be developed.

In the second case, the technique may be useful in the study of enzyme activation or formation. The magnesium effects described by Lodge & Hinshelwood (28) and the sulfhydryl effects described by Morel (29) should probably be attributed to the reactivation of certain enzymes or group of enzymes. However, lag effects are especially interesting in connection with the study of enzymatic adaptation.

Lag and enzymatic adaptation.—Enzymatic adaptation is defined as the formation of a specific enzyme under the influence of its substrate (30). If cells are transferred into a medium containing, as sole source of an essential nutrient, a compound which was not present in the previous medium, growth will be impossible unless and until an enzyme system capable of handling the new substrate is developed. If other potential factors of lag are eliminated, the determination of lag times becomes a means of studying the adaptive properties of the enzyme system involved (Fig. 7).

The technique has proved especially useful for the study of adaptive enzymes attacking organic compounds serving as sole organic source (11, 31). The work of Pollock (32) shows that it can also be applied in the case of adaptive systems specific for certain hydrogen acceptors (nitrate and tetrathionate). A further development of the technique is suggested by the work of Stanier

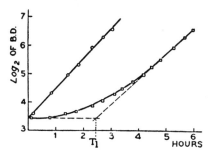

Fig. 7.—Growth of *E. coli* in synthetic medium with glucose (circles) and xylose (squares) as organic source. Culture previously maintained on arabinose medium, temperature 37° C. Growth on glucose proceeds without any lag. Lag time (T_l) on xylose is approximatively 2.5 hours (46).

(33) and Cohen (34) on the possibility of identifying metabolic pathways through a systematic study of cross adaptation.

In general, lag-time measurements may be especially useful in the detection and preliminary identification of adaptive effects, but they could not, of course, replace more direct methods of estimating enzymatic activities.

A broader approach to the problem of relations between lag and enzymatic adaptation should also be considered. As emphasized by Hinshelwood (18), the lag and acceleration phases represent essentially a process of equilibration, the functioning of a regulatory mechanism, by virtue of which a certain enzyme balance inside the cells is attained. That such a mechanism must exist is obvious, since in its absence, the cells could not survive even slight variations of the external environment. However, the nature of the postulated mechanisms is still completely obscure. The kinetic speculations of Hinshelwood, although interesting as empirical formulations of the problem, do not throw any light on

105

the nature of the basic mechanisms involved in the regulation of enzyme formation by the cells.

The most promising hypothesis for the time being appears to be that this regulation is insured through the same mechanism as the formation of adaptive enzymes, which implies the assumption that all the enzymes in a cell are more or less adaptive. The competitive effects observed in enzymatic adaptation (11, 35, 36) agree with the view that the regulation may be the result of a continuous process of selection of mutually interacting enzymes or enzyme-forming systems (30, 37). The kinetics of bacterial growth and, in particular, the lag and acceleration phases certainly constitute the best available material for the study of this fundamental problem.

Division lag.—The largest discrepancies between increase in bacterial density and increase in cell concentration are generally observed during the lag and acceleration phases. This phenomenon has been the subject of much confused discussion (38). Actually, it has been demonstrated by Hershey (39, 40) that a definite lag in cell concentration may occur even when there is no detectable lag in bacterial density. This must mean that cell division mechanisms may be partially inhibited under conditions which do not affect the growth rate and general metabolism of the cell. A number of interesting observations by Hinshelwood *et al.* (18) point to the same conclusion. Further studies on the phenomenon are desirable, as they should throw some light on the factors of cell division in bacteria.

THE INTERPRETATION OF COMPLEX GROWTH CYCLES

Multiple exponential phases.—In many cases, the growth cycle does not conform to the conventional scheme represented in Fig. 1. The interpretation of these complex growth cycles will be briefly discussed here.

One of the most frequently encountered exceptions is the presence of several successive exponential phases, characterized by different values of R and separated by angular transition points. This should in general be interpreted as indicating the addition or removal of one or more rate-determining steps in the steady state system. This type of effect may result from a change in the composition of the medium, for instance from the exhaustion of a compound partially covering an essential nutritional requirement

(34), or from the transitory accumulation of a metabolite, which will eventually serve as a secondary nutritional source (41).

Interpretations are more delicate, and more interesting, when the cause is a change in the composition of the cells themselves. Such effects are frequently encountered with various bacteriostatic agents and have been discussed at length by Hinshelwood (18). But the deliberate confusion entertained by this author between selective and adaptive mechanisms has obscured, rather than clarified, the interpretation of these effects.

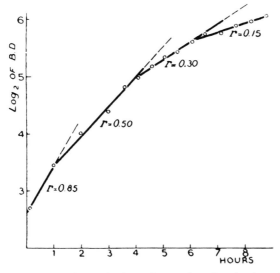

FIG. 8.—Growth of *E. coli* in synthetic medium under suboptimal partial pressure of carbon dioxide (3×10^{-5}). r = growth rate.

In some cases, the phenomenon can be reasonably ascribed to the exhaustion of a reserve metabolite in the cells. An interesting example is afforded by the growth of coli under suboptimal partial pressures of carbon dioxide (26). As seen in Fig. 8 as much as three or four exponential phases can be clearly distinguished suggesting the successive exhaustion of several reserve metabolites, each independently synthesized with the participation of carbon dioxide, a conclusion which is borne out by other lines of evidence.

Diauxie.—This phenomenon is characterized by a double growth cycle consisting of two exponential phases separated by a phase during which the growth rate passes through a minimum, even becoming negative in some cases. It is found to occur in media where the organic source is the limiting factor and is constituted of certain mixtures of two carbohydrates. The evidence indicates that each cycle corresponds to the exclusive utilization of one of the constituents of the mixture, due to an inhibitory effect of one of the compounds on the formation of the enzyme attacking the other (Fig. 9). This striking phenomenon thus reveals the existence

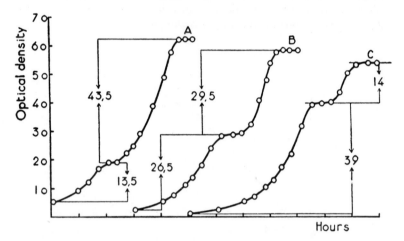

Fig. 9.—Diauxie. Growth of *E. coli* in synthetic medium with glucose+sorbitol as carbon source.

The figures between arrows indicate total growth corresponding to each cycle.
(*a*) Glucose   50 μg. per ml.; sorbitol 150 μg. per ml.
(*b*) Glucose 100 μg. per ml.; sorbitol 100 μg. per ml.
(*c*) Glucose 150 μg. per ml.; sorbitol   50 μg. per ml.
Total growth corresponding to first cycle is proportional to glucose concentration. Total growth of second cycle is proportional to sorbitol concentration (11).

of interactions between closely related compounds in the formation of specific enzymes and has proved valuable in the study of certain aspects of enzymatic adaptations (11, 30, 35). It may perhaps be susceptible of certain technical applications, e.g., for the quantitative analysis of certain mixtures of carbohydrates.

CONCLUDING REMARKS

The time-honored method of looking at a tube, shaking it, and looking again before writing down a + or a 0 in the lab-book has led to many a great discovery. Its gradual replacement by determinations of "turbidity at 16 hours" testifies to technical progress, primarily in the manufacturing and advertising of photelectric instruments. This technique however is not, properly speaking, quantitative, since the quantity measured is not defined. It might be a rate, or a yield, or a combination of both.

In any case, this technique does not take advantage of the fact that the growth of bacterial cultures, despite the immense complexity of the phenomena to which it testifies, generally obeys relatively simple laws, which make it possible to define certain quantitative characteristics of the growth cycle, essentially the three growth constants: total growth (G), exponential growth rate (R), and growth lag (L). That these definitions are not purely arbitrary and do correspond to physiologically distinct elements of the growth cycle is shown by the fact that, under appropriately chosen conditions, the value of any one of the three constants may change widely without the other two being significantly altered. The accuracy, the ease, the reproducibility of bacterial growth constant determinations is remarkable and probably unparallelled, so far as biological quantitative characteristics are concerned.

The general physiological significance of each of the growth constants is clear, provided certain experimental requirements are met. Under certain specific conditions, quantitative interpretations in terms of the primary effect of the agent studied may even be possible. The fallacy of considering certain naive mechanistic schemes, however, as appropriate interpretations of unknown, complex phenomena should be avoided.

There is little doubt that, as further advances are made towards a more integrated picture of cell physiology, the determination of growth constants should and will have a much greater place in the experimental arsenal of microbiology.

LITERATURE CITED

1. DELBRÜCK, M., *Ann. Missouri Botan. Garden*, **32**, 223–33 (1945)
2. LURIA, S. E., *Bact. Revs.*, **11**, 1–40 (1947)
3. HENRICI, A. T., *Morphologic Variation and the Rate of Growth of Bacteria*, 194 pp. (C. C Thomas, Springfield, Ill., 1928)
4. WILSON, G. S., *J. Bact.*, **7**, 405 (1922)
5. KELLY, C. D., AND RAHN, O., *J. Bact.*, **23**, 147 (1932)
6. LODGE, R. M., AND HINSHELWOOD, C. N., *J. Chem. Soc.*, 213–219 (1943)
7. FISHER, K. C., AND ARMSTRONG, F. H., *J. Gen. Physiol.*, **30**, 263 (1947)
8. McILWAIN, H., *Biochem. J.*, **38**, 97–105 (1944)
9. VAN NIEL, C. B., *Bact. Revs.*, **8**, 1–118 (1944)
10. DOGNON, A., in *Techniques de laboratoire*, 197–210 (Masson & Cie, Paris, 1947)

11. MONOD, J., *Recherches sur la croissance des cultures bactériennes*, 211 pp. (Hermann & Cie, Paris, 1942)
12. DUBOS, R. J., *J. Exptl. Med.*, **65**, 873–83 (1937)
13. MONOD, J., *Ann. inst. Pasteur*, **68**, 444 (1942)
14. MONOD, J., *Ann. inst. Pasteur* (In press)
15. JORDAN, R. C., AND JACOBS, S. E., *J. Bact.*, **48**, 579 (1944)
16. LWOFF, A., VAN NIEL, C. B., RYAN, F. J., AND TATUM, E. L., *Cold Spring Harbor Symposia Quant. Biol.*, **11**, 302–3 (1946)
17. BURTON, A. C., *J. Cellular Comp. Physiol.*, **9**, 1 (1936)
18. HINSHELWOOD, C. N., *The Chemical Kinetics of the Bacterial Cell*, 284 pp. (Clarendon Press, Oxford, 1946)
19. JOHNSON, F. H., AND LEWIN, I., *J. Cellular Comp. Physiol.*, **28**, 47 (1946)
20. SCHAEFER, W., *Ann. inst. Pasteur*, **74**, 458–63 (1948)
21. McILWAIN, H., *Biol. Revs.*, **19**, 135 (1944)
22. McILWAIN, H., *Advances in Enzymol.*, **7**, 409–60 (1947)
23. WYSS, O., *Proc. Soc. Exptl. Biol. Med.*, **48**, 122 (1941)
24. KOHN, H. I., AND HARRIS, J. S., *J. Pharmacol. Exptl. Therap.*, **73**, 343 (1941)
25. SCHAEFFER, P., *Compt. rend.*, **228**, 277–79 (1949)
26. LWOFF, A., AND MONOD, J., *Ann. inst. Pasteur*, **73**, 323 (1947)
27. McILWAIN, H., FILDES, P., GLADSTONE, G. P., AND KNIGHT, B. C. J. G., *Biochem. J.*, **33**, 223 (1939)
28. LODGE, R. M., AND HINSHELWOOD, C. N., *J. Chem. Soc.*, 1692–97 (1939)
29. MOREL, M., *Ann. inst. Pasteur*, **67**, 449 (1941)
30. MONOD, J., *Growth*, **11**, 223–89 (1947)
31. MONOD, J., *Ann. inst. Pasteur*, **69**, 179 (1943)
32. POLLOCK, M. R., AND WAINWRIGHT, S. D., *Brit. J. Exptl. Path.*, **29**, 223–40 (1948)
33. STANIER, R. Y., *J. Bact.*, **54**, 339 (1947)
34. COHEN, S. S., *J. Biol. Chem.*, **177**, 607–19 (1949)
35. MONOD, J., *Ann. inst. Pasteur*, **71**, 37 (1945)
36. SPIEGELMAN, S., AND DUNN, R., *J. Gen. Physiol.*, **31**, 153–73 (1947)
37. SPIEGELMAN, S., *Cold Spring Harbor Symposia Quant. Biol.*, **11**, 256–77 (1946)
38. WINSLOW, C. E., AND WALKER, H. H., *Bact. Revs.*, **3**, 147–86 (1939)
39. HERSHEY, A. D., *J. Bact.*, **37**, 290 (1939)
40. HERSHEY, A. D., *Proc. Soc. Exptl. Biol. Med.*, **38**, 127–28 (1938)
41. LWOFF, A., *Cold Spring Harbor Symposia Quant. Biol.*, **11**, 139–55 (1946)
42. LEMON, C. G., *J. Hyg.*, **33**, 495 (1937)
43. TOPLEY, W. W. C., AND WILSON, G. S., *Principles of Bacteriology and Immunity*, 3rd Ed., 2054 pp. (Williams & Wilkins, 1946)
44. LWOFF, A., QUERIDO, A., AND LATASTE, C., *Compt. rend. soc. biol.*, **130**, 1580 (1939)
45. MONOD, J. (Unpublished data)
46. MONOD, J. (Unpublished data)

Editor's Comments on Papers 7, 8, and 9

7   **Schaechter, Maaløe, and Kjeldgaard:** *Dependency on Medium and Temperature of Cell Size and Chemical Composition During Balanced Growth of* Salmonella typhimurium

8   **Harvey, Marr, and Painter:** *Kinetics of Growth of Individual Cells of* Escherichia coli *and* Azotobacter agilis

9   **Bu'lock:** *The Regulation of Secondary Biosynthesis*

The batch culture has wide applications, but it is now used more circumspectly and more incisively. This has revealed new potentialities, as Papers 7 and 8 show, while Bu'lock (Paper 9) considers some of the more traditional problems.

In recent years Maaløe's school in Copenhagen has exploited the exponential phase of batch culture as a condition of "balanced growth" and has used it to address problems of growth in terms relative to the cell. The paper by Schaechter, Maaløe, and Kjeldgaard was an early and very influential landmark of this method [see also Paper 14, and N. O. Kjeldgaard, O. Maaløe, and M. Schaechter, *J. Gen. Microbiol.,* **19,** 607–616 (1958)].

Harvey, Marr, and Painter follow this trend for balanced growth in batch culture, to relate growth in terms of the cell. As a contemporary benchmark, this paper illustrates how the application of mathematics to growth analysis has changed since Müller's and Buchanan's (Paper 2) time; and also indicates how new methods of cell counting and cell sizing by electronic devices have revolutionized the enumeration of cell populations since Wilson's day [G. S. Wilson, *J. Bacteriol.,* **7,** 405–446 (1922)].

The extract from Bu'lock's Squibb Lecture (Paper 9) concerns the contemporary difficulties that surround physiological considerations of growth in batch culture. Bu'lock draws attention to the different qualities associated with development of the cells in a culture and the difficulties of ascribing them to growth. By segregating multiplication from developmental aspects of growth in his "trophophase" and "idiophase" concepts, Bu'lock attempts to reconcile the broad differences in cellular activities, sometimes formerly associated with "proliferating" and "resting" cells. Transient changes, and the heterogeneity of many cultures, are seen to be frustrating obstacles in attempts to elucidate the physiological problems of growth in batch cultures.

Reprinted from *J. Gen. Microbiol.*, **19**(3), 592–606 (1958)

Dependency on Medium and Temperature of Cell Size and Chemical Composition during Balanced Growth of *Salmonella typhimurium*

By M. SCHAECHTER*, O. MAALØE AND N. O. KJELDGAARD

State Serum Institute, Copenhagen, Denmark

SUMMARY: Cell mass, the average number of nuclei/cell and the content of RNA and DNA were studied in *Salmonella typhimurium* during balanced (steady state) growth in different media. These quantities could be described as exponential functions of the growth rates afforded by the various media at a given temperature. The size and chemical composition characteristic of a given medium were not influenced by the temperature of cultivation. Thus, under conditions of balanced growth, this organism exists in one of a large number of possible stable physiological states.

The variations in mass/cell are due to changes in the number of nuclei/cell as well as in mass/nucleus. An increase in the number of ribonucleoprotein particles at higher growth rates could, it appears, largely account for the increase in mass/nucleus. Calculations indicate that the rate of protein synthesis per unit RNA is nearly the same at all growth rates.

It is a classic observation that bacterial cells increase in size during the lag which precedes cell division in a newly-inoculated culture, and become smaller again during the period of declining growth (Henrici, 1928). It is also well known that increase in size and enrichment in ribonucleic acid go hand in hand (Malmgren & Hedén, 1947; Morse & Carter, 1949; Wade, 1952; Gale & Folkes, 1953). Previously, interest has been focused mainly on the striking difference between the small, non-dividing cells of an outgrown culture and the larger forms typical of rapid growth. Hence, cells are often described as 'resting' or 'exponentially growing' and these conditions implicitly considered to be alternative physiological states.

We have studied cells of *Salmonella typhimurium* during unrestricted, balanced growth in a variety of media and at different temperatures. The term 'cell' is used throughout this and the following paper to denote either a colony-forming unit or a microscopically visible rod. In both cases the unit may contain more than one nucleus. The terms 'unrestricted and 'balanced' are defined in the discussion. In each case the growth rate, cell size, and the amounts of ribonucleic acid (RNA) and desoxyribonucleic acid (DNA) and the average number of nuclei/cell were determined.

These experiments show that a large number of physiological states exists, each of which is characterized by a particular size and chemical composition of the cells. *At a given temperature*, average mass, RNA, DNA and number of

* Present address: Department of Microbiology, College of Medicine, University of Florida, Gainesville, Florida, U.S.A.

nuclei/cell can be described as exponential functions of the growth rate. *In a given medium*, cell size and composition are almost independent of the growth temperature. The characteristics of the cells would therefore seem to be determined primarily by the pattern of biochemical activities imposed by the medium.

The figures obtained for mass, RNA and DNA/cell permit estimates to be made of the quantities of protein and nucleic acids synthesized/cell/minute in different media at a given temperature. These calculations suggest that, over a wide range of growth rates, the amount of protein synthesized/minute is roughly proportional to the RNA content of the cell; or that, /unit of RNA, the number of protein molecules synthesized/minute is almost independent of the growth rate.

METHODS

Bacteria. The wild-type strain of *Salmonella typhimurium* used in this work was previously employed in this laboratory (Lark & Maaløe, 1954). The trypto-phan-requiring mutant *try A*-8 of *S. typhimurium*, kindly supplied by Dr M. Demerec, was used in the continuous-culture experiments.

Culture media. The media employed and the growth rates they supported are listed in Table 1. Amino acids and sugars were added after separate steri-lization. All media were adjusted to pH 7·0.

Growth conditions. The organisms were grown in several hundred ml. volumes of medium through which air was constantly bubbled. Balanced growth was maintained by diluting with equal volumes of fresh medium at intervals corresponding to the average generation time. Before sampling, cultures were grown for several hours with frequent checks of the optical density in order to ensure that a constant growth rate had been established. As a rule, the optical density (at 450 mμ, and 1 cm. path) was kept between 0·200 and 0·400. In all media strictly exponential growth is maintained until the optical density reaches 0·800 or more. This optical density corresponds to 1·2–6·2 × 10^8 bacteria/ml. and to *c*. 140 μg./ml. bacterial dry weight in the range of culture media employed.

Continuous culture growth. A continuous culture device using an automatic pipetting machine as feeding pump was employed (Formal, Baron & Spilman, 1956). The culture volume of 600 ml. contained in a cylinder 6 cm. in diameter, was aerated through a fritted glass plate. Efficient stirring was produced by the vigorous aeration; thus an added drop of dye solution became uniformly mixed in the culture liquid well within 1 sec. Excess fluid was continuously removed from the surface by suction. The dilution rate *D* (Monod, 1950) was varied by adjusting both the numbers of strokes delivered by the pipet-ting machine (10–15/min.) and the volume of medium added/stroke (0·5–1·5 ml.). Mutant *try A*-8 was grown with tryptophan as limiting factor in the casamino-acids medium with 1 μg. tryptophan/ml. Feeding was routinely started just after the culture had exhausted the tryptophan present at the time of inoculation. Sampling for the various analyses was carried out after

Table 1. *Culture media employed*

No.	Medium	Concentration	No. of expt.	Average growth rate in doublings/ hr.
1	Brain + heart infusion	Full strength	1	2·80
2	Nutrient broth	Meat extract + 1 % peptone	3	2·75
3	Yeast extract + glucose	Full strength + 0·2 % glucose	2	2·73
4	Placenta broth	Full strength	1	2·70
5	Nutrient broth	Dil. 1:2 with medium no. 14	3	2·60
6	Nutrient broth	Dil. 1:5 with medium no. 14	9	2·40
7	Casamino acids[a]	1·5 % (Difco) + 0·01 % trypto- phan in medium no. 14	2	2·00
8	199 Tissue-culture medium	See[b]	1	1·88
9	20 amino acids	As in medium No. 8 + salt solution[c]	1	1·83
10	Amino acids pool 2[d]	As in medium No. 8 + salt solution[c]	2	1·46
11	Amino acids pool 3[e]	As in medium No. 8 + salt solution[c]	2	1·38
12	Amino acids pool 4[f]	As in medium No. 8 + salt solution[c]	1	1·25
13	Amino acids pool 1[g]	As in medium No. 8 + salt solution[c]	1	1·22
14	Glucose salt (medium K)	0·2 % + Salt solution[c]	9	1·20
15	Succinate salt	0·2 % + Salt solution[c]	2	0·94
16	Lactate salt	0·2 % + Salt solution[c]	2	0·90
17	Dulcitol salt	0·05 % + Salt solution[c]	1	0·83
18	Aspartate salt	0·012 % + Salt solution[c]	1	0·83
19	Methionine salt	0·06 % + Salt solution[c]	1	0·81
20	Histidine salt	0·04 % + Salt solution[c]	1	0·78
21	Threonine salt	0·012 % + Salt solution[c]	1	0·63
22	Lysine salt	0·014 % + Salt solution[c]	1	0·62

(a) This medium, with limiting tryptophan, was employed in the bactostat experiments.

(b) Morgan's medium (Salk, Youngner & Ward, 1954) was employed without anti-biotics, indicator and solutions H–K, I, J, Q, G and P.

(c) Salt solution: $MgSO_4.7H_2O$, 0·1; citric acid, 1·0; $Na_2HPO_4.2H_2O$, 5·0; $Na(NH_4)HPO_4.4H_2O$, 1·74; KCl, 0·74 g./l. Made up as a 50 × concentrate. This solution did not support perceptible growth without the addition of other carbon sources.

(d) Threonine, tyrosine, cysteine, histidine, phenylalanine, isoleucine, hydroxyproline and arginine.

(e) Phenylalanine, isoleucine, hydroxyproline, arginine, leucine, aspartic acid, glycine and tryptophan.

(f) Leucine, aspartic acid, glycine, tryptophan, glutamic acid, alanine, serine and valine.

(g) Glutamic acid, alanine, serine, valine, glutamine, lysine, methionine and proline.

not less than 6 hr. of growth. During this period the optical density remained practically constant at about 0·400.

Mass determination. The values of mass/cell are expressed as the optical density at 450 mμ (1 cm. path) given by a suspension containing 10^7 cells/ ml. The optical density was found to be proportional to the dry weight, irrespective of the cell size. Optical density 0·100 corresponds to 17–18 μg. dry weight/ml.

Plate counts. Samples of the cultures were diluted in steps representing a total dilution of 2 or 4×10^{-4}. The original sampling was done with a 0·025 or 0·050 ml. constriction pipette and the subsequent steps were carried out with

0·1 ml. serological pipettes. For each value of Fig. 1 plating was done from at least six individual dilutions performed within 10 min. and adjusted to give between 300 and 600 colonies/plate. The viable counts were fitted to the growth curve determined by the optical-density measurements. It was found that the number of viable cells/ml. could be measured with an error of less than 10 %.

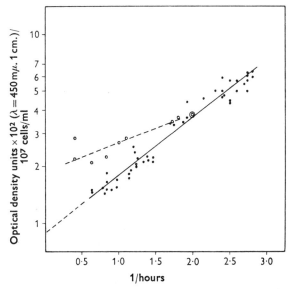

Fig. 1. Dependency of cell mass on growth rate at 37°. From the optical-density (mass) measurements and the viable counts in the different media, values for optical density/10^7 cells/ml. were calculated. The logarithm of these values is plotted against the growth rate expressed as doublings/hr. (●). The stippled line corresponding to the open rings and to the double-ringed point represents results from continuous culture experiments plotted against the dilution rate (○).

Chemical analysis. For nucleic-acid determinations, 40 ml. samples were frozen quickly in a solid CO_2+ethanol bath. They were thawed and centrifuged in a cooled Servall angle centrifuge at 12,000 rev./min. for 20 min., the sediments were resuspended in 2·5 ml. of cold saline and 0·1 ml. of 70 % perchloric acid was added to 2 ml. of this suspension. The material was heated to 70° for 30 min., centrifuged and the supernate collected for colorimetric sugar tests. Deoxyribose was determined by the procedure of Burton (1956) on 1·0 ml. of the acid extract. Ribose determinations following the method of Kerr & Seraidarian (1945) were performed on 0·1 ml. of the extract. All spectrophotometry was done with a Zeiss Model PMQ II spectrophotometer employing 1 cm. cuvettes. Most of the values presented in Figs. 2 and 3 are averages of four independent determinations.

Nuclear staining. Fixation with OsO_4 and staining with thionine were carried out according to Lark, Maaløe & Rostock (1955), except that acid hydrolysis was extended to 6 min. This procedure is not primarily intended to preserve fine structural detail but, in the organism used, it reveals the same

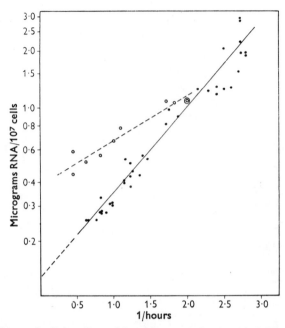

Fig. 2. Dependency of cellular ribonucleic acid on growth rate at 37°. The RNA content of the cultures was calculated from the ribose determinations. (μg. RNA = μg ribose × 4·91). The logarithm of the RNA values (micrograms)/10⁷ viable cells is plotted against the growth rate (●). The stippled line corresponding to the open rings and to the double-ringed point represents results from continuous culture experiments, plotted against the dilution rate (○).

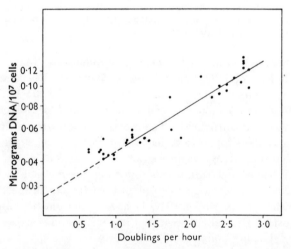

Fig. 3. Dependency of cellular deoxyribonucleic acid on growth rate at 37°. The DNA content of the cultures was calculated from the deoxyribose determinations (μg. DNA = μg. deoxyribose × 2·44). The logarithm of the DNA values μg./10⁷ viable cells is plotted against the growth rate.

number of individual staining bodies in each cell as that obtained by other methods. For reasons of ease of observation the stained preparations were examined under phase contrast microscope (Zeiss). Between 300 and 400 cells per sample were scored as one-, two-, or four-nucleated taking into consideration, when possible, that some of the rods consisted of sister cells in different degrees of separation. A subjective criterion had to be employed in order to score a cell containing two adjacent bodies as one- or two-nucleated. However, repeated counts of the same preparation after an interval of months, or duplicate counts by different observers, always gave results compatible with the sampling error.

RESULTS

Balanced growth was maintained at 37° in the different media listed above (Table 1) and samples analysed for mass (optical density) RNA, DNA, viable counts and number of nuclei/cell. The results are presented in Figs. 1–3 in which logarithms of mass, RNA and DNA per viable cell are plotted against

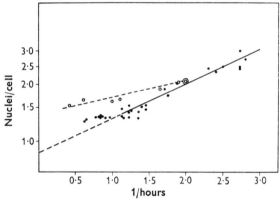

Fig. 4. Dependency of the number of nuclei/cell on growth rate at 37°. The average number of nuclei/cell was calculated from direct counts on stained preparations. The logarithm of the values is plotted against the corresponding growth rate (●). The stippled line corresponding to the open rings and the double-ringed point represents results from continuous culture experiments, plotted against the dilution rate (o).

the growth rate. Figure 4 represents a similar plot for nuclei/cell. Regression analysis showed that the straight lines drawn in Figs. 1–4 adequately represent the observed values. The individual determinations deviate only slightly more from the lines than is to be expected when the combined errors of the chemical analysis and the viable counts are taken into account. For ease of comparison, the increments on the logarithmic scale are the same in all figures.

In Fig. 1, the unbroken line fitted to the solid points shows that cell mass just about doubles/unit increase in doublings/hr. The extrapolation to zero growth rate suggests that the minimal bacterial size would be about half that of a cell growing in glucose-salt medium. This minimal size was actually attained under conditions of nitrogen starvation (see following paper).

Dr K. F. A. Ross kindly carried out measurements on the thickness of formalin-fixed organisms by interference microscopy using a modification of the technique he developed in 1955 (Ross, 1957). He obtained the following figures for the average of ten determinations per sample: $1 \cdot 43 \mu$ for a culture growing in medium no. 2 at $2 \cdot 73$ doublings/hr. (d./hr.); $1 \cdot 22 \mu$ for $1 \cdot 85$ d./hr. (medium no. 7); $0 \cdot 93 \mu$ for $1 \cdot 00$ d./hr. (medium no. 14) and $0 \cdot 87 \mu$ for $0 \cdot 61$ d./hr. (medium no. 22). The cell volumes obtained from these figures and from estimates of cell lengths were found to be proportional to the cell mass, as estimated from optical density measurements.

Figs. 2 and 3 show that, per unit increase in doublings/hr., the amount of RNA/cell increases by $\times 2 \cdot 85$ that of DNA/cell by $\times 1 \cdot 73$. The corresponding factor for the average number of nuclei/cell, derived from Fig. 4, is $\times 1 \cdot 55$. At $37°$, the four parameters examined may thus be described as exponential functions of the growth rate and can be arranged as follows, with regard to the slopes of the semilogarithmic plots: RNA > mass > DNA $\geqslant$ nuclei/cell.

It is to be understood that all 'per cell values' are based, either on viable counts, or, in the case of nuclei, on cytological observations. Simultaneous haemocytometer and colony counts repeatedly showed that, under conditions of balanced growth, viable and total counts did not differ significantly. Counting unstained cells in the phase contrast microscope showed that, in all media, approximately 75 % of the units appeared to be true single cells, the remaining 25 % were 'doublets' representing incompletely separated sister cells. In Figs. 1–3, an approximate correction for doublets can be made by multiplying the ordinate values by $0 \cdot 8$.

The figures for nuclei/cell (Fig. 4) are based on direct counts of stained preparations. It was not always possible after staining to distinguish between single elements and doublets, particularly when the cells are small. Thus, at the lowest growth rates (smallest cells) uninucleated sister cells which remained attached to one another were probably often scored as binucleates. When the number of nuclei/cell is close to unity a bias of this kind may increase the observed value significantly over the true one. This might be the reason why, in Fig. 4, the experimental points deviate from the straight line at low rates, and it may also explain that the values for DNA/nucleus, which at higher growth rates are found to vary in a random manner around the mean value, decrease slightly, but significantly, in this region.

A number of experiments were carried out with cultures grown at lower temperatures. In Table 2 results obtained at $37°$ and at $25°$ with five different media are compared. In all cases, the growth rate at $25°$ was about half that at $37°$; nevertheless, mass, RNA, DNA and number of nuclei/cell remained nearly constant for a given medium. In broth or in the amino-acid medium the figures for nuclei/cell are somewhat higher at $37°$ than at $25°$, but the experiments at $25°$ are few and the observed difference is probably not significant. Moreover, extensive chemical and cytological studies previously carried out with the same organism showed that, in broth, identical values for DNA and for nuclei/cell are obtained at $25°$ and at $37°$ (Lark & Maaløe, 1956; Lark, Maaløe & Rostock, 1955).

In fact, our data suggest that more extensive analyses of 25° cultures would permit graphs to be constructed which would be identical with those of Figs. 1–4 if the growth rate values on the abscissa were reduced to half. Thus, within the temperature range studied, *the size and chemical composition of the cells are related to the growth rate only in so far as it depends on the medium.*

Table 2. *The effect of temperature on cell size and composition*

Medium	No. of expts.	Growth temp. (° C.)	Doublings/ hr.	Optical density* mass	RNA*	DNA*	Nuclei/ cell
No. 6 (broth)	2	25	1·06	5·80	1·64	0·130	2·85
	4	37	2·40	5·00	1·44	0·095	2·50
No. 9 (amino acids)	1	25	0·88	3·66	0·92	0·085	2·05
	2	37	1·83	3·76	0·97	0·056	1·74
No. 14 (glucose)	3	25	0·65	2·32	0·56	0·065	1·46
	5	37	1·20	1·92	0·44	0·048	1·38
No. 15 (succinate)	2	25	0·48	1·47	0·39	0·038	1·31
	2	37	0·93	1·60	0·39	0·042	1·33
No. 16 (lactate)	2	25	0·50	1·50	0·39	0·038	1·30
	2	37	0·90	1·61	0·39	0·039	1·35

* The units for mass, RNA and DNA per cell are the same as in Figs. 1–3, respectively.

The results of the analysis of mass, RNA and nuclei/cell from the continuous culture experiments are presented as open circles in Figs. 1, 2 and 4. The logarithms of cell contents are plotted against the dilution rate D expressed in doublings of culture volume/hr. Monod (1950) and Novick & Szilard (1950) have shown that, in the ideal continuous culture system, D is related to the growth rate μ (in doublings/hr.) by the equation $\mu = D/\ln 2$. As will be seen, this situation was not obtained throughout our experiments. In Figs. 1, 2, and 4 the points marked with a double ring correspond to the maximum rate ($\mu = 2 \cdot 0$) attainable during unrestricted growth of strain *try A*-8 in the Casamino acid medium with excess tryptophan. Thus, for values of D higher than 1·38, corresponding to $\mu = 2 \cdot 0$, the theoretical growth rate calculated from the above formula exceeds the maximum values for unrestricted growth. D values as high as 1·77, corresponding to a theoretical growth rate of 2·6, have been obtained in our experiments. A similar discrepancy between the calculated and the maximum growth rates was reported by Herbert, Elsworth & Telling (1956) and Powell (1956) who attributed it to imperfect mixing.

It will be seen from Figs. 1, 2 and 4, that the values for mass, RNA and nuclei/cell for organisms grown in continuous culture may also be described as exponential functions of the growth rate. The slopes of the stippled lines drawn in Figs. 1, 2, and 4 are considerably less than those obtained for unrestricted growth, indicating that the lower the dilution rate, the more do the values exceed those from batch cultures. The same trend was observed for DNA, but the data are not included in Fig. 3 because of considerable scatter in the values obtained.

DISCUSSION

Our discussion falls into three parts: (1) An account of the physiological states imposed by the medium under unrestricted and restricted, balanced growth. (2) A representation of related and otherwise relevant data from the literature. (3) An analysis of our findings in terms of the major synthetic activities of the bacterial cell.

(1) In liquid cultures of low bacterial concentration, cells continue to grow for a long time in a virtually unchanging environment. This ideal condition leads to the establishment of a steady state of balanced growth, which can be prolonged at will by appropriate dilutions of the culture. In Campbell's apt formulation (Campbell, 1957), growth is said to be *balanced* over a time interval if, during the interval, every 'extensive' property of the system increases by the same factor. Failure to maintain balanced growth throughout an experiment makes it impossible to relate any measured quantity to the growth rate in a direct way.

When the density of a culture approaches saturation the growth rate gradually decreases and it increases upon subsequent dilution with fresh medium. In either case the changes in rates of cell division and of mass synthesis do not run parallel; as a rule, the rate of cell division remains unchanged for some time after the rate of mass synthesis has been lowered or raised in response to a change in the environment. This general phenomenon, which will be analysed in more detail in the next paper (Kjeldgaard, Maaløe & Schaechter, 1958), accounts for the well-known fact that 'exponentially growing cells' are bigger than 'resting cells'. These common, descriptive terms are sometimes taken to mean that only two physiological cell types exist: a small, resting cell and a larger, exponentially growing cell. This is, however, too simple a picture. Under conditions of *balanced growth*, one of a large number of possible physiological states is established. These states are characterized by the size and chemical composition of the cell; they depend on the culture medium, are not grossly influenced by the temperature of cultivation and can probably be maintained as long as the genetic stability of the culture permits. *At a given temperature*, size and composition are found to depend in a simple manner on the growth rate afforded by the medium. This implies that media which give identical growth rates produce identical physiological states, regardless of the actual constituents of the media. The 'resting state' finds a natural place in this system since, in an outgrown broth culture, the size of the cells is reduced to approximately the value expected for zero growth rate (see Fig. 1).

We have so far considered only batch cultures where all the relevant nutrients are present in excess in the medium. Because the growth rate is limited by the *type* of nutrients and not by their *concentration*, we refer to this situation as 'unrestricted growth'. We assume that the growth rate observed under these conditions is the highest which can be attained with the set of nutrients available to the cell. In a continuous culture device of the type we employed, growth is 'restricted' by the rate at which, say, a required amino acid is added, i.e. by the low extra- and intracellular concentration of that component. Under conditions of unrestricted growth the concentration inside the cell of, say, an amino acid might similarly be thought of as being rate-limiting. This need not be the case, however, since, as discussed later, the growth rate may be controlled at the level of protein synthesis without involving limiting intracellular concentrations of amino acids, etc.

In our continuous culture experiments, new medium was added to the culture in pulses each of which momentarily increased the tryptophan concentration by 1–3 μg/l. According to Novick & Szilard (1950) a constant tryptophan concentration of about 3 μg/l. permits growth of *Escherichia coli*, strain B/1, *t* at maximum rate; but below this concentration the growth rate rapidly decreases. Assuming a similar concentration dependency for strain *tryA*-8 of *Salmonella typhimurium*, every pulse of new medium will create conditions for growth at a relatively high rate for a short period. It is thus possible that, in our system, growth is intermittent, and that under such conditions, the growth rate during the pulse of growth afforded by a pulse of new medium, is not related to the dilution rate *D* in the simple manner proposed by Monod (1950) and Novick & Szilard (1950). We could therefore assume that the rate of synthesis during each pulse, rather than the overall rate, determines the size and chemical composition of the cell. However, experiments under more nearly ideal conditions of continuous growth are needed before the discrepancy between the results obtained under conditions of restricted and unrestricted growth can be properly analysed.

(2) Attempts to determine concurrently cell mass, nucleic acid content and the number of nuclei as functions of the growth rate have not previously been reported. Several pertinent studies exist in which one or more of these properties were related to the rate of growth. Thus, Wade (1952) obtained results which suggested a linear relation between RNA phosphorus/mg. N and the growth rate of *Escherichia coli*. Our data (Figs. 1, 2) indicate that this relation may actually be exponential; however, over the range studied by Wade and ourselves (0·6–2·8 generations/hr.), the increase in total RNA/unit mass (unlike RNA/cell) is at most twofold, and a clear distinction between linear and exponential functions therefore cannot be made. On the other hand, the overall increase is significantly less in Wade's experiments than in ours (50–60 % and 100 %, respectively). This difference may be due to the use by Wade of very large inocula (initial culture density: $2\cdot5 \times 10^8$ to 10^9 organisms/ ml.). According to our experience the cells of such a dense culture would not reach, or maintain, the size and RNA concentration characteristic of balanced growth. Wade's careful measurements show that the high growth rate could not always be maintained throughout the experiment; the RNA concentrations, which were related to the *initial* growth rates, may therefore be low compared to the values which would have been obtained had growth been truly balanced.

Caldwell & Hinshelwood (1950) and Caldwell, Mackor & Hinshelwood (1950) determined the RNA concentration (in mg. RNA-phosphorus/mg. N) in different strains of *Aerobacter aerogenes* under various conditions. From the published data it cannot be ascertained whether balanced growth was attained, but their experiments also show that the faster the growth the higher the RNA concentration. In Aerobacter, the observed range of concentration is very high (three- to fourfold) perhaps because the lower growth rates were obtained by adding drugs to the medium or by using selected, slow-growing mutants.

Caldwell & Hinshelwood (1950) also determined the DNA phosphorus in various cultures of *Aerobacter aerogenes*. With increasing values for total nitrogen/cell, a slight, but probably significant rise in DNA/cell was observed. No comparison between these figures and the growth rates can be derived from their data.

Perret (1958), studying *Escherichia coli* strain K12, measured the mean cell length and the number of intracellular structures which appeared as transverse light bands under the phase microscope (probably equivalent to the nuclei described here). He also found that, to the different growth rates obtained in different media and in continuous cultures correspond definite values of cell length and number of 'nuclei'. His figures for nuclei/cell correspond closely to those of Fig. 4.

(3) The existence of a variety of stable physiological states cannot be explained in a simple manner. What follows is an attempt to analyse the significant features of our experimental findings in terms of the major synthetic activities of the cells.

Cytological evidence strongly suggests that the stained bodies referred to as 'nuclei' contain DNA, and we shall assume that most, if not all, the DNA of the cell is located there. A comparison of Figs. 3 and 4 shows that the amount of DNA per stained body is nearly constant, as would be expected if each body represents a nucleus in the physiological and genetic sense. With this in mind it seems opportune to differentiate between variation on the cellular and on the nuclear level. Since at low growth rates the majority of cells are uninucleated it seems natural to consider a single nucleus plus its corresponding cytoplasm, cell wall, etc., as an elementary unit. Multinucleated cells, composed of two or more such units can be thought of as syncytial. We will therefore distinguish between changes in the number of elementary units/cell and in mass/elementary unit. The way in which changes in the number of elementary units/cell come about is treated in the second paper of this series.

The variation in mass per elementary unit will be analysed below in accordance with the following assumptions: (*a*) that a large fraction of the cell's RNA is in the form of ribonucleoprotein particles consisting of about equal parts of RNA and protein and with molecular weight about one million (Schachman, Pardee & Stanier, 1952; Tissières & Watson, 1958); (*b*) that, per nucleus, the cell contains fixed amounts of DNA, cell wall and cell membrane material, varying amounts of free RNA and particles and, finally, a pool of soluble protein and other compounds; and (*c*) that this pool is always made up largely of protein.

The first assumption is supported by data from Wade & Morgan (1957). Comparing resting cells of *Escherichia coli* with cells growing in a complex medium these authors found that the RNA-pentose of particles sedimenting completely in 4 hr. at 100,000 *g* amounted respectively to about 50 and 75 % of the total RNA-pentose. The second assumption is based on estimates of cell diameters and lengths (cf. p. 598) from which it can be estimated that the surface area/nucleus is virtually constant. The actual weight ascribed to cell wall and membrane (see Table 3) is of little importance for our conclusions. The third assumption seems more gratuitous; however, the greater part of the protein of the cell must be in the pool, and since this never exceeds about 60 % of the cell weight (Table 3, column 7) it is fair to assume that the bulk of the pool is protein (free enzymes).

Table 3. *Relative rates of protein synthesis*

1	2	3	4	5	6	7	8 Pool + particle protein synthesized/min.*	9
Doublings/ hr.	Dry weight/ cell*	Nuclei/ cell	Dry weight/ nucleus*	RNA/ nucleus*	Particles/ nucleus†	Pool material/ nucleus*‡	Per nucleus	Per particle (× 10⁴)
0·6	240	1·25§	192	19	11,300	100	1·1	0·98
		1·10	218	22	13,300	120	1·3	1·0
1·2	360	1·45	250	31	22,400	135	3·1	1·4
2·4	840	2·40	350	65	54,000	176	8·8	1·6
2·8	1090	2·90	376	84	81,000	160	10·6	1·3

* All weights in g. × 10¹⁵.

† Calculated on the assumption that, from top to bottom, 50, 60, 70 and 80 % of the RNA is in particles of molecular weight one million, and composed of equal parts of protein and RNA.

‡ Per nucleus, the cell is assumed to consist of: 65 × 10⁻¹⁵ g. of wall, membrane and nuclear material, i.e. about one-third by weight of the smallest cell type; (*b*) particles and free RNA; and (*c*) a pool mainly containing soluble protein.

§ Two values given: above, the one directly observed; below, that obtained by extrapolating the linear part of the curve of Fig. 4 (see p. 597).

Table 3 shows representative figures for mass, average number of nuclei/ cell and mass and RNA/nucleus, taken from Figs. 1, 2, and 4 (columns 2, 3, 4, 5). The right-hand side of the Table contains calculations based on these values and on the assumptions listed above. The number of particles and the pool size per nucleus are presented in columns 6 and 7. The relatively small changes in pool size show that variation in *mass/nucleus is due mainly to variation in number of particles/nucleus*.

In Table 3, column 8 presents maximum values for the synthesis of protein/ nucleus/min. (assuming the pool to be all protein and adding to that the particle protein). It is apparent that the *rate* of protein synthesis is directly proportional to the *amount* of RNA, or the number of particles, and that both increase in rough proportion to the growth rate. Small variations in the soluble protein fraction of the pool or in the particle fraction of RNA do not seriously affect these trends. The rate of protein synthesis/unit RNA, or per particle, thus is nearly independent of the growth rate (column 9). This conclusion is of considerable interest because recent biochemical evidence indicates that these particles, which are analogous to the 'microsomal particles' of animal cells, are directly concerned with protein synthesis (see, for example, review by Crick, 1958).

The constancy of the rate of protein synthesis/particle is most readily interpreted on the simple assumption that all particles participate equally in protein synthesis. This may, however, not be the case. By studying externally induced synthesis of β-galactosidase, it has been possible to show that the enzyme-forming system remains intact during growth in the absence of an inducer (Monod, Pappenheimer & Cohen-Bazire, 1952). If we tentatively identify the system producing β-galactosidase with one class of particles, we have a case where particles which continue to be reproduced remain virtually inactive unless an inducer is present. Addition to a culture of, say, an amino acid may cause repression of enzymes concerned with the synthesis of the added compound (see, for example, Vogel, 1957), and it is sometimes taken for granted that the process of repression also involves a reduction of the size of the enzyme-forming systems. If this is not true, as suggested by the results of Monod *et al.* (1952), cells growing in a complex medium would contain a certain fraction of inactive particles (corresponding to the repressed enzymes), whereas cells grown in minimal medium would contain predominantly active ones (little or no repression).

Without knowing what fraction of the particles is actively synthesizing, the true rate of protein synthesis/particle cannot be estimated. Despite this uncertainty it is clear that, unless a majority of the particles are rendered inactive during growth in complex media, the rate of protein synthesis per particle increases much less than does the growth rate.

It is attractive to imagine that the system responsible for a particularly complex process like protein synthesis perhaps functions with nearly the same efficiency under very different growth conditions (with the reservation, of course, that certain enzyme-forming systems may have *their* function specifically repressed). Addition to a culture growing in minimal medium of compounds like amino acids, purines or pyrimidines increases the growth rate presumably by relieving the cells of the necessity for synthesizing the added compounds. If one assumes that the economy of cell growth is actually based on maintaining a high efficiency of protein synthesis it is evident that an increase in growth rate is possible only if the protein synthesizing system of the individual cell is expanded; i.e. if the number of particles/nucleus is increased. For this to happen, the addition of, say, amino acids to the medium must cause a definite increase in the rate of RNA synthesis which, in turn, brings about the observed, *but smaller* increase in growth rate.

In the next paper we shall see that, in agreement with this hypothesis, the initial effect of adding amino acids or broth to a minimal medium culture is to stimulate RNA synthesis preferentially.

We wish hereby to acknowledge the expert technical assistance of Mr O. Rostock. One of the authors (M.S.) was aided by a post-doctoral fellowship grant from the American Cancer Society and one (N.O.K.) by grants from the Danish National Science Foundation and the Lilly Research Foundation.

REFERENCES

BURTON, K. (1956). A study of the conditions and mechanism of the diphenylamine reaction for the colorimetric estimation of deoxyribonucleic acid. *Biochem. J.* 62, 315.

CALDWELL, P. C. & HINSHELWOOD, C. (1950). The nucleic acid content of *Bact. lactis aerogenes. J. chem. Soc.* 1415.

CALDWELL, P. C., MACKOR, E. L. & HINSHELWOOD, C. (1950). The ribonucleic acid content and cell growth of *Bact. lactis aerogenes. J. chem. Soc.* 3151.

CAMPBELL, A. (1957). Synchronization of cell division. *Bact. Rev.* 21, 263.

CRICK, F. H. C. (1958). On protein synthesis. *Symp. Soc. expl. Biol. N.Y.* (in the Press).

FORMAL, S. B., BARON, L. S. & SPILMAN, W. (1956). The virulence and immunogenicity of *Salmonella typhosa* grown in continuous culture. *J. Bact.* 72, 168.

GALE, E. F. & FOLKES, J. P. (1953). The assimilation of amino-acids by bacteria. 14. Nucleic acid and protein synthesis in *Staphylococcus aureus. Biochem. J.* 53, 483.

HENRICI, A. T. (1928). Morphologic variation and the rate of growth of bacteria. *Microbiology Monographs.* London: Baillière, Tindall and Cox.

HERBERT, D., ELSWORTH, R. & TELLING, R. C. (1956). The continuous culture of bacteria: a theoretical and experimental study. *J. gen. Microbiol.* 14, 601.

KERR, S. E. & SERAIDARIAN, K. (1945). The separation of purine nucleosides from free purines and the determination of the purines and ribose in these fractions. *J. biol. Chem.* 159, 211.

KJELDGAARD, N. O., MAALØE, O. SCHAECHTER, M. (1958). The transition between different physiological states during balanced growth of *Salmonella typhimurium. J. gen. Microbiol.* 19, 607.

LARK, K. G. & MAALØE, O. (1954). The induction of cellular and nuclear division in *Salmonella typhimurium* by means of temperature shifts. *Biochim. biophys. Acta,* 15, 345.

LARK, K. G. & MAALØE, O. (1956). Nucleic acid synthesis and the division cycle of *Salmonella typhimurium. Biochim. biophys. Acta,* 21, 448.

LARK, K. G., MAALØE, O. & ROSTOCK, O. (1955). Cytological studies of nuclear division in *Salmonella typhimurium. J. gen. Microbiol.* 13, 318.

MALMGREN, B. & HEDÉN, C. (1947). Studies of the nucleotide metabolism of bacteria. III. The nucleotide metabolism of Gram negative bacteria. *Acta path. microbiol, scand.* 24, 448.

MONOD, J. (1950). La technique de culture continue, théorie et applications. *Ann. Inst. Pasteur,* 79, 390.

MONOD, J., PAPPENHEIMER, A. M. & COHEN-BAZIRE, G. (1952). La cinétique de la biosynthèse de la β-galactosidase chez *E. coli* considérée comme fonction de la croissance. *Biochim. biophys. Acta,* 9, 648.

MORSE, M. L. & CARTER, C. E. (1949). The synthesis of nucleic acids in cultures of *Escherichia coli* strains B and B/r. *J. Bact.* 58, 317.

NOVICK, A. & SZILARD, L. (1950). Experiments with the chemostat on spontaneous mutations of bacteria. *Proc. nat. Acad. Sci.* 36, 708.

PERRET, C. J. (1958). The effect of growth-rate on the anatomy of *Escherichia coli. J. gen. Microbiol.* 18, vii.

POWELL, E. O. (1956). Growth rate and generation time of bacteria, with special reference to continuous culture. *J. gen. Microbiol.* 15, 492.

ROSS, K. F. A. (1957). The size of living bacteria. *Quart. J. micr. Sci.* 98, 435.

SALK, J. E., YOUNGNER, J. S. & WARD, E. N. (1954). Use of color change of phenol red as the indicator in titrating poliomyelitis virus or its antibody in a tissue-culture system. *Amer. J. Hyg.* 50, 214.

SCHACHMAN, H. K., PARDEE, A. B. & STANIER, R. Y. (1952). Studies on the macromolecular organization of microbial cells. *Arch. Biochem. Biophys.* 38, 245.

TISSIÈRES, A. & WATSON, J. D. (1958). Ribonucleoprotein particles from *Escherichia coli. Nature, Lond.* **182,** 778.

VOGEL, H. J. (1957). Repression and induction as control mechanisms of enzyme biogenesis: the 'adaptive' formation of acetylornithase. In *Chemical Basis of Heredity,* p. 276. Baltimore: The Johns Hopkins Press.

WADE, H. E. (1952). Variation in the phosphorus content of *Escherichia coli* during cultivation. *J. gen. Microbiol.* **7,** 24.

WADE, H. E. & MORGAN, D. M. (1957). The nature of the fluctuating ribonucleic acid in *Escherichia coli. Biochem. J.* **65,** 321.

(*Received* 19 *June* 1958)

8

Reprinted from *J. Bacteriol.*, **93**(2), 605–617 (1967)

Kinetics of Growth of Individual Cells of *Escherichia coli* and *Azotobacter agilis*

R. J. HARVEY, ALLEN G. MARR, AND P. R. PAINTER

Department of Bacteriology, University of California, Davis, California

Received for publication 19 September 1966

ABSTRACT

Escherichia coli and *Azotobacter agilis* were grown in minimal media until a steady state was established. The distribution of cell size was determined electronically. From the equation of Collins and Richmond, the growth rate of individual cells was computed as a function of size. The main features of the growth of individual *E. coli* and *A. agilis* cells revealed by this work were: the specific growth rate decreased at the time of division, and both the absolute and specific growth rates increased between divisions. The frequency function of interdivision times was computed and was found to be positively skewed with a coefficient of variation of approximately 0.3. The results supported the hypothesis of Koch and Schaechter that the size of an individual cell at division is highly regulated.

The kinetics of growth of a bacterial population do not establish the kinetics of growth of the individual. In exponential balanced growth, all extensive properties of the population increase by the same factor during any given time interval (3). However, all extensive properties of an individual cell in this population need not increase in proportion during any arbitrary time interval. It is necessary that the mean value of all extensive properties of dividing cells should be twice the mean value for newly formed cells. The rate of increase in volume of the individual cell can be any function of time which permits the volume to double at the end of the interdivision period.

The method most commonly used for determination of the kinetics of growth of individual bacteria has been direct microscopic measurement of length or volume of cells in microculture as a function of time. Using this method, Adolph and Bayne-Jones (1) found the volume of *Bacillus megaterium* to be an exponential function of time. Growth of cells of *Escherichia coli* was approximately exponential, but the specific growth rate of an individual cell showed random fluctuations in time (2). A similar result was obtained with *B. cereus* by Knaysi (13). The rate of increase of volume of *Streptococcus faecalis* was found to decrease between divisions (14, 16). Using interference microscopy, Mitchison (16) found that the rate of increase in dry weight of *S. faecalis* showed a similar decrease.

In these microscopic measurements, the conditions in the microculture may not have permitted balanced growth. In many experiments it is obvious that growth was not balanced. In the more recent work of Schaechter et al. (20) and of Errington, Powell, and Thompson (6), the state of unrestricted, balanced growth in microculture has been verified. Errington et al. (6) found that the lengths of cells of *Pseudomonas aeruginosa*, *Serratia marcescens*, and *Proteus morganii* were not significantly different from an exponential function of time; however, several significant deviations from exponential increase were observed (6). Schaechter et al. (20) found that the increase in length of *E. coli* and *Salmonella typhimurium* was approximately exponential, but they recognized that microscopic measurement of bacterial cells is subject to a random error of sufficient magnitude to preclude distinguishing linear growth from exponential growth.

The results of direct microscopic measurements can be summarized by stating that the size of individual cells of most bacteria increases continously between divisions, but the exact form of the increase has not been determined.

An alternative to direct microscopic measurement of growth of single cells was developed by Collins and Richmond (4), who demonstrated that the kinetics of growth of individual cells determine the form of the size distribution of the cell population. A more rigorous development of this relationship is given in Appendix A to this paper. From a measurement of the distribution of size of individual cells in the population, the rate of growth of these cells as a function of size may be calculated. The method avoids the limitaions of the method of direct microscopic measurement.

This paper describes the kinetics of growth of *E. coli* and *Azotobacter agilis* determined from electronic measurement of the distribution of volumes of cells in unrestricted balanced growth.

MATERIALS AND METHODS

Symbols. Throughout this paper, the following symbols are used: v, volume of a cell; k, specific growth rate of the population; N, total number of cells in the population; $\lambda(v)$, frequency function of volume of extant cells; $\phi(v)$, frequency function of volume of cells which are dividing; $\psi(v)$, frequency function of volume of cells newly formed by division; $g(v)$, frequency function of volume at division of newly formed cells; $f(\tau)$, frequency function of life length, τ, of newly formed cells; $V(v)$, absolute growth rate of a cell of volume v.

Growth of bacteria. E. coli strain ML30 was grown in glucose-salts medium containing (in grams per liter): glucose, 1.0; NH_4Cl, 0.8; KH_2PO_4, 4.2; K_2HPO_4, 8.5; $MgCl_2 \cdot 6H_2O$, 0.2; Na_2SO_4, 0.2; KCl, 1.0; $CaCl_2$, 0.01; $FeSO_4 \cdot 7H_2O$, 0.0005.

E. coli B/r was grown in glucose-salts medium containing (in grams per liter): glucose, 2.0; NH_4Cl, 2.0; Na_2HPO_4, 6.0; KH_2PO_4, 3.0; NaCl, 3.0; $MgSO_4$, 0.013; Na_2SO_4, 0.011; $CaCl_2$, 0.002; $FeSO_4 \cdot 7H_2O$, 0.0005; $NaMoO_4 \cdot 2H_2O$, 0.000025.

A. agilis (*A. vinelandii* strain 0) was grown in Burks nitrogen-free medium containing (in grams per liter): sucrose, 3.5; KH_2PO_4, 0.41; K_2HPO_4, 0.52; $CaCl_2$, 0.02; $MgCl_2 \cdot 6H_2O$, 0.17; $Na_2MoO_4 \cdot 2H_2O$, 0.00025; $FeSO_4 \cdot 7H_2O$, 0.005; Na_2SO_4, 0.15.

Cultures were grown at 30 C in tubes (30 by 300 mm) containing 100 ml of medium and were aerated by sparging.

Measurement of growth. Periodically the cultures were sampled for measurement of optical density and cell numbers. Optical density was measured at 600 mμ by use of a Beckman model DU spectrophotometer with 1-cm absorption cells. Cell numbers were counted electronically by use of the apparatus previously described (8).

The specific growth rate, k, in hours^{-1} was computed by linear regression according to the equation:

$$\log_e x = \log_e x_0 + kt$$

where x is the optical density or the number of cells per milliliter at time t, and x_0 is the optical density or number of cells per milliliter at time zero.

Measurement of size distributions. Size distributions were measured with the apparatus described by Harvey and Marr (8). Pulses produced by the passage of cells through the Coulter transducer (W. H. Coulter, U.S. Patent 2,656,508, 1953) were differen-

tiated and integrated, and the resulting pulses were measured by a pulse height analyzer. Samples were prepared for measurement by diluting to 10^5 to 2×10^5 cells per milliliter in a solution containing 0.85% NaCl and 0.04% formaldehyde. The size distribution of 2×10^4 to 3×10^4 cells was measured between 1 and 3 hr after sampling.

Measurement of volume by electron microscopy. Formaldehyde (0.04%) was added to a sample of culture, from which the cells were centrifuged, washed once in 0.04% formaldehyde, and resuspended in a solution containing 0.04% formaldehyde and 0.1% serum albumin to give a concentration of 2×10^8 cells per milliliter. The suspension was sprayed onto Formvar-coated grids and was air-dried.

Electron micrographs were made by use of an RCA EMU-3E microscope. Magnification was controlled to within ±0.75%. The micrographs were printed to give a final magnification of about 10,000 ×, and dimensions of cells were estimated from the prints to within ±0.2 mm. Measurement of length is sufficient to estimate relative volume of cells of *E. coli* (15), and was assumed to be sufficient for estimation of relative volumes of *A. agilis*.

RESULTS

Steady-state growth of E. coli and A. agilis. The equation of Collins and Richmond (4) is as follows:

$$V(v) = \frac{k}{\lambda(v)} \left[2 \int_o^v \psi(x)\, dx - \int_o^v \phi(x)\, dx - \int_o^v \lambda(x)\, dx \right] \tag{1}$$

Before the equation can be applied to measured size distributions for calculation of growth rates, it must be established that the culture on which the measurement of size distribution is made is growing in a steady state. In this investigation, the steady state was defined by three criteria. (i) The specific growth rate of the population must be constant. (ii) The distribution of cell size must be independent of time. (iii) As a consequence of (ii), the specific growth rate computed from optical density must be identical with that computed from numbers.

These criteria should be met after growth of a bacterial population for many generations at relatively low cell densities. Both organisms were grown for approximately 20 generations before measurements were commenced. The cultures were periodically diluted with fresh medium so that the culture of *E. coli* did not exceed 5×10^8 cells per milliliter, and the culture of *A. agilis* did not exceed 5×10^6 cells per milliliter.

The population of *E. coli* was sampled at intervals of 0.5 to 0.25 generation over a period of 7 generations for the measurement of optical density, numbers, and size distribution. Measurements of numbers and size distribution of *A. agilis* were made over a period of three generations. Optical density could not be measured at the low densities at which this organism was grown; growth at densities higher than 10^7 cells per milliliter does not approximate a steady state.

Specific growth rates of *E. coli* are compared in Table 1. By use of the *t* test, no significant difference was found between the specific growth rate estimated from numbers and the specific growth rate estimated from optical density. Analysis of variance of the measurements in three sequential periods showed that the specific growth rate did not vary significantly with time. Thus, the criteria of constancy and equality of the specific growth rates are satisfied in cultures of *E. coli*. Numbers of *A. agilis* also increased exponentially, satisfying the condition of constancy of specific growth rate.

The demonstration of the constancy of the frequency function of cell size, $\lambda(v)$, was based on examination of the parameters of this distribution as a function of time. Collins and Richmond (4) reported that the distribution of the logarithm of length of *B. cereus* was Gaussian; thus, the mean and variance of the logarithmic transform sufficiency defined the distribution. The logarithmic transforms of the measured volume distributions of *E. coli* and *A. agilis* were not Gaussian. The mean and variance are not sufficient to define the distribution because of the strong positive skewness. Arbitrarily, we decided to describe the distribution in terms of parameters based on the first four moments.

The mean, variance, g_1, and g_2 statistics (7) were calculated for each distribution measured. From replicate measurements on a single sample, the variance of the estimates of each parameter from measurement error (s^2_M) was calculated.

The variance of the estimates of each parameter over the entire experimental period (s^2_P) was also calculated. If the value of a parameter did not change with time, the variance of its estimates will be due entirely to measurement error. Then the ratio $F = s^2_P/s^2_M$ should satisfy the *F* distribution.

The results of this analysis are shown in Table 2. For both organisms, the variances of mean, variance, g_1, and g_2 were no greater than expected from measurement error; any change with time of the parameters of the distributions was within the limits of measurement error, and the necessary condition of constancy of the size distribution was satisfied.

The measured distributions for each organism were pooled, and the resulting frequency functions are shown in Fig. 1 and 2. The parameters of the pooled distributions were not significantly different from the mean of the estimates of the parameters of the individual distributions. Thus, the pooling of data did not introduce bias.

Parameters of the distribution of size of dividing cells and of newly formed cells. The frequency function of size of dividing cells, $\phi(v)$, must be known or assumed in order to calculate growth rate as a function of size. Since the form of this function is not known, three representative types of functions were chosen, each of which is sufficiently described by the mean and coefficient of variation (CV). These were: (i) the Gaussian function, a symmetrical function; (ii) the Pearson Type III function (5), a positively skewed function; and (iii) a negatively skewed function obtained by a reflection of the variable of the Pearson Type III function. The parameters chosen for these functions are shown in Table 3. The basis for the choice of values for these parameters is given below.

TABLE 1. *Growth of Escherichia coli ML30*

Period of measurement[a]	Optical density		Numbers		*t* for difference	$P(t)$[b]
	Specific growth rate, k, hr^{-1}	Standard error	Specific growth rate, k, hr^{-1}	Standard error		
min						
72–125	0.6886	0.0167	0.6775	0.0132	0.521	>0.1
210–345	0.6700	0.0098	0.6446	0.0096	1.854	>0.05
356–502	0.6754	0.0064	0.6824	0.0079	0.686	>0.1
Value of *F* for constancy of k	0.544		3.951			
$P(F)$	>0.05		>0.025			

[a] The end of each period of measurement was dictated by the necessity for dilution of the culture.

[b] $P(t)$ and $P(F)$ refer to the probability that the observed value of *t* or *F*, respectively, is due to chance.

TABLE 2. *Parameters of size distributions of Escherichia coli and Azotobacter agilis*

Determination	Parameter			
	Mean, μ^3	Variance	g_1	g_2
E. coli ML30[a]				
Mean value	0.8285	0.0456	1.197	1.362
Variance	4.51×10^{-4}	1.02×10^{-5}	4.69×10^{-3}	7.16×10^{-2}
Measured sampling variance	1.60×10^{-4}	5.16×10^{-6}	8.57×10^{-4}	1.41×10^{-2}
F	2.819	1.977	5.472	5.078
$P(F)$	>0.05	>0.05	>0.05	>0.05
A. agilis[b]				
Mean value	5.351	1.961	0.825	0.707
Variance	1.64×10^{-2}	8.01×10^{-3}	2.73×10^{-3}	3.54×10^{-3}
Measured sampling variance	7.91×10^{-3}	5.06×10^{-3}	5.42×10^{-4}	4.63×10^{-3}
F	2.073	1.583	5.037	7.646
$P(F)$	>0.05	>0.05	>0.025	>0.01

[a] Eighteen distributions measured; total number of cells measured $= 4.5 \times 10^5$.
[b] Nine distributions measured; total number of cells measured $= 2.0 \times 10^5$.

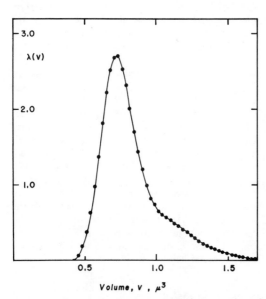

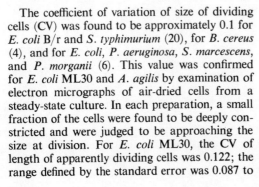

FIG. 1. *Frequency function of volume, $\lambda(v)$, of Escherichia coli ML30. Values below 0.53 μ^3 were obtained by correcting observed values for background.*

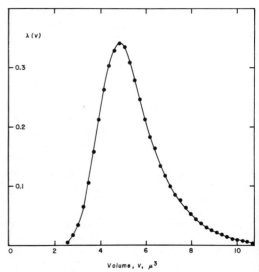

FIG. 2. *Frequency function of volume, $\lambda(v)$, of Azotobacter agilis. Values below 3.3 μ^3 were obtained by correcting observed values for background.*

The coefficient of variation of size of dividing cells (CV) was found to be approximately 0.1 for *E. coli* B/r and *S. typhimurium* (20), for *B. cereus* (4), and for *E. coli, P. aeruginosa, S. marcescens,* and *P. morganii* (6). This value was confirmed for *E. coli* ML30 and *A. agilis* by examination of electron micrographs of air-dried cells from a steady-state culture. In each preparation, a small fraction of the cells were found to be deeply constricted and were judged to be approaching the size at division. For *E. coli* ML30, the CV of length of apparently dividing cells was 0.122; the range defined by the standard error was 0.087 to

TABLE 3. *Parameters of the distribution of size of dividing cells*

Determination	Best estimate	Range
Escherichia coli ML30		
Mean, μ^3	1.306	1.23–1.38
Coefficient of variation	0.12	0.09–0.15
Azotobacter agilis		
Mean, μ^3	8.13	7.68–8.58
Coefficient of variation	0.10	0.09–0.11

0.150. For *A. agilis*, the CV was 0.101, and the range defined by the standard error was 0.090 to 0.110. Since the cells observed had not yet divided but were in different stages of the final division process, the values for the coefficients of variation were overestimated and provided an upper limit to the range of the true value. It was assumed that the CV of size of dividing cells would fall within the range defined by the standard error of the estimate in each case.

The best estimate of the mean (m_ϕ) of $\phi(v)$ was determined from the relationship:

$$m_\phi = v_L/2 + v_S \qquad (2)$$

which is developed in Appendix B. The term v_S is the lower limit and v_L is the upper limit of the range of the distribution of size of cells. This estimate was based on the assumptions that $\phi(v)$ is symmetrical and that division is equal. The probable range of m_ϕ was obtained from the assumption that the function was either extremely negatively or positively skewed. The best estimate and the range of the mean of $\phi(v)$ for *E. coli* ML30 and *A. agilis* are in Table 3.

The validity of these estimates was supported by the distribution of volume of the cells in the effluent from the membrane filter device of Helmstetter and Cummings (9). The majority of such cells had just been formed by division. The frequency function of their volumes should be approximately $\psi(v)$. The value of the mode of the distribution of volume of the effluent cells should provide an estimate of m_ψ. The modal volume of *E. coli* B/r in the effluent was 0.49 μ^3, whereas the estimate obtained from equation 2 was 0.50 μ^3 with a range of 0.465 to 0.538 μ^3.

The ψ-distribution and ϕ-distribution are related; the relationship is a function of the inequality of division (18). The variable p is the ratio of the volume of daughter cell to volume of mother cell at the time of division, and $K(p)$ is the frequency function of the distribution of p. Powell (18) has demonstrated that, if $K(p)$ and $\phi(v)$ are known, $\psi(v)$ can be computed from the following equation:

$$\psi(v) = \int_0^1 \phi(v/p) \cdot \frac{K(p)}{p} \, dp \qquad (3)$$

For equal division, p is not distributed, giving $\psi(v) = 2\phi(2v)$. For *E. coli* ML30, the distribution of $K(p)$ was Gaussian with a CV of 0.04 (15). Similar measurements were made from the electron micrographs of *A. agilis* described above. For *A. agilis*, $K(p)$ was also Gaussian ($p_{\chi^2} = 0.23$) with a CV of 0.02. For most purposes, it can be assumed that division is equal in both organisms.

The act of division referred to throughout this paper is the act of physical separation of daughter cells. Physiological division, whereby the daughter cells become physiologically independent, must precede separation. The temporal relationship between these two processes is not known.

Calculation of growth rates. Growth rates of *E. coli* and *A. agilis* as a function of cell volume were calculated by use of equation 1. Figures 3 and 4 show the results of calculations with the use of the best estimates of the parameters of the distributions of dividing and newly formed cells. The rate of increase in volume in cubic microns per hour was plotted as a function of cell volume in cubic microns. The broken lines show the rates of growth expected if individual cells grew at all times with the same specific growth rate as the population.

Figures 5, 6, and 7 show the effects on the calculated growth rates of *E. coli* of different assumptions regarding the form, coefficient of variation, and mean, respectively, of $\phi(v)$. Results for *A. agilis* showed the same effects. Examination of the results showed that the rate of growth was substantially independent of $\phi(v)$ over a range of

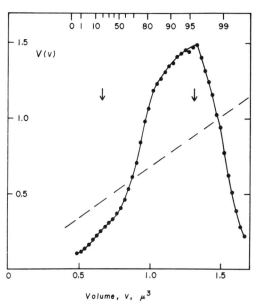

FIG. 3. *Rate of increase of volume, $V(v)$, of Escherichia coli ML30 as a function of cell volume. The distributions of volume of dividing and newly formed cells are assumed to be Gaussian, with parameters as given by the best estimates shown in Table 3. The arrows indicate the best estimates of mean volume of newly formed and dividing cells. The upper scale on the abscissa shows the cumulative probability of the cell size distribution at the corresponding volume on the lower scale.*

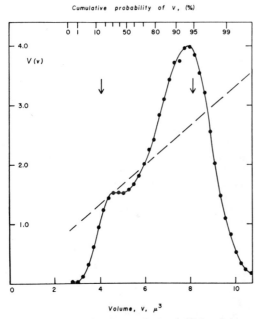

FIG. 4. *Rate of increase of volume, $V(v)$, of Azotobacter agilis as a function of volume. The distributions of volume of dividing and newly formed cells are assumed to be Gaussian, with parameters as given by the best estimates shown in Table 3. The arrows and the upper scale have the same meaning as in Fig. 3.*

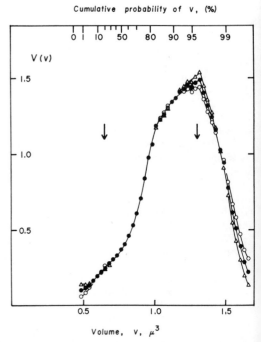

FIG. 5. *Effect of choice of function representing the distributions of volume of dividing and newly formed cells on the calculated values of $V(v)$ for Escherichia coli ML30. The different functions are: ●, Gaussian; ○, Pearson Type III; △, reflected Pearson Type III. Parameters used are the best estimates given in Table 3. The arrows and the upper scale have the same meaning as in Fig. 3.*

cell volume containing approximately 50% of the population. Outside this range the assumptions did influence the calculated growth rates, but did not affect the general form of the functions.

The kinetics of growth computed for *E. coli* and *A. agilis* agreed closely with the kinetics of growth of *B. cereus* determined by Collins and Richmond (4) from the distribution of lengths measured by using light microscopy.

Calculation of distribution of interdivision time. The frequency functions of interdivision times, $f(\tau)$, of *E. coli* and *A. agilis* were calculated from the values of $V(v)$ by use of the equations given in Appendix C. The frequency function of size of dividing cells was assumed to be Gaussian with parameters as given by the best estimates in Table 3. The distribution of volume of newly formed cells was computed from $\phi(v)$, by use of equation 3, and from the experimentally determined parameters of $K(p)$. Hypothetical functions for $f(\tau)$ were calculated by use of the same $\phi(v)$, but assuming either exponential or linear kinetics of growth. Results of these calculations are shown in Fig. 8 and Table 4.

Similar results were obtained for *E. coli* and *A. agilis*. The function $f(\tau)$ calculated from the experimentally determined values of $V(v)$ for either

E. coli or *A. agilis* was positively skewed and had a coefficient of variation of about 0.3. The function $f(\tau)$ calculated assuming either exponential or linear growth was symmetrical.

By direct numerical solution of the functions used in computing $f(\tau)$, the correlation coefficients of life lengths of mother and daughter and of sister and sister cells were computed. For these calculations, it was assumed that a mother divided to produce two sisters of equal size. The correlation coefficients for life lengths of mother and daughter cells were −0.307 and −0.408 for *E. coli* and *A. agilis*, respectively. The correlation coefficients of life lengths of sister and sister cells were 0.866 and 0.722, respectively. The negative correlation of life lengths of mother and daughter cells and the high positive correlation of life length of sister and sister cells were in general agreement with direct experimental measurements (17). The agreement between calculations and experimental results supported the hypothesis of Koch and Schaechter (11) that size controls the timing of cell division.

Cumulative probability of V, (%)

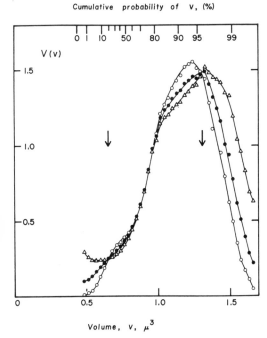

Cumulative probability of V, (%)

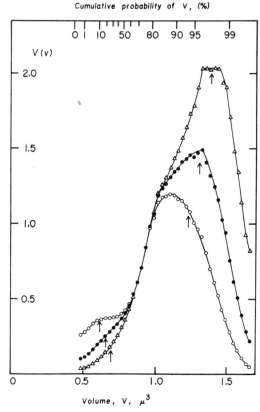

FIG. 6. *Effect of different coefficients on variation of the distributions of volume of dividing and newly formed cells on the calculated values of $V(v)$ for Escherichia coli ML30. The distributions are assumed to be Gaussian, with means as given by the best estimates in Table 3, and coefficients of variation of:* ○, *0.09;* ●, *0.12;* △, *0.15. The arrows and the upper scale have the same meaning as in Fig. 3.*

DISCUSSION

A growth rate, $V(v)$, is the average rate of increase in volume of those cells of volume v. If only size determines the rate of growth, all cells of volume v grow at rate $V(v)$; but, if other factors such as age are significant, the rate $V(v)$ is an average of a distribution of individual rates. The calculation is valid in either case.

From the results in Fig. 3 and 4, one can reconstruct the kinetics of growth of a typical individual if all cells of volume v grow at a rate $V(v)$. After formation at small volume, both the absolute and specific growth rates increased with increasing volume. Just prior to reaching the mean volume of dividing cells, the growth rate was maximal. At larger volumes, the values of $V(v)$ were strongly influenced by the form of the distribution of size of dividing cells, but it appears that individual cells which fail to divide before reaching the mean size of dividing cells grow at a continually decreasing rate.

Continuous exponential growth implies that immediately after division the average growth rate

FIG. 7. *Effect of different means of the distributions of volume of dividing and newly formed cells on the calculated values of $V(v)$ for Escherichia coli ML30. The distributions are assumed to be Gaussian, with coefficients of variation as given by the best estimates in Table 3, and values of:* ○, *1.231 μ^3;* ●, *1.306 μ^3;* △, *1.381 μ^3, for the mean volume of dividing cells. The arrows indicate the mean volumes of newly formed and dividing cells for each curve. The upper scale has the same meaning as in Fig. 3.*

of the two daughter cells will be one-half the growth rate of the mother cell. This was not the case for *E. coli* and *A. agilis*. The growth rate $V(v)$ of cells of volume v was much less than one-half of $V(2v)$ for more than 99% of the population. The specific growth rate of a cell must decrease shortly before, during, or immediately after division. The decrease in growth rate which becomes evident at volumes larger than the mean volume of dividing cells could generally be true, but observable only if cells which are dividing constitute a large fraction of the cells in a given size class.

The kinetics of growth are more complex than predicted by the hypothesis of continuous exponential growth. The validity of this conclusion depends upon the validity of the assumptions regarding the distribution of volume of dividing cells and upon the accuracy of measurement of the population size distributions. The mean and coefficient of variation of the distribution of volume of dividing cells can be determined with

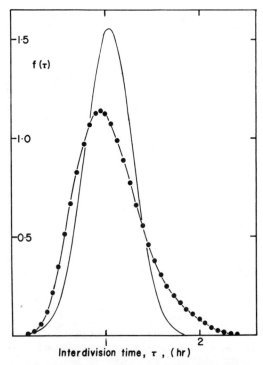

FIG. 8. *Frequency functions, $f(\tau)$, of the distribution of interdivision times. Symbols:* ●, *Escherichia coli ML30; solid line, hypothetical $f(\tau)$, calculated assuming exponential growth. In both cases, the distribution of volume of dividing cells was assumed to be Gaussian, with parameters given by the best estimates for E. coli ML30 in Table 3.*

sufficient accuracy to conclude that reasonable values of these parameters are not consistent with the hypothesis of continuous exponential growth. Over the probable range, the values of these parameters do not affect the general form of $V(v)$.

The form of $\phi(v)$ is assumed, and it might be argued that there exists some frequency function of volume of dividing cells which is consistent with continuous exponential growth. That such a frequency function does not exist can be demonstrated by solving equation 1 for $\phi(v)$ and $\psi(v)$, giving the following equation:

$$2 \int_0^v \psi(x)\,dx - \int_0^v \phi(x)\,dx$$
$$= \frac{V(v) \cdot \lambda(v)}{k} + \int_0^v \lambda(x)\,dx \tag{4}$$

Equation 4 can be solved numerically for $\phi(v)$ if $\lambda(v)$ is known, and the kinetics of growth are assumed. The solution is iterative, $\psi(v)$ being calculated from $\phi(v)$ by equation 3, by use of the known parameters of $K(p)$. Values of $\phi(v)$ were computed from the distributions of volume of E. coli and A. agilis shown in Fig. 1 and 2, with the assumption of continuous exponential growth, i.e., $V(v) = kv$. The result of these calculations for E. coli are in Fig. 9; results for A. agilis were similar. To be consistent with the measured volume distribution and continuous exponential growth, the function $\phi(v)$ must assume both positive and negative values and, hence, cannot be a frequency function. It follows that the measured volume distributions are inconsistent with the hypothesis of continuous exponential growth.

The form of the calculated distribution of interdivision times provides a powerful argument for the validity of the observed kinetics of growth and against the hypothesis of exponential growth.

TABLE 4. *Parameters of the computed distributions of interdivision times, $f(\tau)^a$*

Organism	Kinetics of growth	Parameters of $f(\tau)$			
		Mean	CV	g_1	g_2
		hr			
Escherichia coli[b]	Experimental	1.020	0.336	0.443	0.028
	Exponential	1.044	0.246	−0.034	0.032
	Linear	1.046	0.259	−0.044	−0.080
Azotobacter agilis[c]	Experimental	2.037	0.257	0.901	1.442
	Exponential	2.114	0.201	−0.050	0.027
	Linear	2.074	0.214	−0.069	−0.065

[a] In all cases, the distributions of volume of dividing cells were assumed to be Gaussian, with parameters given by the best estimates in Table 3.
[b] Doubling time was 1.024 hr; coefficient of variation of $K(p)$ was 0.04.
[c] Doubling time was 2.094 hr; coefficient of variation of $K(p)$ was 0.02.

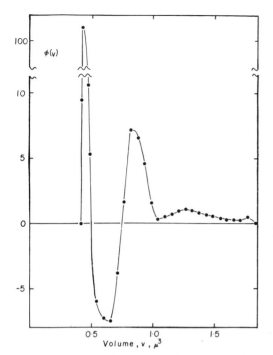

FIG. 9. *Theoretical function* $\phi(v)$ *computed from the measured distribution of volume of Escherichia coli ML30 assuming continuous exponential growth.*

Measurements of the distribution of interdivision times of a number of different organisms have shown that $f(\tau)$ is strongly positively skewed with a coefficient of variation of 0.15 to 0.5 (12, 17). The results in Table 4 are in good agreement with these observations, except that the coefficient of variation of $f(\tau)$ for *E. coli* is somewhat larger than values which have been observed. Powell (14) has asserted that, given exponential growth of individual cells, the function $f(\tau)$ will be symmetrical regardless of the symmetry of $\phi(v)$. We have established this symmetry by numerical analysis using equation C-1 in Appendix C. Since the present calculations of $f(\tau)$ and most previous measurements of this function show a strong positive skewness, it must be concluded that growth of individual cells cannot be strictly exponential. Positive skewness of $f(\tau)$ requires deviations from exponential growth.

Stochastic models based on the assumption that division occurs only after the completion of a number of random and independent cellular processes have been proposed to explain the positive skewness of the distribution of interdivision time (10, 19). Our calculations of $f(\tau)$ are based on the alternative postulate of Koch and Schaechter (11) that the *size* of a cell at division

is under cellular and environmental control. The results support this more deterministic hypothesis.

The precision of determination of the kinetics of growth depends upon the fidelity of measurement of the distribution of size of bacteria. An examination of the accuracy of measurement of volume by the apparatus used in this investigation has been made by use of measurement by electron microscopy as the primary standard (7). The distributions of volumes both of spherical latex particles and of cells of *E. coli* were measured without detectable distortion. However, the necessarily small sample size and the possibility of uncontrolled errors in electron microscopy preclude the detection of minor distortions of the distribution. This applies particularly to errors in estimating moments higher than the second. As Koch (J. Gen. Microbiol., *in press*) has pointed out, accurate measurement of the higher moments of the size distribution is of primary importance in determining the kinetics of growth.

Although systematic errors may have influenced the estimates of $\lambda(v)$, the possibility can be eliminated that the deviations of the observed kinetics from exponential growth are due to random errors in measurement. Equation 1 can be solved for $\lambda(v)$ (*see* Appendix A, equation A-8). Distributions can then be calculated incorporating any assumptions regarding the kinetics of growth and the distributions of size of dividing and newly formed cells. In Table 5 and Fig. 10, size distributions calculated assuming exponential growth are compared with the measured distributions of volume of *E. coli* and *A. agilis*. Distributions calculated assuming a constant growth rate between divisions (linear growth) are also shown. Figure 10 shows marked differences in the form of the distributions. The means, variances, and g_1 and g_2 statistics are compared by use of the t test in Table 5. The means and variances of the measured and theoretical distributions are significantly different in all but a few cases. The g_1 and g_2 statistics are markedly different, and the difference is significant in all instances. Random error in the measurement of the size distribution cannot, therefore, account for the deviations of the observed kinetics from either exponential or linear growth.

The results obtained by direct microscopic measurement of size of cells have been consistent with the hypothesis of continuous exponential growth; however, these results are not inconsistent with the kinetics of growth determined in this investigation. This becomes clear when the data of Fig. 3 and 4 are transformed to yield volume as a function of time, the form of data obtained by direct microscopic measurement. The

TABLE 5. *Comparison of measured size distributions with theoretical distributions calculated assuming exponential or linear growth[a]*

Determination	Assumed		Measured or calculated parameters			
	m_ϕ (μ^3)	CV_ϕ	Mean (μ^3)	Variance	g_1	g_2
Escherichia coli						
Measured			0.8285	0.0456	1.197	1.362
			$\pm 5.01 \times 10^{-3}$	$\pm 2.15 \times 10^{-2}$	$\pm 2.15 \times 10^{-2}$	$\pm 6.31 \times 10^{-2}$
Calculated assuming exponential growth	1.31	0.09	0.9073	0.0412	0.528	−0.376
	1.31	0.12	0.9073	0.0464*	0.569	−0.154
	1.31	0.15	0.9073	0.0532	0.588	0.045
	1.23	0.12	0.8556	0.0412	0.569	−0.154
	1.38	0.12	0.9949	0.0520	0.569	−0.154
Calculated assuming linear growth	1.31	0.09	0.9520	0.0436	0.354	−0.610
	1.31	0.12	0.9581	0.0500	0.409	−0.390
	1.31	0.15	0.9657	0.0576	0.452	−0.196
	1.23	0.12	0.9034	0.0444*	0.409	−0.390
	1.38	0.12	1.0126	0.0556	0.409	−0.390
Azotobacter agilis						
Measured			5.351	1.961	0.825	0.707
			± 0.0427	± 0.0298	± 0.0174	± 0.0628
Calculated assuming exponential growth	8.13	0.09	5.658	1.610	0.528	−0.376
	8.13	0.10	5.658	1.675	0.538	−0.300
	8.13	0.11	5.658	1.746	0.549	−0.221
	7.68	0.10	5.346*	1.500	0.538	−0.300
	8.58	0.10	5.971	1.860	0.538	−0.300
Calculated assuming linear growth	8.13	0.09	5.936	1.716	0.354	−0.610
	8.13	0.10	5.948	1.789	0.354	−0.523
	8.13	0.11	5.961	1.871	0.386	−0.446
	7.68	0.10	5.620	1.603	0.366	−0.523
	8.58	0.10	6.276	1.986*	0.366	−0.523

[a] The distributions of $\phi(v)$ were assumed to be Gaussian, and $\psi(v)$ was computed from equation 3.

* Not significantly different from the measured distribution at the 5% level of probability. All values not so marked are significantly different from the measured distribution.

results of such a transformation, representing the growth of approximately 90% of the population, are shown in Fig. 10 and 11. If the random error in microscopic measurement is only ±5%, the 95% confidence limits of measurement encompass both exponential growth and the observed kinetics for *E. coli*. For *A. agilis*, the confidence limits include linear growth as well.

Our results do not give information about the kinetics of synthesis of cell material. It is possible that the rate of synthesis is independent of the rate of change in volume. Dependence can be tested by pulse-labeling with a radioactive monomer (e.g., an amino acid) and determining the distribution of grains in an autoradiogram with respect to the length of the cells. Goldstein (*unpublished data*) has evidence that the specific rate of protein synthesis decreases at the time of division. A culture of *E. coli* was pulse-labeled with H³-leucine, and the cells were subjected to autoradiography. If the pulse was of long duration, the large, dividing cells had more associated grains than did smaller cells. If the pulse was of short duration, the dividing cells did not have more associated grains.

If the rate of synthesis of protein and other macromolecules is not proportional to the rate of increase in volume of the cell, volume is more critical to cell division than the mass of macromolecules. If the rate of synthesis of protein is proportional to the rate of increase in volume of the cell (i.e., growth is balanced), some mechanism to control the efficiency at which ribosomes function must be envisaged.

Appendix A

Relationship between growth rate of individual cells and the size distribution of the population. The Collins-Richmond equation is a statement of conservation of cells that holds for any culture in exponential balanced growth, for which $\lambda(v)$ and $V(v)$ can be closely approximated by continuous functions and are independent of time or of the number of cells.

If the class T of all cells larger than some v is considered, the number of such cells is

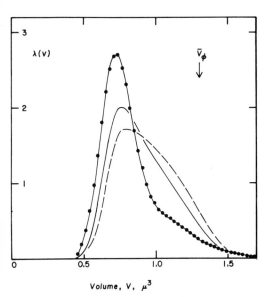

FIG. 10. *Comparison of measured frequency function of Escherichia coli ML30 (●) with theoretical functions calculated assuming continuous exponential growth (solid curve), and linear growth (broken curve) of all cells. In calculating the theoretical functions, the distributions of volume of dividing and newly formed cells were assumed to be Gaussian, with parameters as given by the best estimates for E. coli in Table 3. The arrow indicates the mean volume of dividing cells.*

$$N \int_v^\infty \lambda(x)\, dx$$

and the rate of increase in the number of cells in T is

$$kN \int_v^\infty \lambda(x)\, dx \qquad (A\text{-}1)$$

This rate has three components: cells lost by divisions in T, newly formed cells in T, and cells growing into T. The rate of loss of cells by division is

$$-kN \int_v^\infty \phi(x)\, dx \qquad (A\text{-}2)$$

Since two cells are formed from each dividing cell, the rate of formation of new cells in T is

$$2kN \int_v^\infty \psi(x)\, dx \qquad (A\text{-}3)$$

The rate of growth into T is seen to be

$$NV(v)\lambda(v) \qquad (A\text{-}4)$$

as follows: the number of cells in the interval $(v - dv, v)$ immediately below v is

$$N\lambda(v)\, dv \qquad (A\text{-}5)$$

Only the cells in the interval $(v - dv, v)$ can grow

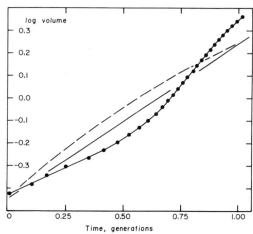

FIG. 11. *Growth of Escherichia coli ML30 expressed as logarithm of volume as a function of time. Symbols: ●, curve obtained by transformation of data shown in Fig. 3; solid line, theoretical curve for exponential growth; broken line, theoretical curve for linear growth.*

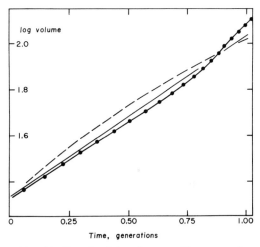

FIG. 12. *Growth of Azotobacter agilis expressed as logarithm of volume as a function of time. Symbols: ●, curve obtained by transformation of data shown in Fig. 4; solid line, theoretical curve for exponential growth; broken line, theoretical curve for linear growth.*

into T in time

$$dt = dv/V(v)$$

Dividing (A-5) by dt gives (A-4) the rate at which cells enter by growth. However, some of the cells in the interval may divide before growing into T. The The rate of divisions in this interval is

$$kN\phi(v)\, dv \qquad (A\text{-}6)$$

but dv can be as small as we like so that these divisions are an infinitesimal part of the rate $NV(v)\lambda(v)$. Thus, we can neglect the dividing cells, and a similar argument shows that we can neglect newly formed cells. The rate at which cells enter T by growth is thus given by the expression in A-4.

From A-1 through A-4, the argument above for the conservation of cells can be stated as

$$k \int_v^\infty \lambda(x)\, dx = V(v)\lambda(v)$$

$$+ 2k \int_v^\infty \psi(x)\, dx - k \int_v^\infty \phi(x)\, dx$$

By use of the fact that λ, ϕ, and ψ are frequency functions, and thus have integrals from zero to infinity equal to one, the familiar form of the equation of Collins and Richmond (4)

$$V(v) = \frac{k}{\lambda(v)} \left[2 \int_o^v \psi(x)\, dx - \int_o^v \phi(x)\, dx - \int_o^v \lambda(x)\, dx \right] \quad \text{(A-7)}$$

can be obtained.

Differentiating A-7 gives a linear equation:

$$V'(v)\lambda(v) + V(v)\lambda'(v) = k[2\psi(v) - \phi(v) - \lambda(v)]$$

The solution of this equation is

$$\lambda(v) = \exp\left[-\int \frac{V'(v) + k}{V(v)}\, dv \right] \int \left(\frac{k}{V(v)} \right.$$

$$\cdot \exp\left[\int \frac{V'(v) + k}{V(v)}\, dv \right] (2\psi(v) - \phi(v)) \bigg)\, dv$$

$$+ C \exp\left[-\int \frac{V(v) + k}{V(v)}\, dv \right] \quad \text{(A-8)}$$

in which C is the constant of integration.

In deriving the Collins-Richmond equation, it is not necessary to assume that all cells of size v have the same growth rate. If, instead, their growth rates are distributed according to the frequency function $P(V)$, there are

$$NP(V)\, dV\lambda(v)\, dv \quad \text{(A-9)}$$

cells in the interval $(v - dv, v)$ that have growth rates between V and $V + dV$. The rate at which these cells enter T is obtained by dividing (A-9) by $dt = dv/V$, giving

$$NP(V)\, dV\lambda(v)V$$

Integrating this expression over V gives the total rate at which cells enter T by growth

$$N\lambda(v) \int_o^\infty VP\,(V)\, dV = N\overline{V}(v)\lambda(v)$$

where $\overline{V}(v)$ is the mean growth rate of cells of size v. Thus, the expression A-4 is valid whether or not the growth rate of cells of size v is distributed; if V is

distributed, expression A-4 is independent of any parameter of the distribution other than the mean.

In the above derivation it is possible to show that, under reasonable conditions on V and λ (each has at most a finite number of discontinuities), V can have a discontinuity at v if and only if λ also has a discontinuity at v. Thus, if λ is continuous, V must also be continuous.

APPENDIX B

Estimation of the mean volumes of dividing and newly formed cells. The method used is a refinement of the method of Collins and Richmond (4). Let m_ψ and m_ϕ represent the means of $\psi(v)$ and $\phi(v)$, respectively, v_L the volume of the largest, and v_S the volume of the smallest cells observed. Since each cell at division forms two daughter cells, then

$$m_\phi = 2m_\psi$$

It is assumed that the probability of occurrence of a newly formed cell at size less than v_S and the probability of occurrence of a division at a size greater than v_L are approximately equal. If $\psi(v)$ and $\phi(v)$ are symmetrical functions, with identical coefficients of variation, C, then

$$\frac{m_\psi - v_S}{Cm_\psi} = \frac{v_L - m_\phi}{Cm_\phi} \quad \text{(B-1)}$$

and, from (B-1),

$$m_\phi = v_L/2 + v_S \quad \text{(B-2)}$$

The relationship in B-2 must be modified if $\psi(v)$ and $\phi(v)$ are not symmetrical functions, to give

$$m_\phi = A(v_L/2 + v_S)$$

The coefficient A is greater than unity if $\phi(v)$ is positively skewed, and less than unity if $\phi(v)$ is negatively skewed. A range of A of 0.67 to 1.5 will encompass most likely cases.

APPENDIX C

Calculation of $f(\tau)$ from growth rate as a function of size. The approach necessary for these calculations was developed for the case of continuous exponential growth by Powell (18). We present this general case.

The frequency function of interdivision time, $f(\tau)$, is the frequency function of the joint distribution of $\psi(v)$ and $g(v)$ after transformation of the variable from volume to time. Let $\psi_T(t)$ and $g_T(t)$ denote the transformed frequency functions such that

$$\psi_T(t)\, dt = \psi(v)\, dv$$

and

$$g_T(t)\, dt = \psi(v)\, dv$$

The time t corresponding to a volume v is

$$t = \int_{v_S}^v \frac{dx}{V(x)}$$

in which v_s is the minimal cell volume. Thus,

$$\psi_T(t) = \psi(v) \cdot V(v)$$

and

$$g_T(t) = g(v) \cdot V(v)$$

Assuming that $g_T(t)$ applies to each cell in $\psi_T(t)$, the frequency function of interdivision time is given by

$$f(\tau) = \int_0^\infty \psi_T(t) g_T(t + \tau) \, dt \qquad \text{(C-1)}$$

Conceptually, the function $g(v)$ determines $\phi(v)$, but in practice is calculated from the assumed $\phi(v)$. The probability of a cell of volume v dividing in the interval $(v, v + dv)$ is

$$\frac{g(v) \, dv}{\displaystyle\int_v^\infty g(x) \, dx}$$

The number of cells subject to this probability is the number growing through volume v in time dt, which from A-4 is

$$N\lambda(v)V(v) \, dt$$

The number of divisions in dt will be

$$\frac{N\lambda(v)V(v)g(v) \, dv \, dt}{\displaystyle\int_v^\infty g(x) \, dx}$$

The number of divisions in dt is also given by A-6; hence,

$$\phi(v) = \frac{\lambda(v)V(v)g(v)}{k \displaystyle\int_v^\infty g(x) \, dx} \qquad \text{(C-2)}$$

Equation C-2 can be rearranged and integrated to give

$$\ln \int_v^\infty g(x) \, dx = - \int_0^v \frac{k\phi(x) \, dx}{V(x)\lambda(x)}$$

or

$$\int_v^\infty g(x) \, dx = \exp\left[- \int_0^v \frac{k\phi(x) \, dx}{V(x)\lambda(x)} \right]$$

Differentiation gives

$$g(v) = \frac{k\phi(v)}{V(v)\lambda(v)} \exp\left[- \int_0^v \frac{k\phi(x) \, dx}{V(x)\lambda(x)} \right] \qquad \text{(C-3)}$$

Acknowledgments

We wish to thank C. L. Squires for technical assistance.

The experimental work was supported by grants ARO-49-092-65-G95 and MD-49-193-65-G160 from the United States Army. Digital computations were financially supported by Public Health Service grant AI-05526 from the National Institute of Allergy and Infectious Diseases.

Literature Cited

1. ADOLPH, E. F., AND S. BAYNE-JONES. 1932. Growth in size of micro-organisms measured from motion pictures. II. *Bacillus megatherium*. J. Cellular Comp. Physiol. **1**:409–427.
2. BAYNE-JONES, S., AND E. F. ADOLPH. 1932. Growth in size of organisms measured from motion pictures. III. *Bacterium coli*. J. Cellular Comp. Physiol. **2**:329–348.
3. CAMPBELL, A. 1957. Synchronization of cell division. Bacteriol. Rev. **21**:263–272.
4. COLLINS, J. F., AND M. H. RICHMOND. 1962. Rate of growth of *Bacillus cereus* between divisions. J. Gen. Microbiol. **28**:15–23.
5. ELDERTON, W. P. 1953. Frequency curves and correlation, 4th ed., p. 65–68. Cambridge Univ. Press, London.
6. ERRINGTON, F. P., E. O. POWELL, AND N. THOMPSON. 1965. Growth characteristics of some gram-negative bacteria. J. Gen. Microbiol. **39**:109–123.
7. FISHER, R. A. 1946. Statistical methods for research workers, 10th ed., p. 70–76. Oliver and Boyd, London.
8. HARVEY, R. J., AND A. G. MARR. 1966. Measurement of size distributions of bacterial cells. J. Bacteriol. **92**:805–811.
9. HELMSTETTER, C. E., AND D. J. CUMMINGS. 1964. An improved method for the selection of bacterial cells at divison. Biochim. Biophys. Acta **82**:608–610.
10. KENDALL, D. G. 1952. On the choice of a model to represent normal bacterial growth. J. Roy. Statist. Soc. Ser. B **14**:41–44.
11. KOCH, A. L., AND M. SCHAECHTER. 1962. A model for the statistics of the cell division process. J. Gen. Microbiol. **29**:435–454.
12. KUBITSCHEK, H. E. 1962. Normal distribution of cell generation rate. Exptl. Cell Res. **26**:439–450.
13. KNAYSI, G. 1940. A photomicrographic study of the rate of growth of some yeasts and bacteria. J. Bacteriol. **40**:247–253.
14. KNAYSI, G. 1941. A morphological study of Streptococcus fecalis. J. Bacteriol. **42**:575–586.
15. MARR, A. G., R. J. HARVEY, AND W. C. TRENTINI. 1966. Growth and division of *Escherichia coli*. J. Bacteriol. **91**:2388–2389.
16. MITCHISON, J. M. 1961. The growth of single cells. III. *Streptococcus faecalis*. Exptl. Cell Res. **22**:208–225.
17. POWELL, E. O. 1955. Some features of the generation times of individual bacteria. Biometrika **42**:16–44.
18. POWELL, E. O. 1964. A note on Koch and Schaechter's hypothesis about growth and fission of bacteria. J. Gen. Microbiol. **37**:231–249.
19. RAHN, O. 1932. A chemical explanation of the variability of the growth rate. J. Gen. Physiol. **15**:257–277.
20. SCHAECHTER, M., J. P. WILLIAMSON, J. R. HOOD, AND A. L. KOCH. 1962. Growth, cell and nuclear divisions in some bacteria. J. Gen. Microbiol. **29**:421–434.

9

Reprinted by permission from *Essays in Biosynthesis and Microbial Development,* John Wiley & Sons, Inc., New York, 1967, pp. 46–49

The Regulation of Secondary Biosynthesis

J. D. BU'LOCK

*   *   *   *   *   *   *

Unfortunately, this commonly used term "growth" is a portmanteau word which is not always easy to define or to relate to what is usually measured. Even the bacteriologist has to distinguish between cell counts, viable-cell counts, and cell weights; but consider filamentous organisms—possibly aseptate and multinuclear—with varying "inclusions" and cell-walls which vary in thickness and often outlast their contents, etc. Only the fact that "growth" is not measured by typical experimental parameters such as dry weight is really clear. Most mycelial cultures have the further disadvantage of heterogeneity, with "young," "old," and "dead" cells present simultaneously. In surface cultures the environmental heterogeneity is also obvious, but even in the most carefully controlled submerged cultures, homogeneity of development and micro-environment is only approximate. Thus we cannot easily observe either "growth" or "development"—yet, to anyone who has worked in the field, it is quite apparent that these imponderables are somehow linked with the circumstances under which secondary metabolism manifests itself.

Much of our own work in this field* has been an attempt to explore and resolve this situation; we have been encouraged by other notable publications on similar topics which I shall refer to. We can begin by looking at a well-known situation typified in Figure 3.2, which represents (say) the production of fat by a yeast. We note that the dry weight of the yeast follows a smooth

*Carried out mainly by D. Hamilton, A. J. Powell, D. Shepherd, and H. M. Smalley.

Editor's Note: A row of asterisks indicates that material has been omitted from the original article.

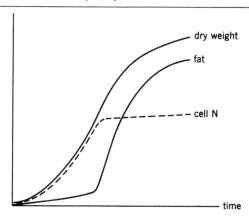

Figure 3.2.   Special biosynthesis—in this case accelerated lipid formation—
following nitrogen exhaustion, at a time not defined by the dry weight curve.

sigmoid curve, yet the first part of this simple "growth" is
accompanied by a low rate of fat synthesis and the second part
by a rapid synthesis. In short, there are two qualitatively different kinds of "growth" concealed in the mathematically simple
sigmoid. When we notice that the second kind of "growth"
occurs after the nitrogen of the medium is exhausted, and indeed involves no net protein synthesis, we realize that (in
terms a bacteriologist would accept) it cannot really be termed
"growth" at all. And if true growth is "what occurs during the
first part of the experiment," we will be content to say that the
rapid synthesis of fat, which is a secondary metabolic phenomenon, occurs *after* growth has ended.

This kind of approach is given far greater refinement in the
work of Borrow et al. (4) on *Gibberella fujikuroi*. Using defined
synthetic media, these authors were able to define experimentally a phase of "balanced" replicatory growth, in which all
the nutrients are taken up in an unchanging ratio, and which
is therefore terminated when any single nutrient is exhausted.

In this phase the gross composition of the mycelium remains uniform. Subsequent events depend on the balance of remaining nutrients, but in general (providing the carbon source is not exhausted), the later "unbalanced" growth is marked by nonreplicatory assimilation, and, in particular, by secondary metabolic reactions such as accelerated lipid synthesis and the formation of gibberellins.

The principles of this work seem to be of general validity, but the ideal of truly balanced growth as the continued exact replication of the inoculum is seldom, if ever, observed in detail. More commonly, the earlier phase is also one of a steadily increasing imbalance, such as the growth of the aminoacid pool discussed later, but here we should note Taber's different characterization based on a study of *Claviceps* fermentations (5). Using phosphate as the exhaustible nutrient, and defining "growth" in terms of polymeric nonextractable mycelial residues (experimentally convenient), Taber distinguishes a phase of true replication, followed by a phase in which "primary shunt-products" such as polyols and oligosaccharides accumulate, followed by a phase in which true secondary metabolites like the ergot alkaloids are formed.

Generally, I believe that the truth lies somewhere between these two views, and that the "growth phase" prior to secondary metabolite formation is not quite so uniform as Borrow et al. imply, nor yet so sharply divisible as Taber's classification suggests. However, before describing some of our own studies, it is worth noting another, rather different, contribution. Most successful laboratory work in this field has been carried out with cultures which have been carefully tended so that they will develop as uniformly and homogeneously as possible. The conditions of industrial fermentations are often fundamentally different, and sharply phased development is seldom observed. Nevertheless, development similar to that in well-phased fermentations does occur. This has been abundantly demonstrated by the work of Becker (6), who applied, not the batch-measurements that are normally used, but cytochemical and morphological criteria of phasing which can be applied to individual mycelial aggregates in a population. Her work shows quite clearly that, in several industrial-type "smooth" fermentations, developmental changes similar to those in "homogeneous" cultures occur, but they spread through the population rapidly or slowly according to the culture conditions.

3. Trophophase and Idiophase

Penicillium urticae is a tolerably well-behaved fungus which we first used for investigations of polyketide synthesis and metabolism, and have since used it as our private equivalent of *E. coli,* i.e., as a basis for sweeping generalizations. To describe its behavior in well-phased submerged cultures we introduced (7) the terms "trophophase" and "idiophase." The trophophase is roughly equivalent to the concept of "balanced growth" used by Borrow et al., but avoids any implication that the inoculum is exactly replicated. The idiophase is the period when idiosyncratic (i.e., species-peculiar) secondary metabolites are produced.

* * * * * * *

References

4. Borrow, A., E. G. Jefferys, R. H. J. Kessel, E. C. Lloyd, P. B. Lloyd, and I. S. Nixon, *Can. J. Microbiol.,* **7,** 227 (1961).
5. Taber, W. A., *Appl. Microbiol.,* **12,** 321 (1964).
6. Becker, Z. E., *Mitt. Versuchssta. Garungsgewerbe, Inst. Angew. Mikrobiol. (Wien),* **18,** 1 (1964).
7. Bu'lock, J. D., D. Hamilton, M. A. Hulme, A. J. Powell, D. Shepard, H. M. Smalley, and G. N. Smith, *Can. J. Microbiol.,* **11,** 765 (1965).

Editor's Comments on Papers 10, 11, and 12

10   **Monod:** *La Technique de culture continue théorie et applications*

11   **Novick and Szilard:** *Description of the Chemostat*

12   **Herbert, Elsworth, and Telling:** *The Continuous Culture of Bacteria: A Theoretical and Experimental Study*

Whereas the Monod theory largely originated the technique of continuous culture from the standpoint of studies of microbial growth, Novick and Szilard developed their method primarily as one for exploring genetical trends in growth. It is doubtful, however, if the continuous culture method would have developed as rapidly without the intervention of the Porton workers and this paper by Herbert, Elsworth, and Telling.

Monod's paper is reproduced here untranslated, as the accompanying articles (Papers 11 and 12) render a translation largely superfluous. Novick and Szilard's apparatus, rather than Monod's, now lends its name to the technique, as it more closely reflects the mode of operation in the method. Another paper by Novick and Szilard [*Proc. Natl. Acad. Sci. (U.S.)*, **36**, 708–719 (1950)] outlined the early experiments made with the chemostat.

The paper by Herbert et al. has served, since its introduction, as a guide and reference to continuous (chemostat) culture. The paper has served to translate and interpret the Monod growth theory and was the first of many contributions from the Porton establishment that have subsequently maintained the lead and the high standard set by this paper.

10

Copyright © 1950 by L'Institute Pasteur, Paris

Reprinted from *Ann. Inst. Pasteur,* **79**(4), 390–410 (1950)

LA TECHNIQUE DE CULTURE CONTINUE
THÉORIE ET APPLICATIONS

par Jacques MONOD.

(Institut Pasteur.
Service de Physiologie microbienne.)

I. — Introduction.

Repiquer une culture bactérienne, c'est, dans l'acception courante, diluer un petit volume de culture dans un grand volume de milieu neuf. Cette opération introduit dans la croissance une discontinuité et dans l'expérience un élément d'incertitude que connaissent bien les bactériologistes. Discontinuité et incertitude seront d'autant moindres que les repiquages seront plus fréquents et pratiqués à dilution moins grande. A la limite on aurait une culture maintenue par dilution continue calculée de façon que la croissance des germes soit exactement compensée. Une culture ainsi entretenue croîtrait indéfiniment, à vitesse constante, dans des conditions constantes. La discontinuité aurait disparu ainsi que l'élément d'incertitude qu'elle comporte. Il est évident que par la constance des conditions de milieu, du taux de croissance, donc de l'état physiologique des germes, une telle culture serait un objet d'expérience extrêmement favorable.

Les recherches poursuivies dans ce laboratoire nous ont amenés à mettre ce principe en œuvre (Monod, Torriani et Doudoroff, 1950) [**2**] et à étudier les propriétés des cultures continues. On verra par ce qui suit, je l'espère, que l'intérêt de cette technique ne se borne pas à l'obtention de cultures permanentes et stables. Les cultures continues constituent, dans certaines conditions, des systèmes en équilibre, de sorte que l'étude d'un phénomène en fonction du temps peut y être souvent remplacée par sa mesure à l'état d'équilibre, ce qui présente de grands avantages, théoriques autant que pratiques.

J'essaierai ici d'exposer les propriétés générales de ces systèmes continus, et d'indiquer les moyens de les utiliser dans différents types d'expériences. Ce n'est là qu'une première approximation. Le problème mériterait d'être traité, du point de vue théorique, d'une façon plus rigoureuse et plus approfondie que je ne saurais le faire. Quant à la réalisation technique, on ne trouvera ici que la description d'un montage assez primitif, dont le seul mérite est la simplicité.

II. — THÉORIE.

A. CROISSANCE EXPONENTIELLE CONTINUE. CONDITIONS D'ÉQUILIBRE. — Considérons un récipient B contenant un volume donné V_B de culture bactérienne. Supposons que, les conditions de milieu étant favorables, cette culture se développe à taux constant. Supposons que du milieu neuf, en réserve dans une nourrice N, soit amené de façon continue dans le récipient B par une tubulure *ad hoc* (T_1), tandis que, grâce à un artifice quelconque, une quantité égale de milieu est retirée à chaque instant par une seconde tubulure (T_2) aboutissant à un second récipient (P). Supposons que les bactéries tombant dans le récipient P cessent immédiatement de se multiplier (soit qu'elles soient congelées, soit que le récipient P contienne une substance antiseptique ou bactériostatique). Supposons enfin qu'en dépit du milieu neuf constamment admis dans la culture, l'homogénéité de la suspension bactérienne et des substances nutritives dissoutes soit assurée par un brassage efficace du liquide dans le récipient B. Ce brassage est supposé assurer également l'équilibre

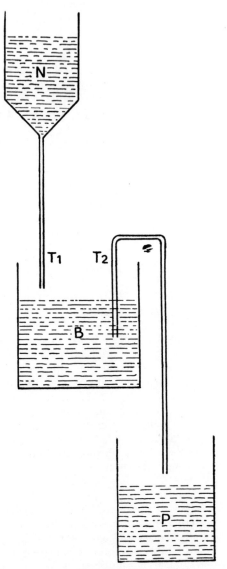

FIG. 1. — Schéma d'un appareil à culture continue.

entre le milieu liquide et l'atmosphère gazeuse du récipient.

Soit : x_{B} la masse bactérienne contenue dans le récipient B ;

$\qquad x_{\text{P}}$ la masse bactérienne contenue dans le récipient P ;

$\qquad x_{\text{B}}$ la masse bactérienne totale.

On a par définition :

$$x_{\text{B}} = x_{\text{T}} - x_{\text{P}}$$

et en dérivant par rapport au temps :

$$\frac{dx_{\text{B}}}{dt} = \frac{dx_{\text{T}}}{dt} - \frac{dx_{\text{P}}}{dt}. \tag{1}$$

Etant donné que seules les bactéries contenues dans B se multiplient, on voit que :

$$\frac{dx_{\text{T}}}{dt} = \mu x_{\text{B}} \tag{2}$$

μ est une constante que nous appellerons « taux de croissance népérien » (1). L'accroissement de la masse bactérienne en P est à chaque instant proportionnel à x_{B} et au débit du système. Si nous définissons ce débit (D) comme le rapport du volume débité par unité de temps, au volume V_{B}. nous pourrons écrire :

$$\frac{dx_{\text{P}}}{dt} = D . x_{\text{B}}. \tag{3}$$

En combinant (1), (2) et (3) on obtient :

$$\frac{dx_{\text{B}}}{dt} = (\mu - D)\, x_{\text{B}}. \tag{4}$$

et en intégrant :

$$\text{Log}_e\, \frac{x_{\text{B}}}{x_0} = (\mu - D)\, t \tag{5}$$

x_0 est ici une constante exprimant la masse bactérienne au temps $t = 0$.

Si nous considérons un intervalle de temps quelconque, $t_2 - t_1$. et l'accroissement correspondant, $x_2 - x_1$, de la masse bactérienne dans B, nous pouvons écrire :

$$\frac{\text{Log}_e x_2 - \text{Log}_e x_1}{t_2 - t_1} + D = \mu. \tag{5 bis}$$

Cette équation permet de calculer le taux de croissance, connaissant l'accroissement de la culture et le taux de dilution.

(1) Le taux de croissance est généralement défini comme le nombre de divisions (ou de doublements de la masse) par unité de temps. On passe du « taux de croissance népérien » (μ) au taux de croissance usuel (μ_2) en divisant par $\text{Log}_e 2$:

$$\mu_2 = \frac{\mu}{0{,}69}.$$

Le système est en équilibre lorsque $\mu = D$. Dans cette condition, la masse bactérienne ou, ce qui revient au même, la densité de la culture en B est constante, tandis que le « produit » s'accumule en P proportionnellement au temps.

Si l'on dispose d'un moyen de régler le débit, il est donc possible, en principe, de maintenir indéfiniment la culture à toute densité compatible avec la croissance au taux maximum, tandis que la multiplication se poursuit au taux correspondant à la phase exponentielle.

B. Croissance continue a taux limité. — La croissance d'une culture dans un milieu non renouvelé modifie la composition de ce milieu et crée ainsi des conditions qui ralentiront et, en définitive, arrêteront la croissance. Dans le milieu constamment renouvelé du récipient « B », ceci ne se produira pas si le taux de dilution équilibre *exactement* le taux de croissance maximum. Pour peu que le taux de dilution soit *inférieur* au taux de croissance, il en ira autrement. La densité de la culture dans B s'accroîtra à un taux « apparent » égal à μ-D, tant que les conditions de milieu demeureront optimales. Tôt ou tard l'apport de milieu neuf ne suffira plus à maintenir ces conditions pour une population accrue, de sorte que le taux de croissance diminuera. Il est clair qu'il atteindra ainsi, nécessairement, la valeur d'équilibre ($\mu = D$), mais qu'il ne tombera pas sensiblement *au-dessous* de cette valeur, car alors la population diminuerait et l'apport désormais en excès de milieu neuf tendrait à restituer les conditions primitives, par conséquent à augmenter le taux de croissance. *Celui-ci s'ajustera donc automatiquement à la valeur d'équilibre et y demeurera indéfiniment.*

Ces conditions de fonctionnement pour lesquelles le système tend vers un équilibre stable sont particulièrement intéressantes et nous allons chercher à les définir d'une façon plus rigoureuse.

Considérons une culture se développant dans un milieu de composition définie, et telle que le seul « facteur limitant » de la croissance soit la concentration de l'un des aliments essentiels, par exemple l'aliment carboné. On sait (*Cf.* Monod, 1942-1949) [1] que dans un tel milieu (non renouvelé) la croissance totale (2) est proportionnelle à la concentration initiale de l'aliment carboné. On sait aussi que le taux de croissance varie avec la concentration de l'aliment carboné suivant une loi assez bien exprimée par la relation hyperbolique :

$$\mu = \nu_0 \frac{S}{S_K + S} \tag{6}$$

(2) Différence entre la densité initiale et la densité maximum. La densité de la culture est définie comme le poids sec de substance bactérienne par unité de volume.

dans laquelle μ est le taux de croissance correspondant à une concentration S d'aliment limitant, μ_0 le taux de croissance maximum, qui prévaut quand S est grand, S_K, une constante caractéristique de l'organisme et de la substance considérée.

On sait enfin que, dans la plupart des cas connus, la valeur de la constante d' « affinité » S_K est très petite par rapport aux valeurs de la concentration S permettant un développement assez abondant des cultures. Ceci revient à dire que le taux de croissance est, en pratique, indépendant de la concentration de l'aliment carboné, sauf lorsque celle-ci est excessivement faible.

Considérons maintenant une culture se développant dans le récipient B (fig. 1) alimenté par une nourrice contenant un milieu de composition telle que la concentration de l'aliment carboné constitue le seul facteur limitant de la croissance. Soit S_0 la concentration de l'aliment limitant dans le milieu neuf ; S, la concentration de cet aliment dans B ; R, la constante de rendement (*Cf.* Monod, *loc. cit.*). On peut écrire :

$$\frac{dS}{dt} = D(S_0 - S) - \frac{1}{R}\frac{dx_T}{dt}.$$ (7)

En effet : 1° la concentration S de l'aliment dans B tend à se rapprocher de la concentration S_0 dans le milieu neuf d'autant plus vite que la différence de ces concentrations est plus grande et le débit plus rapide ; 2° la concentration S tend à diminuer d'autant plus vite que l'accroissement de la masse bactérienne totale est plus rapide. Ce second terme

$$\left(-\frac{1}{R}\frac{dx_T}{dt}\right)$$

exprime l'hypothèse que le rendement de la croissance est constant, indépendamment de son taux (3). En remplaçant

$$\frac{dx_T}{dt}$$

dans l'équation (7) par sa valeur tirée de l'équation (2), on obtient :

$$\frac{dS}{dt} = D(S_0 - S) - \frac{x_B}{R}\mu$$

et en remplaçant μ par sa valeur, donnée par (6) :

$$\frac{dS}{dt} = D(S_0 - S) - \frac{x_B}{R}\mu_0\frac{S}{S_K + S}.$$ (8)

(3) Voir à ce sujet Monod (1942) [1], Teissier (1942) [3] et plus loin, p. 398.

Une culture ne peut être dite à l'équilibre que si S et x_B sont constants, c'est-à-dire lorsque l'on a à la fois :

$$\frac{dS}{dt} = 0$$

et

$$\frac{dx_B}{dt} = 0.$$

D'après les équations (4), (6) et (8), ces deux conditions peuvent s'écrire respectivement :

$$D\,(S_0 - S) = \frac{\mu_B}{R}\,\mu_0\,\frac{S}{S_K + S} \qquad (9)$$

et

$$\mu_0\,\frac{S}{S_K + S} = D. \qquad (10)$$

Il est évident que l'équilibre n'est pas réalisable si $D > \mu_0$ puisque l'équation (10) ne serait vérifiée pour aucune valeur de S ou de D. En revanche, un équilibre stable est nécessairement atteint si $D < \mu_0$. En effet si, à un moment quelconque x_B est inférieur à la valeur satisfaisant l'équation (9), alors

$$\frac{dS}{dt}$$

est positif, S s'accroît, de sorte que

$$\mu_0\,\frac{S}{S_K + S} - D$$

prend une valeur positive, et la culture s'accroît. L'inverse se produit si x_B est supérieur à la valeur d'équilibre. De même, si S est plus grand que la valeur satisfaisant l'équation (10), alors

$$\frac{dx_B}{dt}$$

est positif [équation (4)], la culture croît, de sorte que l'égalité (9) n'est plus satisfaite, et

$$\frac{dS}{dt}$$

devient négatif. L'inverse se produit si S est inférieur à la valeur satisfaisant l'équation (10). Si, l'équilibre étant atteint, on modifie le débit (tout en le maintenant inférieur à μ_0), le système évolue, pour les mêmes raisons, vers un nouvel équilibre ; la densité bactérienne et la concentration de l'aliment limitant se stabilisent à de nouvelles valeurs, telles que le taux de croissance soit de nouveau égal au taux de dilution. *L'expérimentateur dispose* *donc là d'un moyen de modifier à son gré le taux de croissance et de le régler à une valeur quelconque inférieure à μ_0.*

On voit, d'après (10), qu'à l'équilibre l'équation (9) se simplifie en :

$$x_B = R \, (S_0 - S). \qquad (9 \; bis)$$

De plus, comme l'équilibre stable exige

$$D = \mu_0 \, \frac{S}{S_K + S} < \mu_0.$$

comme, d'autre part, la valeur de la fraction

$$\frac{S}{S_K + S}$$

est pratiquement indépendante de S tant que S est d'un ordre de grandeur supérieur à S_K ; comme enfin les valeurs expérimentalement déterminées (v. ci-dessus) de S_K sont très petites, l'équilibre ne sera atteint que pour de faibles valeurs de S.

Cette remarque est importante car, en pratique, les valeurs choisies pour la concentration initiale, S_0, seront presque invariablement beaucoup plus grandes que les valeurs d'équilibre de S. A l'équilibre $S_0 - S$ sera donc très peu différent de S_0, et l'équation (9 *bis*) se simplifiera encore en :

$$x_B = RS_0.$$

Autrement dit, tout se passera comme si, malgré la dilution continue du milieu, l'aliment limitant était entièrement consommé par les bactéries. Eclairons cette conclusion d'un exemple. Pour *E. coli* se développant en milieu défini, avec du glucose comme aliment limitant, la constante S_0 est de l'ordre de 10^{-5} (Monod, 1942). Or, la concentration de glucose qui donnerait une « bonne culture » aux yeux d'un bactériologiste serait de 10^{-3} à 5×10^{-3}, soit 100 à 500 fois plus grande. Avec $D = 0,5 \, \mu_0$ l'équation (10) s'écrirait :

$$0,5\mu_0 = \mu_0 \, \frac{S}{10^{-5} + S}$$

d'où $S = 10^{-5}$. L'équation (9 *bis*) nous donnerait alors, avec $S_0 = 2 \times 10^{-3}$.

$$x_B = R \, (2.10^{-3} - 10^{-5})$$

et l'on voit que x_B ne différerait que de 1/200 de la valeur maximum correspondant à l'utilisation intégrale de l'aliment carboné. Avec $D = 0,9 \, \mu_0$, la densité de la culture, à l'équilibre, ne serait inférieure que de 5 p. 100 à la valeur maximum.

Il apparaît donc que dans ces conditions, non seulement le système est auto-régulateur, mais encore le taux de croissance

151

peut être fixé à toute valeur voulue (inférieure à 0,9 μ_0 environ) et modifié en cours même d'expérience, sans que cependant la densité de la culture subisse de variations appréciables. Ce sont surtout ces propriétés singulières du « régime auto-régulateur » qui font l'intérêt des cultures continues ; dans les paragraphes qui suivent on verra comment ces propriétés peuvent être mises à profit pour l'étude de quelques problèmes-types de physiologie microbienne.

C. APPLICATIONS A QUELQUES PROBLÈMES DE PHYSIOLOGIE MICROBIENNE. — *Relation entre le taux de croissance et la concentration d'un aliment limitant.* — L'étude de cette relation constitue évidemment l'une des applications les plus immédiates de la technique de culture continue. Lorsque pour une telle étude on utilise des cultures en milieu non renouvelé, on rencontre de graves difficultés qui tiennent aux faibles valeurs des constantes d'affinité : dans la plupart des cas, les concentrations d'aliment limitant donnant des cultures visibles se trouvent dans la zone de saturation. Le problème consiste donc à maintenir la concentration de l'aliment limitant à des valeurs constantes et très faibles. Or, c'est là le résultat obtenu automatiquement avec une culture continue lorsque le débit est inférieur au taux de croissance maximum, c'est-à-dire en régime autorégulateur.

Il n'est pas nécessaire d'insister sur le fait évident que l'emploi de cette technique n'est justifié que si la constitution du milieu est telle que l'aliment étudié constitue bien le *seul* facteur limitant de la croissance (*Cf.* à ce sujet, Monod, 1942, p. 33). Ceci posé, il n'est en revanche nullement nécessaire que la forme générale de la relation soit connue. Il faut et il suffit qu'un équilibre *stable* soit atteint pour qu'on puisse écrire :

$$\mu = D = f(S).$$

Or, un équilibre stable ne peut manquer d'être atteint, pour peu que la relation présente la forme générale d'une courbe de saturation, ce qui, *a priori*, semble devoir être toujours le cas. Le dosage direct de l'aliment limitant, dans le liquide de la culture, permet alors d'établir la forme de la relation $\mu = f(S)$.

Il n'est pas non plus nécessaire, pour que la technique soit applicable, que les variations de x_B (densité de la culture) avec le taux de dilution soient négligeables. En fait, la seule limite à l'application de cette technique sera la précision du dosage chimique. On pourrait, en principe, éviter le dosage direct de S, en déduisant sa valeur de la relation (9 *bis*) :

$$x_B = R(S_0 - S).$$

Mais comme les variations de x_B à différents équilibres seront toujours faibles (ou même insensibles), les erreurs seraient considérables. En outre, comme nous allons le voir, la relation (9 *bis*) représente une approximation qui pourrait être en défaut dans certains cas.

Rendement de la croissance. — En effet, considérer l'équation (9 *bis*) comme exacte, revient à admettre l'hypothèse [exprimée par le second terme de l'équation (7)] que le rendement de la croissance est indépendant du taux de croissance. Lorsque cette hypothèse est acceptable, l'équation (9 *bis*) permet de déterminer la constante de rendement R. Le dosage de l'aliment limitant à l'équilibre est inutile. En principe, il suffit de déterminer x_B pour deux ou plusieurs valeurs de S_0. Ce n'est pas là une application particulièrement intéressante de la méthode puisque, dans la mesure même où l'hypothèse d'indépendance est exacte, on peut utiliser tout aussi bien les techniques habituelles. On sait que cette hypothèse se vérifie avec précision lorsque la croissance est limitée par l'épuisement de l'aliment carboné (*Cf.* Monod, *loc. cit.*). Il est probable qu'elle représente une bonne approximation dans la plupart des cas, mais qu'elle cesse d'être exacte au delà de certaines limites, ou pour certains aliments. Par exemple, toute dépense nutritive affectée d'un « coefficient d'entretien » appréciable ne saurait être indépendante du taux de croissance. Si l'on manque de données à cet égard, c'est en partie parce qu'on ne disposait pas d'une technique adéquate. Les « cultures continues » dont on peut faire varier le taux de croissance sans pour cela modifier ni la température, ni la composition du milieu (si ce n'est la concentration d'un aliment) apportent une solution à ce problème expérimental. Si nous admettons que le rendement, R, est une fonction, φ (μ), du taux de croissance, l'équation (9 *bis*) devient :

$$\frac{x_B}{S_0 - S} = \varphi\,(\nu) = \varphi\,(\mathbf{D}). \qquad (9\ ter)$$

La détermination de S et de x_B pour différentes vitesses de dilution permet d'établir la forme de la relation entre le taux de croissance et le rendement.

Cependant, il ne faudrait pas l'oublier, cette méthode n'est justifiée que si le rendement est indépendant de la *concentration* de l'aliment. C'est là une seconde « hypothèse d'indépendance » qui pourrait se trouver en défaut lorsque la concentration de l'aliment varie beaucoup, ou lorsqu'elle devient très petite, ce qui est précisément le cas pour l'aliment limitant à l'équilibre. On conçoit, en effet, qu'une substance métabolisée par deux systèmes enzymatiques distincts avec des constantes d'affinités assez différentes pourrait ne pas donner les mêmes rendements à forte et à faible concentration. La solution de cette difficulté consiste à composer le milieu de façon que le facteur limitant soit une source alimentaire *autre que celle dont on veut déterminer le rendement*. L'équation (7), donc l'équation (9 *ter*), n'en reste pas moins valable pour cela. Il faut encore cependant que la concentration S_0 de l'aliment étudié ne soit pas choisie trop grande afin que S_0-S soit mesurable avec précision, ni trop petite, de façon que S reste assez grand pour que ses variations puissent être considérées comme sans effet.

Les variations du rendement en fonction de la concentration d'un aliment peuvent avoir, dans certains cas, un intérêt propre. Ici encore l'artifice qui consiste à limiter le taux de croissance par un aliment autre que celui dont on étudie le rendement est applicable. D étant maintenu constant, on déterminerait S par dosages, pour différentes valeurs de S_0.

On sait enfin que le rendement de la croissance bactérienne est fonction de la température (*Cf*. Monod, 1942, p. 106). Mais il n'est pas possible, par les techniques usuelles, d'étudier cet effet indépendamment des variations concomitantes du taux de croissance. La culture continue à taux limité permet d'atteindre ce résultat dans certaines conditions. Supposons que, par réglage du débit, le taux de croissance d'une culture soit fixé à une valeur légèrement inférieure au taux maximum correspondant à une température donnée. Supposons que cette température soit assez loin de l'optimum. Cette culture peut être portée à toute température pour laquelle le taux maximum est supérieur au débit fixé, sans que soit modifié le taux de croissance. Il est évident, cependant, que l'accroissement de la température se traduira, dans de telles conditions, par une variation (en général une diminution) de la concentration d'équilibre de l'aliment limitant. On devra donc éventuellement tenir compte de l'effet de dilution discuté dans les paragraphes précédents, et l'éviter en faisant en sorte que l'aliment limitant ne soit pas celui dont on cherche à déterminer le rendement en fonction de la température.

La possibilité de dissocier, dans une certaine mesure, deux phénomènes physiologiques, tous deux fonction de la température, constitue sans doute l'une des applications les plus intéressantes de la méthode. Nous y reviendrons tout à l'heure.

Vitesses de synthèses. — L'étude de la cinétique d'un processus de synthèse spécifique, tel que la formation d'un enzyme ou autre constituant cellulaire, comporte de graves difficultés, et de toutes sortes, la plupart inhérentes au phénomène lui-même. Certaines de ces difficultés cependant tiennent aux techniques de culture. Les suspensions dites « non proliférantes » ne peuvent être employées s'il s'agit d'un phénomène lié, fût-ce indirectement, à la croissance. D'ailleurs, les propriétés de ces suspensions qui contiennent un nombre variable, généralement indéterminé, de germes non viables, se modifient rapidement avec le temps. Une culture en voie de croissance ne peut être considérée comme physiologiquement stable qu'au cours de la phase exponentielle, souvent trop courte pour les besoins de l'expérience. Encore la composition du milieu se modifie-t-elle très rapidement au cours de cette phase. Enfin, la variation continue de la densité bactérienne au cours de l'expérience introduit une difficulté supplémentaire. L'emploi de cultures continues que l'on peut maintenir indéfiniment dans un milieu constant, à taux de croissance constant, à densité constante, est donc tout indiqué.

Les cultures continues présentent pour ce type d'expériences d'autres avantages remarquables. En premier lieu, mesurer le taux d'une réaction de synthèse dans une culture continue revient à déterminer un état d'équilibre, au lieu de mesurer à intervalles successifs l'accumulation d'une substance. En second lieu, le contrôle du taux de croissance permet d'étudier le degré de dépendance (ou d'indépendance) de la réaction de synthèse considérée, à l'égard des processus de synthèse dans leur ensemble. Il permet en somme de distinguer, dans une certaine mesure, les effets d'un agent actif à la fois sur le phénomène envisagé et sur d'autres qui pourraient le masquer. Afin de préciser les propriétés de ces systèmes, nous allons considérer maintenant quelques modèles théoriques de réactions de synthèse.

Supposons une culture continue, en train de se multiplier dans les conditions du régime auto-régulateur. Supposons un constituant cellulaire Z, un enzyme par exemple, synthétisé par les bactéries en train de proliférer. Soit Z_B la quantité de Z dans le récipient B, Z_P la quantité de Z dans le récipient P, Z_T la quantité totale.

On peut écrire :

$$\frac{dZ_B}{dt} = \frac{dZ_T}{dt} - \frac{dZ_P}{dt} \qquad (11)$$

l'accroissement de Z dans le récipient P est donné par :

$$\frac{dZ_P}{dt} = Z_B D \qquad (12)$$

D représentant, comme ci-dessus, la vitesse de dilution.

Pour avoir un système complet d'équations [homologues des équations (1), (2) et (3)], il reste à exprimer l'accroissement de Z_T en fonction des conditions de milieu, de Z_B, éventuellement du taux de croissance, de la température, etc., c'est-à-dire à faire une hypothèse sur le mécanisme ou la nature du processus de synthèse par lequel s'élabore la substance Z. Cette équation, que nous appellerons « hypothétique » pourra, suivant les mécanismes envisagés, prendre des formes très différentes. Cependant.

mis à part certains cas qu'il faudrait qualifier de pathologiques, l'équation hypothétique devra exprimer le fait que la concentration de tout constituant cellulaire ne saurait excéder une certaine limite. Soit X cette limite, qui s'exprimera par exemple comme fraction de la masse bactérienne x_B contenue dans le récipient B. L'équation hypothétique générale sera de la forme :

$$\frac{dZ_T}{dt} = (X - Z_B) \psi \tag{13}$$

dans laquelle ψ représente la fonction hypothétique proprement dite. En comparant les équations (11), (12) et (13), on voit que le système tend nécessairement vers un équilibre puisque

$$\frac{dZ_T}{dt}$$

diminue tandis que

$$\frac{dZ_P}{dt}$$

augmente lorsque Z_B augmente, et inversement ;

$$\frac{dZ_B}{dt}$$

tendra donc toujours vers zéro. Ceci à la condition que ψ ne prenne pas de valeurs infinies, hypothèse que l'on peut exclure *a priori*. Soit Z_E la valeur de Z_B à l'équilibre. La détermination de Z_E donne une mesure de la vitesse de la réaction de synthèse dans les conditions choisies, puisque à l'équilibre on peut écrire, d'après (11) et (12) :

$$\frac{dZ_T}{dt} = \frac{dZ_B}{dt} = Z_E D \cdot$$

Voyons maintenant comment varie Z_E suivant la forme de la fonction ψ, c'est-à-dire suivant l'hypothèse faite sur le mécanisme de la réaction (ou du système de réactions) de synthèse. Il est clair que toutes sortes de fonctions hypothétiques pourraient être envisagées suivant les variables et les phénomènes considérés. Nous nous bornerons ici à quelques cas simples et typiques, de nature à mettre en lumière les propriétés du système.

Supposons d'abord qu'on veuille étudier l'effet sur la synthèse de Z, d'une substance présente dans le milieu. Soit S la concentration de cette substance. Faisons, sur le mécanisme d'action de cette substance, l'hypothèse la plus simple possible : à savoir que la vitesse de la réaction de synthèse est proportionnelle à la concentration S. La fonction ψ devient alors :

$$\psi = KS$$

K étant une constante de proportionnalité.

156

L'équation hypothétique (13) s'écrit donc :

$$\frac{dZ_T}{dt} = (X - Z_B) KS. \qquad (13\ a)$$

Comme à l'équilibre on a :

$$\frac{dZ_T}{dt} = \frac{dZ_P}{dt} = Z_E D$$

et, par définition,

$$Z_B = Z_E.$$

On peut écrire en remplaçant

$$\frac{dZ_T}{dt}$$

et Z_B par leurs valeurs :

$$Z_E D = (X - Z_E) KS$$

d'où :

$$Z_E = X \frac{KS}{D + KS}. \qquad (A)$$

Dans ce cas, comme on le voit, Z_E serait une fonction hyperbolique de D et de S. Sa valeur tendrait vers X lorsque S deviendrait grand, ou lorsque D tendrait vers O. Z_E ne serait nul que pour $D = \infty$ ou $S = O$. Les courbes figuratives de Z_E en fonction de S et de D sont données par les figures 2 et 3. Remarquons que dans ce calcul nous admettons implicitement que S est indépendant de D. C'est dire que la substance « active » dont il s'agit de déterminer l'effet ne saurait être l'aliment limitant. Il n'y a pas là, d'ailleurs, de difficulté. Il faut cependant prendre garde que la concentration S à l'équilibre n'est pas égale à la concentration S_0 dans le milieu neuf, si la substance active est métabolisée à un taux appréciable. Compte tenu de ces observations, la vérification expérimentale de l'hypothèse ainsi que la détermination de la valeur des constantes n'offrent pas de difficulté de principe.

Supposons maintenant que, conservant la même hypothèse quant à l'effet de la substance activante, nous fassions l'hypothèse supplémentaire que la réaction est *autocatalytique*, autrement dit que sa vitesse est à chaque instant proportionnelle à la concentration de la substance Z dans les cellules. L'équation hypothétique devient :

$$\frac{dZ_T}{dt} = (X - Z_B) KSZ_B. \qquad (13\ b)$$

Le même raisonnement que ci-dessus conduit à la solution suivante pour l'état d'équilibre :

$$Z_E = X - \frac{D}{KS}. \qquad (B)$$

Les courbes figuratives de Z_E en fonction de D et de S, d'après l'équation (B) sont données par les figures 2 et 3. On voit que ces courbes sont bien différentes de celles auxquelles conduisait la première hypothèse [réaction non autocatalytique, équation (A)]. En particulier, Z_E est une fonction *linéaire* de D, et s'annule pour une valeur finie de la variable. Autrement dit, lorsque le débit augmente au delà de certaines valeurs, la sub-

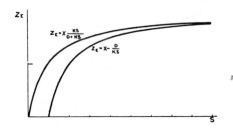

Fɪɢ. 2. — Courbes théoriques exprimant la variation de la concentration d'équilibre d'un constituant Z, en fonction de la concentration de l'inducteur S. Au-dessus : réaction non autocatalytique (équation A). Au-dessous : réaction autocatalytique (équation B).

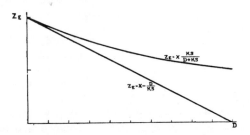

Fɪɢ. 3. — Courbes théoriques exprimant la variation de la concentration d'équilibre du constituant Z en fonction du taux de dilution D. Au-dessus : réaction non autocatalytique (équation A). Au-dessous : réaction autocatalytique (équation B).

stance Z n'est plus formée. De même Z_E devient nul lorsque la concentration de la substance active tombe au-dessous d'une valeur limite. A supposer que l'expérience ne puisse être faite pour ces valeurs critiques de D et de S, l'extrapolation doit permettre de distinguer entre des résultats expérimentaux qui vérifieraient soit l'équation A, soit l'équation B. La mesure de Z_E pour différentes valeurs du débit et de la concentration de la substance active permettrait donc de déterminer si le processus de synthèse du constituant Z est, ou n'est pas, doué des propriétés d'une réaction autocatalytique.

Les hypothèses exprimées par les équations (13 a) et (13 b) sup-

posent implicitement l'indépendance entre le taux de croissance
et le taux de la réaction de synthèse. Si l'on supposait, au
contraire, que le taux de cette réaction soit directement proportionnel au taux de croissance, μ, les termes D et μ disparaî-
traient de la solution puisque au régime autorégulateur D = μ.
L'équation A deviendrait :

$$Z_E = X \frac{KS}{C + KS} \qquad (A')$$

et l'équation B

$$Z_E = X - \frac{C}{KS} \qquad (B')$$

C représente une constante. On voit que dans ce cas Z_E devient
indépendant de D, quelle que soit d'ailleurs la forme de la fonction hypothétique. La détermination de Z_E en fonction de D
permettrait donc d'établir éventuellement les limites de validité
de « l'hypothèse d'indépendance ».

Nous avons supposé jusqu'ici, pour plus de simplicité, que
la vitesse de la réaction de synthèse était *directement* proportionnelle à la concentration de la « substance active » S. On peut
envisager aussi le cas, plus probable, où la réaction serait de
nature enzymatique. Sa vitesse ne serait plus alors proportionnelle à S, mais à une fonction de S qui exprimerait le phénomène
de saturation caractéristique des réactions enzymatiques. Si, par
exemple, on adopte, pour cette fonction, la forme de l'équation
de Michaelis :

$$v = V \frac{S}{S_K + S}$$

dans laquelle V représente la vitesse maximum et S_K la constante
d'affinité, les équations (A) et (B) deviennent respectivement :

$$Z_E = VX \frac{S}{DS_K + DS + VS} \qquad (A'')$$

et

$$Z_E = X - \frac{DS_K + DS}{VS}. \qquad (B'')$$

On verra sans peine que les équations (A'') et (B'') conduisent à
des prévisions expérimentales qui ne sauraient se confondre entre
elles, ni avec les prévisions des équations homologues (A) et (B).
L'expérience, fondée sur ces prévisions, peut donc en principe
départager ces différentes hypothèses.

Si, au lieu de choisir comme variable la concentration d'une
substance supposée « activante », nous avions introduit une
constante de vitesse, supposée fonction d'une variable indépendante quelconque, les résultats eussent été les mêmes. Une telle
variable indépendante serait par exemple la température. Nous

avons vu tout à l'heure que les conditions du régime autorégulateur pouvaient être ainsi fixées, que le taux de croissance soit indépendant de la température dans un intervalle assez large. On voit maintenant comment des variations de température, à taux constant, pourraient être utilisées pour l'analyse du mécanisme d'une réaction de synthèse.

On pourrait envisager encore beaucoup d'autres formes de l'équation hypothétique, destinées à exprimer toutes sortes de mécanismes. Mais il ne s'agit ici que d'exposer les principes. Les exemples que nous venons de discuter suffisent à montrer comment la mesure de la concentration d'un constituant cellulaire dans une culture continue parvenue à l'équilibre permet d'étudier, en fonction de diverses variables, les propriétés de la réaction de synthèse.

Les avantages, si considérables qu'ils soient en théorie comme en pratique, de cette méthode d'équilibre, ne doivent pas nous faire oublier qu'une culture continue se prête également, et bien mieux qu'une culture normale, à la mesure de l'accumulation (ou de la disparition) d'un constituant cellulaire en fonction du temps. Pour que l'emploi de cette seconde méthode soit justifié, il faut et il suffit que les conditions de culture (taux de croissance, débit, composition du milieu) ne varient pas au cours de l'essai. Ceci acquis, on dispose là d'un second moyen de vérifier par l'expérience les prévisions de l'équation hypothétique. En remplaçant

$$\frac{dZ_T}{dt} \quad \text{et} \quad \frac{dZ_P}{dt}$$

dans l'équation (11) par leurs valeurs tirées de (12) et de l'équation hypothétique (13) on obtient :

$$\frac{dZ_B}{dt} = K (X - Z_B) \psi - Z_B D$$

ou, sous forme intégrale :

$$\int \frac{dZ_B}{KX \psi - (K \psi + D) Z_B} = \int dt + Cte.$$

L'intégration donne Z_B en fonction du temps. Pour que l'expérience de vérification soit significative, il faut évidemment qu'au temps zéro la concentration Z_B soit différente de la concentration d'équilibre Z_E correspondant aux conditions choisies. Mais, et ceci est fort important, comme l'équilibre est réversible, il est indifférent que Z_B au temps initial soit plus petit ou plus grand que Z_E. Pratiquement, l'expérience consistera, une fois l'équilibre obtenu dans des conditions données, à modifier les conditions d'équilibre, puis à mesurer Z_B à intervalles adéquats, jusqu'à ce que le nouvel équilibre soit réalisé. En principe, on peut imposer à une même culture une succession indéfinie de ces ruptures d'équilibre. Inutile d'insister sur la richesse des combinaisons expérimentales qu'offre cette technique.

Taux de mutation. — On ne peut ici que mentionner l'application possible de la technique de culture continue à l'étude des mutations bactériennes. Une véritable discussion de ce problème nous entraînerait trop loin. On sait que les théories relatives à la sélection des formes mutantes dans les populations bactériennes ne sont en général valables que pour des cultures en train de se multiplier à taux constant. De plus, l'hypothèse est presque toujours faite, mais n'a jamais été démontrée, que la probabilité de mutation est proportionnelle au taux de croissance. La vérification de cette hypothèse fondamentale semble possible avec les cultures continues à taux limité, de même que la détermination, à taux de croissance constant, des coefficients de température des fréquences de mutation.

III. — Réalisation.

A. *Appareillage.* — Les développements théoriques qui précèdent sont fondés sur des définitions assez rigoureuses des conditions de culture. Il serait vain de chercher à appliquer la théorie si les conditions expérimentales s'écartaient trop sensiblement de ces définitions. Divers types fort différents d'appareils peuvent être imaginés, qui permettraient sans doute de réaliser des conditions assez proches des exigences théoriques. Je n'en décrirai ici qu'un seul. Ce montage est schématisé par la figure 4.

La condition la plus importante est l'homogénéité de la culture, hors quoi la notion de « concentration d'équilibre » de l'aliment limitant devient illusoire. Il faut donc que la culture soit brassée continuellement, intégralement et rapidement. Il faut encore que l'équilibre avec la phase gazeuse soit assuré, ce qui entraîne que le rapport de la surface au volume soit grand. Dans le montage adopté, le récipient de culture est un ballon B, fixé et centré sur un support rotatif. La capacité du ballon est de deux litres pour un volume de culture de 100 à 400 cm³. Un moteur M imprime au ballon une vitesse de rotation de l'ordre de 200 à 400 tours/minute. Le brassage obtenu est très énergique : à cette vitesse, la dispersion d'une goutte de colorant introduite dans le liquide est pratiquement instantanée. L'aération est assurée par le film liquide qui recouvre la plus grande partie de la surface intérieure du ballon. La meilleure répartition de ce film liquide est obtenue quand l'axe de rotation fait un angle de 3 à 4 degrés avec l'horizontale.

Le milieu neuf est en réserve dans une nourrice, en l'occurrence une fiole à toxine, placée 1,50 m. environ au-dessus du niveau du ballon. Le liquide passe dans une tubulure en caoutchouc et un serpentin (Srp) constitué par un tube capillaire en verre de 2 m. de long environ, pour aboutir à une pipette compte-gouttes. Le capillaire est choisi de diamètre tel que le débit moyen désiré soit obtenu lorsque l'extrémité du compte-gouttes est environ à 1 m. au-dessous du niveau de la nourrice. Le réglage se fait par déplacement vertical du compte-gouttes. On obtient ainsi un débit

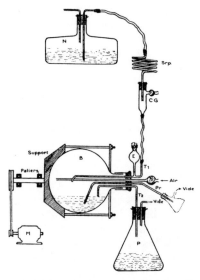

Fig. 4. — Montage d'un appareil à croissance continue. N, nourrice; Srp, serpentin capillaire; C.G., compte-gouttes; B, ballon rotatif; T$_1$, tubulure d'arrivée; E, tubulure d'ensemencement; Pr, tubulure de prélèvement (en pointillé, fiole de prélèvement); T$_2$, tubulure de niveau; P, produit; M, moteur.

suffisamment stable, sans qu'il soit indispensable de maintenir rigoureusement constant le niveau du liquide dans la nourrice. Le compte-gouttes, calibré par pesées, permet une mesure précise du débit à l'aide d'un chronographe.

Le volume du liquide en B est maintenu constant, quel que soit le débit, grâce à une tubulure de niveau (T 2) aboutissant à un récipient P, dans lequel on entretient un vide suffisant et où s'accumule le « produit ». Le volume constant ne dépend pas seulement du niveau de la tubulure T 2 mais aussi de la vitesse de rotation qui détermine l'épaisseur, nullement négligeable, du film liquide. Un accroissement de vitesse non compensé par un changement du niveau de la tubulure T 2 se traduit par un accroissement de volume et inversement. Pour tenir compte de cet effet on détermine le volume par pesée, après essai dans les conditions choisies.

Le système comporte en outre une tubulure munie d'un filtre en coton, par où de l'air (ou tout autre mélange gazeux approprié) est envoyé dans le ballon, après avoir passé dans un flacon laveur contenant de l'eau. Une tubulure d'ensemencement (E), et une tubulure de prélèvement (Pr) peuvent utilement compléter le dispositif.

Le col du ballon est protégé par une pièce annulaire en verre, portée par un bouchon de caoutchouc à travers lequel passent les tubulures.

L'appareil est monté dans une chambre-étuve portée à la température voulue.

B. *Conduite des expériences.* — L'ensemble nourrice — compte-gouttes — tubulures est stérilisé d'un part, le ballon de l'autre. Le montage est fait stérilement. Le ballon est rempli par la tubulure E d'une « culture de départ » à densité convenable. Le débit est amorcé en créant une dépression temporaire dans l'ampoule du compte-gouttes. Les prélèvements directs sont faits en abouchant une fiole à vide à l'extrémité de la tubulure « Pr ».

Ceci dit, tout dépendra du type d'expérience envisagé. Quel qu'il soit, cependant, certaines précautions devront être respectées :

1° La densité bactérienne à l'équilibre ne doit jamais être telle que des facteurs inconnus ou incontrôlés (vitesse de dissolution de l'oxygène, par exemple) ne deviennent limitants. Aussi est-il sage de ne pas dépasser 0,3 à 0,5 µg de poids sec par centimètre cube, et d'utiliser des cultures plus diluées chaque fois que c'est possible. Lorsqu'on peut se contenter de cultures très diluées (0,05 µg par centimètre cube par exemple), il est facile de composer les milieux de façon que la « concentration d'équilibre » des aliments autres que l'aliment limitant soit pratiquement la même que dans le milieu neuf. Ce qui simplifie l'expérience, les calculs et les interprétations.

2° Les prélèvements directs modifient le volume en B, par conséquent le taux de dilution, d'où rupture d'équilibre. Il faut donc qu'ils soient assez petits pour que l'écart soit négligeable, ou assez espacés pour que le système soit entre-temps revenu à l'équilibre. Lorsqu'il n'est pas nécessaire d'effectuer le prélèvement à un moment bien défini (détermination d'un état d'équilibre), on peut avec avantage recueillir le « produit » au lieu de prélever directement.

3° Les conditions dans l'échantillon prélevé changent radicalement et *très rapidement* : ainsi la disparition de l'aliment limitant peut n'être l'affaire que de quelques secondes. Le traitement adéquat doit donc être appliqué *immédiatement*. Le plus efficace et le plus généralement applicable consiste à recevoir l'échantillon sur de la glace pilée.

4° **Les risques de contamination**, avec cet appareillage, ne sont pas négligeables. Nous n'en avons pas observé cependant au cours de nombreuses séries d'expériences d'une durée **maximum** de sept à huit heures. Le montage décrit n'est pas conçu pour un fonctionnement permanent.

IV. — Conclusions et commentaires.

1° Le maintien d'une culture bactérienne, à densité constante, à taux de croissance constant, dans un milieu de composition constante est possible en théorie et réalisable en pratique, grâce au principe de « dilution continue à volume constant ».

Pour qu'un tel système soit en équilibre, il faut et il suffit que le taux de dilution (rapport du volume débité par unité de temps au volume de la culture) soit égal au taux de croissance (nombre de divisions par heure) multiplié par le coefficient 0,69 (logarithme népérien de 2). Ce résultat peut être atteint de deux façons différentes : a) en ajustant le taux de dilution de façon qu'il équilibre le taux de croissance maximum ; cet équilibre est instable ; il ne peut être maintenu que par ajustements continuels ; b) en laissant croître la culture jusqu'à ce qu'une condition devienne limitante (concentration d'un aliment essentiel), le taux de dilution étant fixé à une valeur inférieure à celle qui équilibre le taux de croissance maximum. Cet équilibre est stable. Le taux de croissance est alors limité par la concentration d'un aliment, concentration déterminée elle-même par le taux de dilution (régime autorégulateur).

2° Avec une culture continue en régime autorégulateur, l'expérimentateur dispose d'un moyen de fixer le taux de croissance à toute valeur voulue inférieure au maximum. *Autrement dit, le taux de croissance devient, dans une certaine mesure, une variable indépendante.* C'est là une ressource expérimentale précieuse ; inutile d'insister, dans ce résumé, sur ses nombreuses applications possibles. En revanche, il faut souligner le danger qu'il y aurait à s'en prévaloir, dans l'interprétation des expériences, pour traiter le taux de croissance comme une variable univoque et abstraite. Le « réglage » du taux de croissance, dans une culture continue, est obtenu par l'intervention d'un facteur limitant. Beaucoup d'éléments d'interprétation dépendent de la nature du facteur limitant, de ses effets sur la composition et le métabolisme des bactéries, etc...

3° Les cultures continues se prêtent particulièrement bien à l'étude de la cinétique des processus de synthèse. La composition d'une cellule en train de croître, entendue comme le rapport de chaque constituant cellulaire à la masse totale, tend vers un équilibre lorsque le taux de croissance est constant. La concentration de chaque constituant cellulaire, à l'équilibre, dépend des constantes de vitesse du processus de synthèse intéressé. La détermination de l'équilibre obtenu (c'est-à-dire la mesure de la concentration stable d'un constituant cellulaire), en fonction de diverses variables, revient à déterminer l'effet de ces variables sur la vitesse du système de réactions par quoi s'élabore ce constituant. Cette méthode d'équilibre présente évidemment de grands avantages théoriques et pratiques. Elle permet de comparer des modèles théoriques à des résultats expérimentaux dans des conditions telles que beaucoup de facteurs « étrangers » soient annulés, ou plutôt maintenus constants. En somme, la technique de culture continue appliquée à l'étude de ces difficiles problèmes permet de gagner quelques degrés de liberté dans l'expérimentation. Mais ici encore la simplification du problème expérimental ne doit pas créer l'illusion que les phénomènes eux-mêmes soient simplifiés. Fût-il entièrement satisfaisant, le modèle théorique d'un processus de synthèse n'est qu'une représentation partielle d'une somme de phénomènes. Il n'y a pas à craindre, d'ailleurs, que les résultats ne s'accommodent trop aisément d'interprétations naïvement mécanicistes. Ces quelques degrés de liberté supplémentaires donnent au contraire le moyen de mettre les schémas théoriques à plus rude et plus sévère épreuve.

4° La technique de culture continue trouverait sans doute des applications intéressantes dans l'analyse de la mutabilité. Mais il faut à ce propos mentionner une difficulté qui n'a pas été envisagée dans la discussion théorique. On a, pour cette discussion, supposé implicitement que les cultures étaient génétiquement homogènes. Homogènes tout au moins en ce qui concerne les caractères ou propriétés en cause. Dans l'expérimentation, au contraire, on ne peut négliger *a priori* les facteurs de sélection. L'hypothèse que la sélection intervient doit être considérée dans tous les cas, ne fût-ce que pour l'éliminer par des contrôles adéquats (Monod, Torriani et Doudoroff, 1950). Savoir comment en tenir compte, ou l'éliminer, quels contrôles faire, dépend du problème envisagé et ne peut être discuté ici.

BIBLIOGRAPHIE

[1] Monod (J.). Recherches sur la croissance des cultures bactériennes. Actualités scientifiques et industrielles. Hermann, éd., Paris, 1942. *Ann. Rev. Microb.*, 1949, **3**, 371.

[2] Monod (J.), Torriani (A. M.) et Doudoroff (M.). Ces *Annales* (sous presse).

[3] Teissier (G.). *La Revue Scientifique*, 1942, 209-214.

Reprinted from *Science*, **112**(2920), 715–716 (1950)

Description of the Chemostat

Aaron Novick and Leo Szilard

*Institute of Radiobiology and Biophysics,
University of Chicago*

We have developed a device for keeping a bacterial population growing at a reduced rate over an indefinite period of time. In this device, which we shall refer to as the Chemostat, we have a vessel (which we shall call the growth tube) containing V ml of a suspension of bacteria. A steady stream of nutrient flows from a storage tank at the rate of w ml/sec into the tube. The contents of the tube are stirred by bubbling air through it, and the bacteria are kept homogeneously dispersed throughout the tube at all times. An overflow sets the level of the liquid in the growth tube, and through that overflow the bacterial suspension leaves the tube at the same rate at which fresh nutrient enters it.

The chemical composition of the nutrient is such that it contains a high concentration of all growth factors required by the bacterium, with the exception of one, the controlling growth factor, the concentration of which is kept relatively low. The concentration of the controlling growth factor, a, in the storage tank will determine the density, n, of the bacterial population in the growth tube in the stationary state, and it can be shown that, except for very low values of n, we have

$n = \dfrac{a}{A}$, where A is the amount of the controlling growth

factor needed for the production of one bacterium.

The growth rate $\alpha = \dfrac{1}{n}\dfrac{dn}{dt}$ of a strain of bacteria is a

function of the concentration, c, of the controlling growth factor in the medium, and in general we may expect the growth rate, at low concentrations c, first to increase rapidly with increasing concentration and then slowly to approach its highest attainable value, α_{max}.

The Chemostat must be so operated that the washing-

out time, $\dfrac{w}{V}$, should be lower than the growth rate α_{max}

for high concentrations of the controlling growth factor. It can be shown that in that case a stationary state will become established in which the growth rate, α, will be

just equal to the washing-out rate, $\dfrac{w}{V}$.

What happens is that n will increase until it becomes so large that the bacteria will take up the controlling growth factor from the tube just as fast as it is necessary in order to reduce c to the point where the growth rate

$\alpha(c)$ becomes equal to the washing-out rate, $\dfrac{w}{V}$.

Using a tryptophane-requiring strain of coli and a simple lactate medium with tryptophane added, we have used both lactate and tryptophane as the controlling growth factor. Using tryptophane, we have kept bac-

terial populations growing over long periods of time at rates up to ten times lower than normal. We are thus able to force protein synthesis to proceed very slowly while certain other biochemical processes may continue at an undiminished rate.

A study of this slow-growth phase by means of the Chemostat promises to yield information of some value on metabolism, regulatory processes, adaptations, and mutations of microorganisms. A study of the spontaneous mutations of bacteria growing in the Chemostat has been made and is being published elsewhere.

Because for most investigations a number of such Chemostats will be needed, we attempted to perfect a simple yet adequate design. Of various possible designs, we eliminated those in which changes in the barometric pressure affect the rate of flow of the nutrient from the storage tank into the growth tube. We also discarded designs that permit growth of the bacteria on the inner walls of the growth tube, or permit growth of bacteria in the Chemostat anywhere except homogeneously dispersed in the liquid nutrient in the tube. After trying out several designs, we found the one shown in Fig. 1 satisfactory.

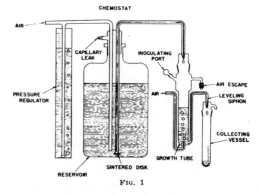

FIG. 1

A tube leading to the bottom of the storage tank is connected to a small air compressor (for example, an air pump such as is used for aerating aquaria). When the compressor is first started, the air rises rapidly in bubbles through the nutrient liquid in the storage tank and accumulates in the space above the liquid level until the pressure in the nutrient at the bottom of the tank becomes equal to the air pressure in the tube. The air space in the storage tank above the liquid level communicates through a narrow capillary with the outside air, and therefore the air will continue indefinitely to bubble through the nutrient liquid in the storage tank, but at a very slow rate (of perhaps one bubble per minute).

The pressure of the air entering the tube is regulated by a simple pressure regulator consisting of an air outlet located at the bottom of a glass cylinder filled with water

2

up to a certain level. Above this level, the air communicates freely with the outside air. By changing the water level in the pressure regulator, the air pressure can be adjusted to any value required for the operation of the Chemostat.

In this arrangement, the pressure at the bottom of the storage tank will always be greater than the pressure of the outside air by the height of the water column in the pressure regulator, and hence will be independent of the height of the level of the nutrient liquid. This is important because the level of the nutrient will gradually fall during the operation of the Chemostat.

From the storage tank the nutrient liquid is forced through a sintered glass filter into the growth tube, where it is mixed drop by drop with the bacterial suspension. The content of the growth tube is continuously stirred by aeration.

The level of the liquid in the tube is set by a siphon, and the volume of the bacterial suspension is thus maintained constant. The nutrient liquid and the bacteria suspended in it leave the tube through the syphon at the same rate at which fresh nutrient enters. The air space above the nutrient liquid in the growth tube communicates with the outside air, hence the pressure which forces the nutrient liquid through the sintered disk is at all times equal to the height of the water column in the pressure regulator.

If, after the Chemostat has been in operation for some time, the barometric pressure falls very suddenly, the pressure of the air entering into the storage tank also falls suddenly, and the nutrient liquid will rise in the air pressure tube to a certain height. If this happens, the pressure at the bottom of the storage tank will no longer exceed the outside pressure by the height of the water column in the regulator, but rather by a greater amount, and the flow of the nutrient liquid into the growth tube increases. Because of the capillary communication between the air space above the nutrient liquid and the outside air, this condition will be quickly corrected. As air flows out of the storage tank through the capillary outlet, the pressure diminishes, and the liquid which had risen into the air pressure tube in the tank is pushed out. Thus, within a short period of time, the pressure at the bottom of the storage tank is restored to its former value.

In this manner the Chemostat keeps the rate of flow of the nutrient liquid into the growth tube constant, independent of changes in barometric pressure and in the liquid level in the tank. The flow rate can be changed as desired by changing the water level in the pressure regulator.

12

Reprinted from *J. Gen. Microbiol.*, **14**(8), 601–622 (1956)

The Continuous Culture of Bacteria; a Theoretical and Experimental Study

By D. HERBERT, R. ELSWORTH AND R. C. TELLING

Microbiological Research Department (Ministry of Supply), Porton, Wiltshire

SUMMARY: A theoretical treatment of continuous culture is given, which allows quantitative prediction of the steady-state concentrations of bacteria and substrate in the culture, and how these may be expected to vary with change of medium, concentration and flow-rate. The layout and operation of a small pilot plant for the continuous culture of bacteria are described. This plant has been operated continuously for periods of up to 4 months without breakdown or contamination of the culture. No alterations in the properties of the organisms studied have occurred during such periods of continuous culture. Results are given of a series of experiments on the continuous culture of *Aerobacter cloacae* in a chemically defined medium, designed to allow quantitative comparison with the results predicted by the theory. The relative advantages of batch and continuous culture as production processes are discussed, and it is concluded that continuous culture may usually be expected to show a five to tenfold increase in output as compared with a batch process.

The continuous culture of micro-organisms is a technique of increasing importance in microbiology. The essential feature of this technique is that microbial growth in a continuous culture takes place under steady-state conditions; that is, growth occurs at a constant rate and in a constant environment. Such factors as pH value, concentrations of nutrients, metabolic products and oxygen, which inevitably change during the 'growth cycle' of a batch culture, are all maintained constant in a continuous culture; moreover, they may be independently controlled by the experimenter. These features of the continuous culture technique make it a valuable research tool, while it offers many advantages, in the form of more economical production techniques, to the industrial microbiologist. Nevertheless, the technique has so far been comparatively little used. (The review of Novick (1955) lists nearly all the work on the subject that has yet appeared.) The reasons for this relative neglect are, we believe, twofold.

The first reason is the lack of a generally accepted theoretical background. Continuous culture presents theoretical problems of an essentially kinetic nature which must be solved before the technique can be intelligently applied. The basis of a correct theoretical treatment has, in fact, already been laid in important papers by Monod (1950) and Novick & Szilard (1950), but later writers on the subject (Golle, 1953; Finn & Wilson, 1954; Northrop, 1954) have either disagreed with, or failed to understand, this work; at least their theoretical treatments are quite different (and in our view erroneous). This may have been because the earlier writers presented little experimental data in support of their theoretical conclusions. One object of the present work was to obtain quantitative data to test the general validity of Monod's theory.

A second reason for the neglect of continuous culture techniques, in particular by industrial microbiologists, is the apparently widespread belief that they are so difficult as to be impracticable. Difficulty in maintaining sterility during long runs and the probability of mutations are two objections which have frequently been raised (e.g. Warner, Cook & Train, 1954*a, b*). These objections, which have been answered by Dawson & Pirt (1954), are based solely on conjecture since no serious attempts have hitherto been made (in Great Britain at least) to apply continuous techniques in industrial microbiology. Our own experience shows that these purely practical difficulties have been greatly exaggerated.

The present paper attempts to deal with both of the aspects of continuous culture mentioned above. In the first part a mathematical theory of continuous culture is presented; the second part describes the operation of a small pilot-scale continuous culture apparatus and gives data obtained during continuous growth of *Aerobacter cloacae*, for comparison with results predicted by theory.

THEORY

General principles of continuous flow systems

All continuous flow systems consist essentially of some form of reactor into which reactants flow at a steady rate and from which products emerge. The factors governing their operation are: (i) the way in which material passes through the reactor (which depends upon its design); (ii) the kinetics of the reaction taking place. As Danckwerts (1954) stated, (i) may be characterized by the *distribution of residence-times* of molecules or minute particles passing through the system. Most reactors lie between the extremes of the completely-mixed tank and the ideal tubular type with 'piston flow' and no mixing. In ideal piston flow all particles have the same residence-time, equal to the *mean residence-time*, $\bar{t}$, while complete mixing produces a wide spread of residence-times about the mean.

In a piston-flow reactor the extent of the reaction (whatever its kinetics) will be the same as for a batch reactor operated for a period equal to $\bar{t}$, while in a completely mixed reactor some of the material will have reacted for a considerably longer and some for a considerably shorter period than $\bar{t}$. Danckwerts (1954) pointed out that the piston flow reactor will be the more efficient for chemical reactions whose rates fall off as the reaction proceeds, but the completely-mixed reactor will be more efficient for reactions of the 'autocatalytic' type whose rate increases with time. Since bacterial growth is an autocatalytic process, the completely-mixed reactor will therefore be the most efficient type for continuous bacterial culture (a fact which has not always been appreciated), and only this type will be considered here.

Completely-mixed continuous culture apparatus. In the type of apparatus to be considered the reactor consists of some form of culture vessel in which the organism can be grown under suitable conditions. Sterile growth medium is fed into the culture vessel at a steady flow-rate f and culture emerges from it at the same rate, a constant-level or similar device keeping the volume of culture in the vessel (v) constant. The contents of the vessel are sufficiently well stirred to approximate to the ideal of complete mixing, so that the entering growth medium is instantaneously and uniformly dispersed throughout the vessel.

Residence-times in such a culture vessel will be determined not by the absolute values of the flow-rate and culture volume but by their ratio which we call the *dilution rate, D*, defined as $D = f/v$, i.e. the number of complete volume-changes/hr. The mean residence-time of a particle in the culture vessel is evidently equal to $1/D$.

Assume for the moment that the bacteria in the culture vessel are not growing or dividing. With complete mixing, every organism in the vessel has an equal probability of leaving it within a given time. It can easily be shown to follow that the fraction of the total organisms in the vessel having a residence-time $\geq t$ is e^{-Dt}. The *wash-out rate*, i.e. the rate at which organisms initially present in the vessel would be washed out if growth ceased but flow continued is therefore:

$$-\frac{dx}{dt} = Dx, \tag{1}$$

where x is the concentration of organisms in the vessel. The distribution of residence-times and the wash-out rate in a completely-mixed continuous culture vessel of this sort can thus be adequately described in fairly simple terms. Their application to continuous culture requires, in addition, a knowledge of the basic kinetics of the growth process.

Kinetics of bacterial growth

Theoretical discussions of bacterial growth usually start from the familiar 'exponential growth' equation

$$\frac{1}{x}\frac{dx}{dt} = \frac{d(\log_e x)}{dt} = \mu = \frac{\log_e 2}{t_d}, \tag{2}$$

where x is the concentration of organisms (dry weight of organisms/unit volume) at time t, μ is the *specific growth rate** and t_d is the *doubling time*,† i.e. the time required for the concentration of organisms to double. In this equation, μ and t_d are usually assumed to be constants; it is insufficiently appreciated, however, that this assumption is correct only when all substrates necessary for growth are present in excess.

Monod (1942) first showed that there is a simple relationship between the specific growth rate and the concentration of an essential growth substrate, μ being proportional to the substrate concentration when this is low but reaching a limiting saturation value at high substrate concentrations according to the equation

$$\mu = \mu_m\left(\frac{s}{K_s + s}\right), \tag{3}$$

* The actual rate of increase of concentration of organisms (dx/dt) is called the *growth rate*; the rate of increase/unit of organism concentration $\left(\frac{1}{x}\frac{dx}{dt}\right)$ is called the *specific growth rate*.

† The doubling time is sometimes confused with the *mean generation time*, but strictly the two could only be identical if all organisms had identical individual generation times.

where s is the substrate concentration, μ_m is the *growth rate constant* (i.e. the maximum value of μ at saturation levels of substrate) and K_s is a *saturation constant* numerically equal to the substrate concentration at which $\mu = \frac{1}{2}\mu_m$. It follows from equation (3) that exponential growth can occur at specific growth rates having any value between zero and μ_m, provided the substrate concentration can be held constant at the appropriate value—a fact of major importance in continuous culture (Monod, 1950; Novick & Szilard, 1950).

Monod (1942) also showed that there is a simple relationship between growth and utilization of substrate. This is shown in its simplest form in growth media containing a single organic substrate (e.g. glucose, ammonia and salts); under these conditions the growth rate is a constant fraction, Y, of the substrate utilization rate:

$$\frac{dx}{dt} = -Y\frac{ds}{dt},\tag{4}$$

where Y is known as the *yield constant*. Thus over any finite period of growth

$$\frac{\text{weight of bacteria formed}}{\text{weight of substrate used}} = Y.$$

If the values of the three growth constants μ_m, K_s and Y are known, equations (2) to (4) provide a complete quantitative description of the 'growth cycle' of a batch culture (Monod, 1942). The same equations and constants are equally applicable to the theoretical treatment of continuous culture.

Growth in continuous culture

Experimental arrangement. Consider bacteria growing in a completely-mixed type of continuous culture vessel as described above, the inflowing medium containing a single organic substrate (e.g. glucose) at a concentration s_R. It is assumed that all other substrates are present in excess, and the culture vessel is so efficiently aerated that the oxygen supply is always adequate; the supply of organic substrate is therefore the sole growth-limiting factor. (Alternatively, conditions may be assumed to be completely anaerobic, when the same theoretical treatment will apply, though the actual values of μ_m and Y will be different.) The variables within the control of the experimenter are the substrate concentration and flow rate of the incoming culture medium and a complete theory must describe how variation of these affects the growth rate and the concentrations of organisms and of substrate in the growth vessel.

Changes in concentration of organisms. In the culture vessel the organisms are growing at a rate described by equation (2) and simultaneously being washed away at a rate determined by equation (1). The net rate of increase of concentration of organisms is given by the simple balance equation (individual terms referring to *rates* in each case):

$$\text{increase} = \text{growth} - \text{output},$$

or

$$\frac{dx}{dt} = \mu x - Dx.\tag{5}$$

Hence if $\mu > D$, dx/dt is positive and the concentration of organisms will increase, while if $D > \mu$, dx/dt is negative and the concentration of organisms will decrease, eventually to zero; i.e. the culture will be 'washed out' of the culture vessel. When $\mu = D$, $dx/dt = 0$ and x is constant; i.e. we have a steady state in which the concentration of organisms does not change with time. Under such steady-state conditions, the specific growth rate, μ, of the organisms in the culture vessel is exactly equal to the dilution rate D. This equation has been derived by several workers, but it has not always been realized that by itself the equation gives no information on what dilution rates make a steady state possible. To know this, we must also know how dilution rate affects the concentration of substrate in the culture vessel, since as already mentioned (equation 3) the value of μ depends on s.

Changes in substrate concentration. In the culture vessel, substrate is entering at a concentration s_R, being consumed by the organisms and flowing out at a concentration s. The net rate of change of substrate concentration is obtained by another balance equation (individual terms again referring to rates):

$$\text{increase} = \text{input} - \text{output} - \text{consumption}$$

$$= \text{input} - \text{output} - \frac{\text{growth}}{\text{yield constant}} \quad \text{(from eqn. 4),}$$

$$\frac{ds}{dt} = Ds_R - Ds - \frac{\mu x}{Y}. \tag{6}$$

Fundamental equations of continuous culture. Equations (5) and (6) both contain μ, which is itself a function of s (equation 3). Substituting (3) in these equations we have:

from (5)
$$\frac{dx}{dt} = x\left\{\mu_m\left(\frac{s}{K_s + s}\right) - D\right\}, \tag{7}$$

and from (6)
$$\frac{ds}{dt} = D(s_R - s) - \frac{\mu_m x}{Y}\left(\frac{s}{K_s + s}\right). \tag{8}$$

These are virtually identical with equations given by Monod (1950), though the above derivation is different from (and, we think, simpler than) Monod's. They define completely the behaviour of a continuous culture in which the fundamental growth relations are given by equations (2) to (4).

The steady state. It is apparent from consideration of equations (7) and (8) that if s_R and D are held constant and D does not exceed a certain critical value (see equation 12), then unique values of x and s exist for which both dx/dt and ds/dt are zero; i.e. the system is in a steady state. Solving (7) and (8) for $dx/dt = ds/dt = 0$, these steady-state values of x and s, which will be designated $\tilde{x}$ and $\tilde{s}$ are given as

$$\tilde{s} = K_s\left(\frac{D}{\mu_m - D}\right), \tag{9}$$

$$\tilde{x} = Y(s_R - \tilde{s}) = Y\left\{s_R - K_s\left(\frac{D}{\mu_m - D}\right)\right\}. \tag{10}$$

From these equations the steady-state concentrations of bacteria and substrate in the culture vessel can be predicted for any value of the dilution rate and concentration of inflowing substrate, provided the values of the growth constants μ_m, K_s and Y are known. They may be said to describe completely the behaviour of a continuous culture running under steady-state conditions. These equations also were first derived by Monod (1950).

While these equations describe accurately the situation existing once a steady state has become established, no proof was given by Monod that, starting from non-steady state conditions (such as exist at inoculation for example), a steady state must inevitably be reached. Rigorous proof of this has lately been provided by our colleague E. O. Powell who has shown (unpublished work) that, starting from any initial values of x and s, the system inevitably adjusts itself to the steady state defined by equations (9) and (10), and that this is the only stable state of the system. For example, consider a system which has just been inoculated, when x is very small, s is nearly equal to s_R and $\mu > D$. The concentration of organisms will increase, but owing to the resulting fall in substrate concentration the specific growth rate will decrease, until eventually μ becomes equal to D. At this point the combined rates of substrate consumption and loss just balance the rate of substrate addition and the system shows no further tendency to change. The system is stable in the sense that small accidental fluctuations from the steady-state values will set up opposing reactions which will restore the *status quo*. It is this automatic self-adjusting property of the system that makes continuous culture a readily feasible possibility.

As previously mentioned, in the steady-state the specific growth rate is equal to the dilution rate

$$\mu = \frac{\log_e 2}{t_d} = \mu_m \cdot \left(\frac{\tilde{s}}{K_s + \tilde{s}}\right) = D. \tag{11}$$

The doubling time t_d is therefore equal to $0 \cdot 693/D$; e.g. if one volume per hour is flowing through the culture vessel, the mass of organisms will be doubling every 42 min.

As is evident from equations (9) and (10), the steady-state values of the concentrations of organisms and substrate depend solely on the values of s_R and D (since μ_m, K_s and Y are constant for a given organism and growth medium). By varying s_R and D an infinite number of steady states can be obtained.

Effect of varying dilution rate. Fig. 1 shows how the mean generation time and the steady-state concentrations of bacteria and substrate in a continuous culture may be expected to vary when the dilution rate is varied, the inflowing substrate concentration being held constant; the curves are plotted from equations (9) and (10). The concentration of organisms has a maximum value when the dilution rate is zero, the substrate concentration then also being zero; i.e. the situation corresponds to the final stage of a batch culture. As the dilution rate increases the substrate concentration increases and the concentration of organisms falls, until at a critical value of D the concentration of organisms becomes zero and the substrate concentration becomes equal to s_R;

if plotted on appropriate scales the curves for bacterial and substrate concentrations are mirror-images.

The critical value of the dilution rate, above which complete 'wash-out' occurs, is obviously of great practical importance. This critical value, which is designated D_c, will be seen from equation (5) to be equal to the highest possible value of μ, which is the value attained when $\tilde{s}$ has its highest possible value s_R, and is given by

$$D_c = \mu_m \left(\frac{s_R}{K_s + s_R} \right). \tag{12}$$

When $s_R \gg K_s$, which is usually the case, then $D_c \sim \mu_m$.

At all dilution rates greater than D_c, dx/dt will be negative (from equation 5), bacteria will be washed out of the culture vessel faster than they can grow, and no steady state with $\tilde{x} > 0$ is possible.

Effect of inflowing substrate concentration. In Fig. 2 the variation with the dilution rate of the steady-state concentrations of organisms and substrate is shown for a number of different values of the inflowing substrate concentration s_R; the curves are plotted from equation (9). It will be seen that at a given dilution rate below the critical the concentration of organisms is nearly proportional to s_R (see equation 9), but the concentration of substrate is independent of s_R, i.e. the curve relating dilution rate to the concentration of substrate in the culture vessel is the same whatever the concentration of substrate in the inflowing culture medium (cf. Novick & Szilard, 1950). In other words, when the dilution rate is fixed, the substrate concentration must come to a level (determined by equation 9) that makes μ equal to D, and this level is independent of s_R.

The curve relating concentration of organisms to dilution rate is seen (Fig. 2) to be displaced vertically as s_R increases, the drop in $\tilde{x}$ at high dilution rates being steeper for higher values of s_R. Consideration of equation (3) shows that the important factor here is not the absolute value of s_R but the value of the ratio s_R/K_s. The higher this ratio, the greater the fraction of total substrate that can be consumed without an appreciable decrease in the specific growth rate. Hence as s_R/K_s is increased the concentration of organisms is maintained at nearly the maximum value up to higher values of D and the critical dilution rate D_c approaches more closely to μ_m (cf. equation 12). The value of K_s is very low (10^{-4} M or less) for most substrates, so that s_R/K_s ratios will usually be high in practice.

Performance criteria: output and yield. When a continuous culture system is viewed as a production process, its performance may be judged by two criteria: (i) the quantity of bacteria produced in unit time, which will be called the *output rate*; (ii) the quantity of bacteria produced from unit weight of substrate, which will be called the *effective yield*.

The total output from a continuous culture unit in the steady state is obviously equal to the product of flow-rate and concentration of organisms, or $f\tilde{x}$; the output/unit volume of culture is therefore $D\tilde{x}$, and from equation (10):

$$\text{Output} = D\tilde{x} = DY \left\{ s_R - K_s \left(\frac{D}{\mu_m - D} \right) \right\}. \tag{13}$$

Now as D is increased from 0 to D_c, $\tilde{x}$ decreases from Ys_R to 0, and it can be shown that there is a value of D for which the product $D\tilde{x}$ is a maximum; in other words, for any system there is a particular dilution rate, D_M, which gives the maximum output of organisms in unit time. This maximum output value of D is obtained by differentiating equation (13) with respect to D and equating to zero, and is

$$D_M = \mu_m \left\{ 1 - \sqrt{\frac{K_s}{K_s + s_R}} \right\}. \tag{14}$$

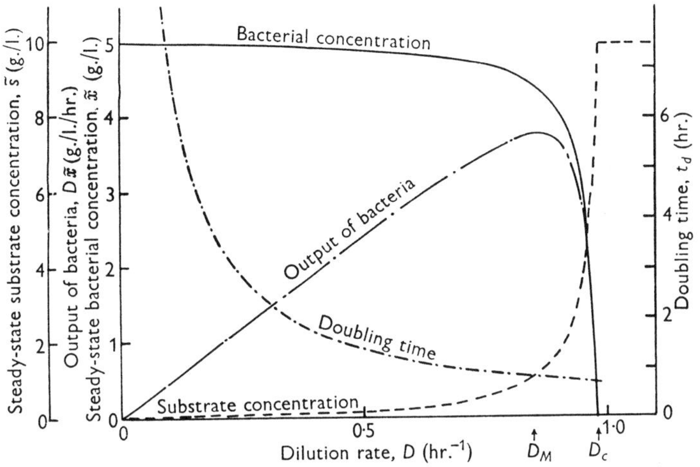

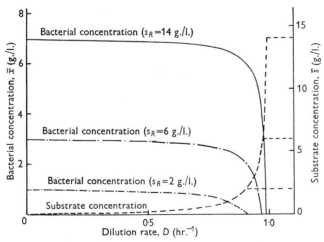

Fig. 1. Steady-state relationships in a continuous culture (theoretical). The steady-state values of substrate concentration, bacterial concentration and output at different dilution rates are calculated from equations (9) and (10), for an organism with the following growth constants: $\mu_m = 1\cdot0$ hr.$^{-1}$, $Y = 0\cdot5$ and $K_s = 0\cdot2$ g./l.; and a substrate concentration in the inflowing medium of $s_R = 10$ g./l.

Fig. 2. Effect of varying the concentration of substrate in the inflowing medium (s_R) on the steady-state relationships in a continuous culture (theoretical). The curves are calculated from equations (9) and (10) for an organism with $\mu_m = 1\cdot0$ hr.$^{-1}$, $Y = 0\cdot5$, and $K_s = 0\cdot1$ g./l., for media of three different substrate concentrations.

The steady-state concentration of organisms at this dilution rate is obtained by substituting the above value of D_M in equation (10) and is

$$\tilde{x}_M = Y\{(s_R + K_s) - \sqrt{(K_s(s_R + K_s))}\}. \tag{15}$$

The *maximum output rate*, $D_M\tilde{x}_M$, is the product of equations (14) and (15) and is given below (equation 23).

Output of organisms is plotted against dilution rate in Fig. 1. The output is nearly proportional to D at low D values (up to about $0.5\ D_c$); as D increases the output curve flattens, reaches a maximum and falls rapidly to zero at the critical dilution rate.

The yield constant Y, defined as in equation (4), is related to the steady-state concentrations of organisms and substrate in a continuous culture by the expression

$$Y = \frac{\text{output of bacteria}}{\text{substrate utilized}} = \frac{\tilde{x}}{s_R - \tilde{s}}. \tag{16}$$

The *effective yield*, Y_E, is defined as the ratio of bacteria formed to substrate supplied in the inflowing culture medium and is given by

$$Y_E = \frac{\text{output of bacteria}}{\text{input of substrate}} = \frac{\tilde{x}}{s_R} = \frac{Y(s_R - \tilde{s})}{s_R}. \tag{17}$$

At all flow-rates > 0, the effective yield is less than the yield constant owing to the substrate wasted in the outflow, s, which increases with the dilution rate (Fig. 1).

The efficiency of utilization of the substrate supplied in the inflowing growth medium is given by equations (9) and (17) as

$$\frac{Y_E}{Y} = \frac{s_R - \tilde{s}}{s_R} = \frac{s_R - K_s\left(\dfrac{D}{\mu_m - D}\right)}{s_R}. \tag{18}$$

It follows that for maximum utilization of substrate the dilution rate should be as low as possible; this is, strictly speaking, incompatible with a maximum output of bacteria, which requires the dilution rate to be high (D_M being close to D_c). However, the shape of the curve relating $\tilde{s}$ to D (Fig. 1) is such that loss of substrate in the outflow is in practice negligible up to quite high dilution rates; in the example plotted utilization of substrate is $> 95\%$ complete at all dilution rates up to $0.7\ D_c$, and is still 90% complete at the maximum output rate D_M.

The efficiency of utilization of substrate at the maximum output rate D_M is given by equations (14) and (18) as

$$\frac{s_R - \tilde{s}_M}{s_R} = \frac{(s_R/K_s + 1) - \sqrt{(s_R/K_s - 1)}}{s_R/K_s}. \tag{19}$$

The efficiency of utilization will be seen to depend solely on the ratio s_R/K_s and approaches 100% if this ratio is made sufficiently high; in other words, high substrate concentrations are advantageous for efficient utilization of substrate.

To summarize, the conditions for maximum production efficiency, combining a high output with efficient utilization of substrate, will be obtained with a flow-rate at or a little below the maximum output rate D_M and the highest practicable substrate concentration. It must be emphasized, however, that these are the optimum conditions when the object is to produce micro-organisms. When the desired product is a fermentation product whose formation is proportional to the amount of substrate breakdown (e.g. ethanol, lactic acid), optimum conditions should be much the same, but they may be widely different for the production of complex metabolic products such as antibiotics or exotoxins.

Comparison of continuous and batch culture. The relative outputs of continuous and batch cultures are of interest from the production standpoint. The output of a batch culture of course varies throughout the growth cycle, but a mean output can be calculated as follows.

Consider a batch of medium of initial substrate concentration s_0, inoculated initially with organisms to a concentration x_0; the maximum growth attained when all substrate has been utilized is x_m. Then the total time of one production cycle is

$$t = \frac{1}{\mu_m} \log_e \frac{x_m}{x_0} + t_L, \tag{20}$$

where the first term is the time which would be occupied if the organisms grew exponentially at maximum rate from start to finish, and the second term t_L is an overall 'delay time' which includes the initial lag and final retardation phases of growth and the 'turnaround time' necessary to take down, sterilize and re-assemble the plant preparatory to a second cycle.

The total amount of organisms produced (from equation 3) is

$$x_m - x_0 = Y s_0 \left(\frac{x_m - x_0}{x_m} \right). \tag{21}$$

The mean output is therefore

$$\frac{\text{total organisms produced}}{\text{total cycle time}} = \frac{\mu_m Y s_0 \left(\frac{x_m - x_0}{x_m} \right)}{\log_e \frac{x_m}{x_0} + \mu_m t_L} \tag{22}$$

The maximum output of a continuous culture is from equations (14) and (15):

$$D_M \tilde{x}_M = \mu_m Y s_R \left\{ \sqrt{\frac{K_s + s_R}{s_R}} - \sqrt{\frac{K_s}{s_R}} \right\}^2, \tag{23}$$

where the term in brackets has a value close to 1·0 for high s_R/K_s ratios such as obtain in practice and may therefore be neglected. Hence, when the same growth medium is used in both cases, so that $s_0 = s_R$, we have from equations (22) and (23)

$$\frac{\text{continuous output}}{\text{batch output}} = \frac{\log_e \frac{x_m}{x_0} + \mu_m t_L}{(x_m - x_0)/x_m} \tag{24}$$

The value of this ratio is affected mainly by the growth rate of the organism and the total delay time, inoculum size having a lesser effect. Assuming a '5 % inoculum' ($x_0/x_m = 0.05$), and a delay time t_L of 6 hr., calculations from (24) give the following results:

Doubling time of organism hr.	Ratio: continuous output / batch output
0·5	11·9
1·0	7·6
2·0	5·3
4·0	4·3

The delay time of 6 hr. assumed in these calculations is fairly optimistic; both plant turn-around time and growth lag would exceed this in many instances, which would further increase the above ratios. In the majority of cases, therefore, a continuous process would be expected to show at least a five- to tenfold advantage over the corresponding batch process.

APPARATUS AND METHODS

Continuous culture apparatus. The apparatus used is of conventional small pilot plant type and is fabricated wholly in stainless steel; the layout is shown in Fig. 3. It consists of two sterilizing tanks S_1, S_2 (each of 300 l. working volume) connected in parallel and used alternately to feed growth medium into the culture vessel, *CV*. This has a working capacity of 20 l. and is of conventional design; mixing and aeration are effected by a vane-disk impeller mounted on a central shaft passing through a stuffing-box in the lid, with wall baffles and injection of sterile air immediately under the impeller, similar to the arrangement of Chain *et al.* (1954). The oxygen transfer rate, measured by the method of Cooper, Fernstrom & Miller (1944), is 135 mmole O_2/l./hr. with an air flow of 1·0 vol./vol. culture/min. The level of culture in the vessel is maintained constant by a side overflow tube through which the culture flows by gravity into two calibrated collecting tanks M_1, M_2, which are connected in parallel, and thence to a large holding tank *H*. Effluent air leaves the culture vessel by a separate outlet; both entering and exit air is sterilized by heat. Air and medium flow-rates are indicated by Rotameters R_1, R_2, and are controlled by manual adjustment of cocks; a second check on medium flow-rate is provided by the calibrated tanks M_1, M_2. Sampling points SP_{1-2}, SP_3, are provided on the sterilizers and culture vessel respectively. On the culture vessel there is a point AP_1 for the addition of inoculum and anti-foam ('Anti-foam A', Midland Silicones Ltd.). Temperature is controlled by circulating thermostatically controlled water through an internal coil (not shown in diagram) in the culture vessel and was 37° for all experiments described in this paper.

Operation. The whole equipment is sterilized empty with internal steam at 20 lb./sq.in., followed by filling and sterilization of the medium reservoirs. The culture vessel is then charged with 20 l. of sterile medium and inoculated with 1 l. of a shake-flask culture of the selected organism grown on the same medium. Growth is allowed to proceed batchwise until the concentration of organisms reaches 90 % of the expected peak value; medium flow through the culture vessel is then started at a selected value and the culture from then on run continuously. Samples of sterile medium are taken from each sterilizing tank for analysis and samples of the culture are usually taken hourly; samples of the exit air are also taken for gas analysis.

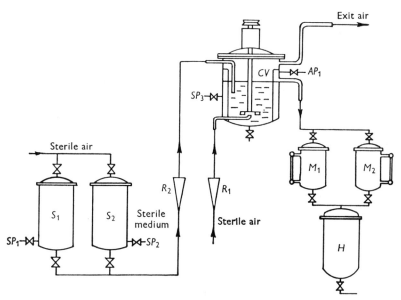

Fig. 3. Pilot plant for continuous culture of bacteria (schematic). S_1, S_2, sterilizing tanks (300 l. working volume); *CV* culture vessel (20 l. working volume); M_1, M_2, measuring tanks (30 l.); *H*, holding tank; R_1, R_2, rotameters; SP_1, SP_2, SP_3, sampling points; AP_1, anti-foam addition point.

Observations and measurements. Each batch of medium was examined or analysed for sterility, pH value, ammonia-N content and concentration of organic substrate, which in the present instance was glycerol, determined according to Neish (1952). Samples from the culture vessel were tested for pH value, concentration of glycerol, total count (by Helber counting chamber), viable count (by the Miles & Misra (1938) technique on plates of Hartley's digest agar) and purity of the microbial species being grown.

Organisms and culture medium. Aerobacter cloacae (*Cloaca cloacae* strain NCTC 8197) was used throughout and grown in medium of the following composition: 0·09 M-$(NH_4)_2HPO_4$, 0·01 M-NaH_2PO_4, 0·01 M-K_2SO_4, 0·001 M-$MgSO_4$, 0·0001 M-$CaCl_2$, 0·00002 M-$FeSO_4$, and 0·0272 M (0·25 %, w/v) glycerol; the pH value after sterilizing was 7·3–7·4. In this medium glycerol (selected in preference to glucose for its stability on sterilization) was the sole carbon source and was also the growth-limiting component, all other components being present in excess. A rather low concentration was used to ensure fairly low bacterial concentrations so that aeration was always adequate.

RESULTS

Growth in batch culture

As a preliminary to continuous culture studies, data were obtained on the growth of the organism in batch culture, using the same apparatus and growth medium. Typical experiments are shown in Fig. 4. After a variable lag period, which depends on the age and size of the inoculum, growth proceeds exponentially until almost all the substrate has been exhausted (a sign that aeration is not rate-limiting). Under these conditions the organisms are growing for most of the time in substrate concentrations which are high compared with K_s, and equation (3) becomes

$$\frac{1}{x}\frac{dx}{dt} = \frac{d\,(\log_e x)}{dt} \sim \mu_m. \tag{25}$$

The value of μ_m is given by the slope of the straight line obtained by plotting $\log_e x$ against t, the best straight line being fitted by the 'method of least squares'. The mean value obtained from a number of experiments was $\mu_m = 0.85$ hr.$^{-1}$ (doubling time $t_d = 49$ min.).

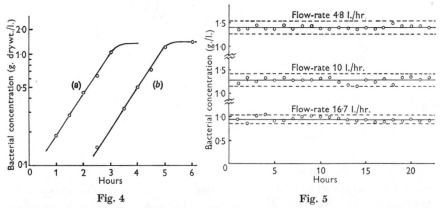

Fig. 4   Fig. 5

Fig. 4. Growth curves of *Aerobacter cloacae* in batch culture. Details of growth conditions and culture medium are given in the text. Curve (*a*) was obtained from a seed culture at the start of a continuous culture run; curve (*b*) was obtained after the culture had been operated continuously for 6 weeks (the points on curve (*b*) have been displaced 2 hr. to the right to avoid overlapping with curve (*a*)).

Fig. 5. Steady states in continuous culture. *Aerobacter cloacae* was grown in continuous culture as described in the text; the data plotted are hourly measurements of bacterial concentration at three different flow-rates. The continuous lines are the lines of best fit, calculated statistically; the dotted lines are the 95 % fiducial limits.

In the medium used, the total growth $(x_m - x_0)$ obtained in a batch culture was strictly proportional to the initial concentration of glycerol in the medium (cf. Monod, 1942). The mean of several determinations of the yield constant was $Y = 0.53$ g. dry weight of organism/g. glycerol used; this corresponds to a 61 % conversion of the glycerol carbon to bacterial carbon. (The carbon content of the bacteria was found to be 45 % of the dry weight.)

Growth in continuous culture

Steady-state operation. In accordance with theory, steady-state operation was found to be possible over a wide range of flow-rates, the range actually tested being from 4 to 22 l./hr., corresponding to dilution rates of 0·2–1·1 hr.$^{-1}$. The critical flow-rate for complete wash-out (obtained by extrapolation of the curves of Fig. 6) was about 24 l./hr., corresponding to a critical dilution rate of about 1·2 hr.$^{-1}$.

Over the stable range of flow-rates the culture was self-adjusting; i.e. on setting the flow-rate to a given value, the concentrations of organisms and substrate would move towards and settle down at steady levels which were maintained indefinitely so long as the flow-rate remained unaltered; on changing the flow-rate, new steady-state levels were automatically attained. After a change of flow-rate, some hours might elapse before the culture had stabilized at the new steady state, particularly when the change in D was large. The culture was therefore always run for at least 24 hr. after a flow-rate change, before measurements at the new steady state were begun.

Examples of steady states at three different flow-rates are given in Fig. 5; measurements of bacterial concentration were taken hourly for 24 hr., after a preliminary 24 hr. stabilization period. Statistical analysis of the results showed that the plots of bacterial concentration against time did not differ significantly from straight lines of zero slope; the apparent small fluctuations in bacterial concentration were purely random and within the errors of measurement (95 % fiducial limits for each curve are shown by the dotted lines).

Maintenance of purity of culture. Under this heading we include both contamination of the culture with foreign micro-organisms and mutation of the parent organism to an extent sufficient to alter the characteristics of the culture. To detect such occurrences, samples taken directly from the culture vessel were repeatedly examined (*a*) microscopically, (*b*) by plating out and examination of colonies, (*c*) by regular subculturing and examination by biochemical tests of typical colonies (and any atypical ones, when observed).

Such tests readily detect gross contamination, but statistical considerations show that they are not very efficient in revealing small degrees of contamination. Suppose that in the whole culture there is an average of z contaminants to every n total organisms. If z is small its distribution in samples of n will be Poissonian; i.e. the probability (P) of finding r contaminants in a single sample of n cells is $e^{-z}z^r/r!$, and the probability of finding no contaminant at all is e^{-z}. Hence in 5 % of cases no contaminants will be observed when the expected number is three ($P = e^{-3} = 0·05$). For example, even if no contaminants are found in a sample of 1000 colonies examined, there is still a 5 % chance that the degree of contamination is really as high as 0·3 %. The same considerations apply in any form of sterility testing and are discussed in more detail by Elsworth, Telling & Ford (1955).

It is therefore practically impossible to show that a culture is completely free from contaminants by a single test. Repeated tests over long periods are more significant, however, since the permanent existence of a low equilibrium level of contaminants is improbable; foreign contaminants (or mutants) are most likely either to displace the parent culture altogether, or to disappear.

With the above reservations it can be said that over the past two years no difficulty has been found in operating the continuous culture apparatus for long periods without any detectable contamination. The longest individual run lasted 108 days and was still free from contamination when terminated voluntarily; 2–3 months is the average period for most runs. The construction of a leak-free apparatus did not prove too difficult, and contaminations experienced in the earlier stages of the work could usually be ascribed to defective initial sterilization or faulty aseptic technique during sample-taking or addition of anti-foam. If due attention be paid to these points, we see no reason why continuous cultures should not be maintained free from contamination more or less indefinitely.

So far as mutations are concerned, we have never during the longest periods of continuous operation been able to detect any organisms differing from the parent strain in microscopic or colonial morphology, biochemical reactions or growth characteristics. In view of the possibility of faster-growing though morphologically identical mutants displacing the parent strain more or less completely, we have on numerous occasions re-determined the value of the growth-rate constant μ_m at intervals during the course of a continuous culture. This can easily be done by stopping the medium flow, draining off 95 % or more of the culture (the remainder serving as inoculum), re-filling the culture vessel with fresh medium and growing up as a batch culture, the growth rate and yield constant being determined as in Fig. 4. No significant change in μ_m or Y was ever observed, even after long periods of continuous culture (cf. curves (a) and (b) of Fig. 4).

Quantitative tests of continuous culture theory. Table 1 summarizes quantitative data on steady-state bacterial and substrate concentrations at twenty-one different flow-rates. These were all obtained during a single run, over a period of 65 days continuous operation. In Fig. 6 (a)–(d) the steady-state bacterial concentration and output, substrate concentration and yield constant are plotted against flow-rate, for comparison with the results to be expected according to Monod's theory.

The theoretical curves shown were plotted from equations (9) and (10), using the values $\mu_m = 0.85$ hr.$^{-1}$ and $Y = 0.53$ obtained in the batch culture experiments. Determination of K_s was more difficult. Accurate values are not readily obtained from batch culture experiments, since at the low substrate concentrations necessary, s is continually decreasing during the experiment. Theoretically, K_s is most easily determined from continuous culture experiments, for a single measurement of the steady-state substrate concentration at any dilution rate allows K_s to be calculated if μ_m is known, as is shown by re-arranging equation (9) in the form

$$K_s = \tilde{s} \left(\frac{\mu_m - D}{D} \right). \tag{26}$$

In particular, when $D = \mu_m/2$, we have $\tilde{s} = K_s$, and this is probably the most accurate method for determining K_s, provided the analytical method for

40

determining $\tilde{s}$ is sufficiently sensitive. In the present case, unfortunately, the value of K_s was so small that over most of the lower range of dilution rates the substrate concentration was too low to measure with the available methods for glycerol estimation (Table 1). K_s was therefore determined from equation (14), using experimentally determined values of μ_m and the maximum output

Table 1. *Quantitative data on the continuous culture of* Aerobacter *cloacae*

The apparatus and general plan of the experiment are described in the text. Each row of figures in the table refers to a different steady state. Glycerol concentration in the inflowing culture medium (s_R) was 2·5 g./l.

Flow rate (l./hr.)	Duration* of test	Bacterial concentration (g. dry wt./l.)	Output of bacteria (g./hr.)	Glycerol concentration in culture (g./l.)	Yield constant $\left(\dfrac{\text{g. bacteria}}{\text{g. glycerol}}\right)$
f	(hr.)	$\tilde{x}$	$f\tilde{x}$	$\tilde{s}$	$\dfrac{\tilde{x}}{s_R - \tilde{s}}$
0	(Batch culture)	1·32†	—	0	0·53†
4·6	22	1·33	6·12	<0·03‡	0·55
4·8	26	1·39	6·66	<0·03	0·57
5·0	25	1·30	6·48	<0·03	0·53
7·2	25	1·26	9·05	<0·03	0·52
8·4	14	1·25	10·47	—	—
10·0	25	1·29	12·88	<0·03	0·53
11·5	22	1·31	15·04	<0·03	0·54
11·9	24	1·27	15·08	—	—
13·8	23	1·23	16·92	—	—
13·9	34	1·22	17·0	<0·03	0·50
14·3	16	1·04	14·9	<0·03	0·43
14·5	17	1·07	15·5	—	—
15·7	25	0·96	15·1	<0·03	0·40
15·8	23	1·26	19·0	<0·03	0·52
15·9	20	0·91	14·3	<0·03	0·36
16·6	22	1·00	16·5	—	—
16·7	24	0·98	16·3	0·26	0·44
17·3	12	0·95	16·4	0·28	0·43
18·3	8	0·72	13·1	0·88	0·45
19·5	6	0·35	6·84	1·66	0·45
20·0	9	0·48	9·68	1·37	0·42
22·4	8	0·24	5·31	1·93	0·34

* The culture was allowed to stabilize for 18–24 hr. at each flow-rate before observations commenced; it was then sampled hourly during the test period. The figures for bacterial and glycerol concentrations are the means of the hourly samples.

† Means of a number of batch culture experiments.

‡ This figure represents the lowest concentration of glycerol which can be measured with accuracy by the method used.

dilution rate D_M and solving for K_s. The value obtained, which may be subject to considerable error, was $K_s = 1·35 \times 10^{-4}$ M, or 12·3 μg. glycerol/ml. This is comparable with the values of K_s found by Monod (1942) for other carbohydrate substrates (namely 4, 2 and 20 μg./ml. for glucose, mannitol and lactose respectively), but considerably higher than the value of 10^{-3} μg./ml. found by Novick & Szilard (1950) with tryptophan as substrate. The above values of μ_m, K_s and Y were used in calculating the 'theoretical' curves shown in Fig. 6.

Qualitatively, the results plotted in Fig. 6 may be seen to be in general agreement with the theory. The steady-state concentration of organisms falls only very slowly as the dilution rate is increased from zero to about 0·8 hr.$^{-1}$ (Fig. 6a); as the flow-rate is increased still further the bacterial concentration drops fairly steeply, the curve extrapolating to complete wash-out at $D = D_c = 1·2$ hr.$^{-1}$ (approx.). The output of bacteria (Fig. 6b) is nearly proportional to the flow-rate up to a maximum at $D_M = 0·79$ hr.$^{-1}$, thereafter falling fairly sharply to zero. The curve for steady-state glycerol concentration (Fig. 6c) should be a mirror-image of that for bacterial concentration (cf. Figs. 1

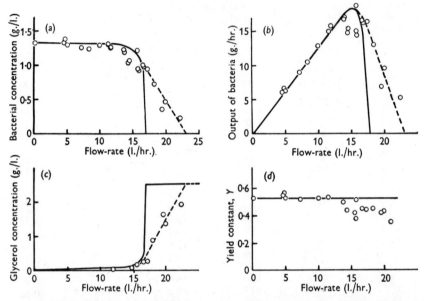

Fig. 6. Quantitative relationships in continuous culture of *Aerobacter cloacae*; comparison of experiment with theory. The data of Table 1 are plotted for comparison with the theoretical relationships calculated from equations (9) and (10). The continuous lines are the theoretical curves; the dotted lines are drawn to fit the experimental points.

and 2), and this is seen to be approximately the case; at all flow-rates up to about 16 l./hr., however, utilization of substrate was so nearly complete that the residual glycerol was too low for accurate measurement.

Quantitatively, agreement with theory is good at all the lower flow-rates (below $D = 0·8$ hr.$^{-1}$), discrepancies being less than the experimental errors of bacterial dry weight and glycerol determinations. At the higher flow-rates, however, there are definite discrepancies; the curves for bacterial concentration and output and for glycerol concentration, though of the right general shape, show definite deviations from the predicted curves, bacterial concentrations being higher and substrate concentrations being lower than the values expected. In particular, the observed value of the critical dilution rate is definitely higher than that calculated from equation (12), and genuine steady-state conditions appear to have been achieved at dilution rates with $D > \mu_m$,

40-2

when theory predicts that complete 'wash-out' should occur. Another discrepancy is that the yield constant Y, which at low dilution rates has the same value as was found in batch culture experiments, shows a definite tendency to decrease at higher dilution rates (Fig. 6d). Possible reasons for these divergences are discussed below.

DISCUSSION

One object of the present work was to compare the results obtained during the operation of a continuous culture apparatus with those predicted by the theory given earlier. It is also relevant to compare them with the results predicted by the theories of other workers.

A common feature of all theories of continuous culture is the elementary one that net growth must be the resultant of exponential growth minus exponential 'wash-out', or re-writing equation (5):

$$\frac{d\,(\log_e x)}{dt} = \mu - D. \tag{27}$$

Hence when a steady state $(d\,(\log_e x)/dt = 0)$ exists, then $\mu = D$, i.e. the specific growth rate must be equal to the dilution rate. An error made by a number of authors in mathematical discussions of continuous culture is to confuse, in the above equation, the specific growth rate μ (which varies with the substrate concentration) and the growth rate constant μ_m. Obviously, if one puts in equation (27) $\mu = \mu_m = $ constant, it follows that a steady state is possible only at one particular flow-rate, when $D = \mu_m$. This is the assumption made by Golle (1953) who states that 'there is only one rate of medium flow...at which steady-state conditions will be maintained'. Finn & Wilson (1954) also stress this point, and the same idea appears to be implicit in the writings of Adams & Hungate (1950) and Northrop (1954). If this were so, a continuous culture would be an inherently unstable system and very difficult to operate; moreover, restriction to a single flow-rate would greatly limit its usefulness. The experimental data presented in this paper show conclusively that a continuous culture is an inherently stable system adjusting itself automatically to changes in dilution rate. Any number of steady states can be obtained at different dilution rates anywhere between zero and the critical value, as the theories of Monod (1950) and Novick & Szilard (1950) predict.

The essential features of the Monod–Novick & Szilard theories and the theoretical treatment given in this paper* is that they take into account the observed facts that bacteria can grow only at the expense of the substrate utilized, and that their specific growth rate is a function of the substrate concentration. It then becomes apparent that a continuous culture apparatus is a device for controlling growth through control of the substrate concentration; each dilution rate fixes the substrate concentration at that value which makes μ equal to D.

---

* We wish to emphasize that the theoretical treatment given in this paper is merely an expansion and development of those of Monod and of Novick & Szilard (particularly the former), and is based on the fundamental principles which they originated.

This important role of the substrate is not considered in the mathematical papers of Golle (1953), Finn & Wilson (1954) and the other authors mentioned above. This has led to some incorrect conclusions. For example, Golle (1953) discusses at some length the mathematics of a series of two or more culture vessels run in cascade, and concludes that there are distinct advantages in this procedure. Now it is apparent from the theory given earlier that over the useful range of flow-rates in a continuous culture, the substrate is nearly completely utilized and the issuing medium virtually exhausted; hence negligible growth could occur in any subsequent culture vessels in series with the first. Our experimental results confirm that this almost complete utilization of substrate does in fact occur.

It is to be emphasized that the theoretical treatment of continuous culture given in this paper is based on the minimum number of extremely simple postulates. It might well be objected that even under ideal conditions the growth behaviour of bacteria cannot be completely represented by such simple equations as (2), (3) and (4). Further refinement of the theory will undoubtedly become necessary as knowledge of the subject increases. The gratifying degree of agreement between experimental and predicted results shows, however, that the basic principles of the theory must be sound. Qualitatively, it provides a good general picture of the behaviour of a continuous culture, and quantitative agreement is good over most of the range. It is necessary to discuss, however, the quantitative deviations from theory found at high flow-rates, and in particular the apparent existence of steady states at dilution rates higher than the maximum specific growth rate μ_m.

According to equation (27) the specific growth rate must be equal to the dilution rate; when the latter exceeds the maximum possible growth rate, complete 'wash-out' should occur and no steady state should be possible. Since steady states were in fact found with $D > \mu_m$, this must mean that either (*a*) the maximum growth rate in a continuous culture is higher than in a batch culture (the value of μ_m was determined from batch culture experiments), or (*b*) the wash-out rate is less than would be predicted from equation (1). We believe the latter explanation to be correct.

A possible reason for an organism growing faster in continuous culture than in batch culture might be the selection of faster-growing mutants. However, as previously mentioned, no evidence for any permanent selection of this kind was found, the value of μ_m being unchanged after long periods of continuous culture. Another explanation could be based on the normal variation in the growth-rates (or generation times) of individual bacteria; in a continuous culture the faster-growing bacteria might be selected. This possibility seems to be eliminated by the work of Powell (1955), who found that there is zero correlation between the generation time of a bacterium and the generation times of the two daughter-cells into which it divides. In other words, an unusually fast-growing organism is just as likely as not to have unusually slow-growing progeny; selection would therefore not affect the mean generation time.

The alternative explanation, suggested by our colleague Mr E. O. Powell, is that the actual wash-out rate in the culture vessel is effectively less than that given by equation (1). Though this equation has been accepted by all previous writers on the subject, it is based on the unverified assumption of 'perfect mixing' within the culture vessel. This implies that each drop of liquid entering the vessel is uniformly distributed throughout its contents in an infinitesimal time, a condition difficult to attain in practice even with vigorous stirring. Incomplete mixing means that the dilution rates in different regions of the culture vessel will not be uniform; there will be a distribution of local dilution rates about the mean or overall dilution rate $\bar{D} = f/v$. Hence when $\bar{D}$ is greater than the critical value D_c (calculated from equation 12) there might still be regions within the culture vessel where the local dilution rate was less than D_c, and in these regions organisms will continue to be produced, so that complete wash-out will not occur. Rough calculations show that quite small deviations from perfect mixing could have surprisingly large effects, owing to the steep descent of the D against x curve in the critical region (Fig. 1). A theoretical and experimental study of this subject is being made by Mr E. O. Powell, and it will not be further discussed here, but we believe that the greater part of the deviations from theory found in our experiments can be explained in this way. This view is reinforced by the fact that in similar experiments with the same organism in a laboratory scale continuous culture apparatus fitted with a highly efficient mixing system, the deviations from theory at high flow rates are very small (Herbert, D., to be published).

Another 'apparatus effect' can produce results similar to those of imperfect mixing. During long experiments, a solid film of bacteria of considerable thickness builds up on the walls of the culture vessel above the liquid level, portions of which are continually becoming detached by splashes or condensed liquid running down the walls. This continued re-inoculation of the culture from the walls also leads to a continued production of bacteria at flow-rates which should theoretically cause complete wash-out. It is difficult to estimate the magnitude of this inoculation rate, but suppose it to be w g. bacteria/l./hr., then it can be shown that

$$\tilde{x} = \frac{w}{D - D_c}. \tag{28}$$

Hence if D is only a little above D_c, a quite small value of w can produce a quite high value of $\tilde{x}$.

Other imperfections in the apparatus (e.g. short-term fluctuations of flow-rate or liquid level about the mean values) can also be shown to have effects in the same direction. On the whole, it appears that most of the observed discrepancies can be attributed to 'apparatus effects' of this sort, rather than to inadequacy of the biological side of the theory. Such effects, however, cannot account for the apparent decrease in the yield constant Y at high flow-rates, and it is possible that the independence of yield constant and growth rate assumed in equation (4) may need to be modified; this point is now under investigation.

On the practical side, we consider that the results reported show that the continuous culture of bacteria on a fairly large scale is a readily feasible proposition with apparatus of quite simple design. Complexity was deliberately avoided and the apparatus was designed for production rather than research purposes; nevertheless, the degree of control available permits of useful quantitative investigations. Results are comparable with those obtained in smaller and more complicated types of laboratory continuous culture apparatus in use in this Department. The high production rate may be emphasized; though the working capacity of the culture vessel was only 20 l., no difficulty was found in producing culture at the rate of 300 l./day and higher rates could be achieved with faster-growing organisms. The plant could be scaled up considerably without any major changes in design.

Contamination, often alleged to be a major obstacle to the operation of continuous culture on a plant scale, was not found to be a real difficulty, and runs of several months' duration are now routine. Another alleged difficulty, that of mutation, has not troubled us at all, no changes in the culture having been observed after long periods of continuous operation. In this respect our results are at variance with those of other workers (summarized by Novick, 1955) who have used continuous culture techniques as a means of studying mutation rates. It might be argued that this was due to the type of growth medium used in our experiments, since almost every mutation (of those affecting nutrition at least) would be at a disadvantage compared with the parent type. However, we have had similar experiences with other types of chemically defined and complex growth media, and it could also be argued that in much of the work done on mutations in continuous culture the conditions have been favourable for mutant survival. For a given organism and growth medium, we believe that mutants are somewhat less likely to build up in a continuous culture than in a batch culture. For mutations are rare events arising singly, and a single organism with a generation time τ has a probability of $1 - e^{-D\tau}$ of being washed out of the culture vessel before it has divided once. Hence an appreciable fraction of the mutants arising in a continuous culture will be removed before they have progeny, while in a batch culture all will remain.

In both the theoretical and experimental parts of this paper, continuous cultivation has been regarded as a process for converting substrate into bacteria—a deliberately one-sided approach. Obviously it has many other aspects; for example, as a research tool for elucidating problems of biosynthesis or as a process for the production of metabolic products. It is hoped to make these the subjects of future investigations.

We are grateful to Dr D. W. Henderson for continued encouragement in this work, to Mr E. O. Powell for invaluable discussions and advice on theoretical aspects and to Mr S. Peto for assistance with statistical problems.

REFERENCES

ADAMS, S. L. & HUNGATE, R. E. (1950). Continuous fermentation cycle times; prediction from growth curve analysis. *Industr. Engng Chem.* **42**, 1815.

CHAIN, E. B., PALADINO, S., UGOLINI, F. & CALLOW, D. S. (1954). Pilot plant for fermentation in submerged culture. *R.C. Ist. sup. Sanit.* **17**, 132.

COOPER, C. M., FERNSTROM, G. A. & MILLER, S. A. (1944). Performance of agitated gas-liquid contractors. *Industr. Engng Chem.* **36**, 504.

DANCKWERTS, P. V. (1954). Continuous flow of materials through processing units. *Industr. Chem. Mfr.* **30**, 102.

DAWSON, P. S. S. & PIRT, S. J. (1954). Engineering and microbiological processes. *Chem. & Ind. (Rev.)* p. 282.

ELSWORTH, R., TELLING, R. C. & FORD, J. W. S. (1955). Sterilization of bacteria by heat. *J. Hyg., Camb.* **55**, 445.

FINN, R. K. & WILSON, R. E. (1954). Population dynamics of a continuous propagator for micro-organisms. *Agric. Fd. Chem.* **2**, 66.

GOLLE, H. A. (1953). Theoretical considerations of a continuous culture system. *Agric. Fd. Chem.* **1**, 789.

MILES, A. A. & MISRA, S. S. (1938). The estimation of the bactericidal power of the blood. *J. Hyg., Camb.* **38**, 732.

MONOD, J. (1942). *Recherces sur la croissance des cultures bactériennes.* Paris: Hermann & Cie.:

MONOD, J. (1950). La technique de culture continue; théorie et applications. *Ann. Inst. Pasteur.* **79**, 390.

NEISH, A. C. (1952). *Analytical Methods for Bacterial Fermentations.* Nat. Res. Coun. Canada, Rep. no. 46–8–3 (2nd revision), Saskatoon.

NORTHROP, J. H. (1954). Apparatus for maintaining bacterial cultures in the steady state. *J. gen. Physiol.* **38**, 105.

NOVICK, A. (1955). Growth of bacteria. *Annu. Rev. Microbiol.* **9**, 97.

NOVICK, A. & SZILARD, L. (1950). Experiments with the Chemostat on spontaneous mutations of bacteria. *Proc. nat. Acad. Sci., Wash.* **36**, 708.

POWELL, E. O. (1955). Some features of the generation times of individual bacteria. *Biometrika*, **42**, 16.

WARNER, F. E., COOK, A. M. & TRAIN, D. (1954a). Engineering and microbiological processes. *Chem. & Ind. (Rev.)* p. 114.

WARNER, F. E., COOK, A. M. & TRAIN, D. (1954b). Engineering and microbiological processes. *Chem. & Ind. (Rev.)* p. 283.

(*Received* 15 *November* 1955)

Editor's Comments on Paper 13

13　**Pirt and Callow:** *Studies of the Growth of* Penicillium chrysogenum *in Continuous Flow Culture with Reference to Penicillin Production*

Pirt and Callow had previously demonstrated the advantage of two-stage continuous culture for the production of 2,3-butanediol by *Aerobacter aerogenes* (*Selected Scientific Papers*, **2**, 292–313, Rome: Instituto Superiore di Sanita, 1959). The paper here is chosen instead for several reasons: (1) it demonstrated the ability to grow mycelial organisms continuously in steady states; (2) it confirmed that mycelial organisms conformed to the Monod and Novick–Szilard formulations; and (3) it indicated the need for two-stage continuous culture for the production of secondary metabolites (e.g., penicillin). Later progress has been described by S. J. Pirt and D. S. Callow [*Sci. Rept. Inst. Super. Sanita*, **1**, 250–259 (1961)] and S. J. Pirt [*Chem. & Ind. (London)*, **1968**, 601–603].

It is of interest to consider this paper in relation to Bu'lock's comments in Paper 9; and it is surprising that the lead given by Pirt and Callow has not been followed up to any extent by mycologists.

13

Copyright © 1960 by the Society for Applied Bacteriology

Reprinted from *J. Appl. Bacteriol.*, **23**(1), 87–98 (1960)

STUDIES OF THE GROWTH OF *PENICILLIUM CHRYSOGENUM* IN CONTINUOUS FLOW CULTURE WITH REFERENCE TO PENICILLIN PRODUCTION

By S. J. PIRT and D. S. CALLOW

*Microbiological Research Establishment,
Porton, Wiltshire*

SUMMARY: A filamentous mould was cultured by the continuous flow method in which medium is supplied at a constant rate and the culture volume is kept constant. Flow rates up to 0·1 culture volumes/hr were used. The mycelial dry weight concentration and the yield of mycelium/g of carbon source used were equal to or slightly greater than the maximum obtained in batch culture. With glucose concentrations up to 80 g/l. at a flow rate of 0·05 culture volumes/hr, about 45% of the substrate carbon was converted into mycelial carbon and the remainder oxidized to CO_2.

With unlimited amounts of all nutrients available growth of the mould followed the exponential law, as does bacterial growth, and therefore the mould had a constant doubling time.

The oxygen demand of the mould as a function of growth rate was determined.

Conditions were found under which the rate of penicillin production/g of mycelium remained at its maximum value for 1000 hr.

MOULD fermentations are carried out either by the submerged batch culture method, as in penicillin production, or by the more primitive surface culture method, as in citric acid production. In the reported attempts to develop the continuous flow culture of filamentous moulds (Kolachov & Schneider, 1952; Bartlett & Gerhardt, 1958) the organism used was *Penicillium chrysogenum* and attention was concentrated on the penicillin production aspect to the exclusion of growth studies. These investigations did not show that the continuous process had any marked advantage over the batch process.

In the last decade, the continuous flow culture of bacteria has become firmly established as a research tool because of the much greater control of the environment which it makes possible (for example, see Málek, 1958; Herbert, 1958; Pirt & Callow, 1959c). A characteristic of these studies of continuous flow bacterial culture was the attention paid to the fundamental aspects. So far, no fundamental study of the continuous flow culture of a filamentous mould has been reported; it is the aim of this work to furnish such a study and to demonstrate that continuous flow culture according to the Monod-Novick-Szilard principles (Herbert, 1958) is applicable to a filamentous mould.

The process of penicillin production has two aspects; the first is growth of the mould, and the second is synthesis of penicillin. In fact, the process may be a two-stage one in which the conditions required for growth are different from the conditions required for penicillin formation. The work described here, in contrast to the previous investigations, was primarily on the growth of the mould in continuous flow culture. A preliminary account of this work has been given (Pirt & Callow, 1959a).

METHODS

Apparatus. The culture apparatus was an improved form of that described by Elsworth, Meakin, Pirt & Capell (1956). The automatic pH control system described by Callow & Pirt (1956) was used.

Aeration and agitation. Vortex aeration (Chain *et al.* 1952) was used for most of the work because the absence of baffles ensured that the walls of the vessel below the liquid level were swept clean and the accretion of mycelium on the walls prevented. Stirrer speeds in the vortex system were in the range 1000–1400 rev/min. The 63 mm diameter impeller was fitted the reverse way to that used in the baffled system, that is, with the greater height of vane above the disc. The culture volume was 1·5–2·0 l.

The culture was supplied with an excess of oxygen. This was verified by varying the oxygen demand of the culture and showing that the oxygen uptake rate was limited by the oxygen demand of the culture and not by the oxygen solution rate. This condition was in general realized when the oxygen solution rate (σ_s), determined by the sulphite method with a copper catalyst, was three times the oxygen demand of the culture (σ_c). The minimum possible ratio of $\sigma_s : \sigma_c$ was not determined.

Media. The media were based on that of Jarvis & Johnson (1947). Glucose was autoclaved separately and added at a concentration 2·3 times the dry weight required. All quantities given are in g/l.

Medium M3, used to produce mycelial dry weight up to 10 g/l., contained: $MgSO_4.7H_2O$, 0·25; $FeSO_4.7H_2O$, 0·10; $CuSO_4.5H_2O$, 0·005; $ZnSO_4.7H_2O$, 0·02; Na_2SO_4, 0·50; $MnSO_4.4H_2O$, 0·024; $CaCl_2.6H_2O$, 0·075; EDTA, 0·566; phenylacetic acid, 1·0; KH_2PO_4, 2·0; $(NH_4)_2SO_4$, 6·15. The phosphate and ammonium sulphate were mixed, adjusted to pH 6·4 with NaOH and autoclaved separately. The other salts were adjusted to pH 7·4 with NaOH and sterilized by autoclaving. There was no salt precipitation in this medium.

Medium M16, used to produce mycelial dry weight from 10 to 20 g/l., contained: $MgSO_4.7H_2O$, 0·40; EDTA, 0·744; $(NH_4)_2SO_4$, 16·95; and the other constituents of M3.

Medium M18, used to produce 20 to 30 g of dry mycelium/l., contained: $MgSO_4.7H_2O$, 0·80; EDTA, 0·982; $(NH_4)_2SO_4$, 17·8; $(NH_4)_2HPO_4$, 2·0; and the other constituents of M3 except the Na_2SO_4.

Medium M22, for dry weights between 20 and 30 g/l., contained: $MgSO_4.7H_2O$, 0·80; EDTA, 0·982; $(NH_4)_2HPO_4$, 2·0; $(NH_4)_2SO_4$, 17·8; Mo, V, Ga and Sc, each at a concentration of 0·02 mg of metal/l., the Mo being added as ammonium molybdate, V as vanadium pentoxide, Ga as the metal dissolved in HCl, Sc as ScO dissolved in H_2SO_4; and the other constituents of M3, with the exception of the Na_2SO_4.

Either 2N or 4N NaOH and 0·5N or N H_2SO_4 were used in the pH control.

The medium for growing the inoculum differed from M3 in that the phosphate was replaced by 0·1M potassium phosphate buffer, pH 6·8; the glucose concentration was 10 g/l. and phenylacetate was omitted.

192

Alkaterge-C (Commercial Solvents Corp., Terre Haute, U.S.A.) 30% (v/v) in liquid paraffin was added periodically to the culture to inhibit foaming (Pirt & Callow, 1958*a*).

Analytical methods. Mycelial dry weight was determined by filtering 10 ml of culture in a tared sintered glass crucible of porosity 2; the mycelial mat was washed with water and dried to constant weight at 100–105°.

Mycelial carbon was estimated from the dry weight by multiplying by the factor 0·471 (a series of carbon analyses by dry combustion of mycelium grown at different pH values gave carbon percentages varying progressively from 46·1 at pH 6·0 to 48·3 at pH 7·5).

Glucose was determined by ferricyanide reduction. Initially the samples were simply diluted to contain less than 2·5 mg of glucose/5 ml. Latterly, however, the ion exchange resin treatment of Strange, Dark & Ness (1955) was applied to samples from the culture to remove interfering substances; this treatment, it was found, removed reducing material corresponding to about 0·5 mg of glucose/ml. Samples of the culture fluid (5 ml) were shaken with 0·5 g each of acidic and basic resins.

Penicillin was determined by the biological assay method of Humphrey & Lightbown (1952); the amounts are expressed as Oxford units.

Mould strains. The strains of *Penicillium chrysogenum* used were: Wis 47-1564, Wis 49-133, Wis 54-1255, and strain D, a variant derived from Wis 51-20.

Preparation of inoculum. Stock cultures were maintained on ground maize moistened with asparagine and glycerol (Whiffen & Savage, 1947). The inoculum medium in 100 ml amounts contained in 1 l. conical flasks was given a heavy inoculum of spores, enough to ensure filamentous growth, and incubated at 25° on a shaker which gave an oxygen solution rate in sodium sulphite + copper ions of about 40 m-mole O_2/l./hr. After 40 hr growth the inoculum (200 ml) was transferred to the continuous culture vessel. At this stage the inoculum had a dry weight of 3–4 mg/ml. If phenylacetate was present in the medium to which the inoculum was transferred the pH was adjusted to 7·0; in the absence of phenylacetate the pH value could be lower.

Tests for contamination. Apart from direct microscopical examination of the culture, contaminating organisms were detected by the following tests: by plating on meat tryptic digest agar incubated at 37°; by plating on the medium of Backus & Stauffer (1955) incubated at 25°; by adding 1 ml of culture to 10 ml of meat tryptic digest broth incubated at 25°. Contamination has not been a problem in this work and no more than weekly tests were required.

Temperature. The culture temperature, unless otherwise stated, was 25°.

Other methods. Methods other than those reported here have been previously described by Pirt & Callow (1958*b*).

RESULTS

Steady state dry weight and glucose concentrations

Fig. 1 shows the dry weight at various times in cultures in which glucose supply was the growth limiting factor. The medium flow was started after an initial period of batchwise growth. The dry weight obtained in a continuous culture with a dilution rate of about 0·05 hr⁻¹ was slightly higher than the maximum obtained in batch culture with the same medium. When the steady state had been reached the dry weight generally showed little variation. Factors which can cause large variation in the dry weight are: accretion of mycelium on the walls of the vessel, the dry weight increasing should mycelium fall back into the culture; formation of the pellet type of growth (Pirt & Callow, 1959b), causing erratic variation in dry weight; and accretion of mycelium at the effluent point, forming a filter which causes mycelium to accumulate in the vessel. This may occur if the overflow pipe is horizontal rather than vertical.

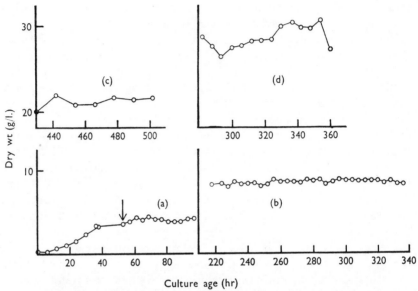

Fig. 1. Mycelial dry weights in the steady states of continuous cultures of *P. chrysogenum*. (a) Batchwise growth followed by continuous flow culture of strain Wis 47-1564 in Medium M3 (without phenylacetate) at pH 6·5, dilution rate (D), 0·086 hr⁻¹; the arrow indicates when flow began: (b) strain Wis 47-1564 in medium M3 with 19·5 g of glucose/l., pH 7·2, D, 0·079 hr⁻¹: (c) strain Wis 54-1255 in medium M16 with 54·5 g of glucose/l., pH 6·9, D, 0·055 hr⁻¹: (d) strain Wis 54-1255 in medium M18 with 82·7 g of glucose/l., pH 6·7, D, 0·057 hr⁻¹.

The concentration of residual glucose in the steady state was about 0·5 g/l., when the initial glucose concentration was 20 g/l. The residual glucose concentration increased to 1·2 g/l. when the initial glucose concentration was 85 g/l. The actual value of the residual glucose concentration may have been lower, since it was near the lower limit of the range of glucose concentration which could be estimated by the method used and also the reducing power may not have been entirely due to glucose. The results show that in the steady state of a continuous flow culture the glucose utilization may be practically complete.

The phenylacetate added did not affect the dry weight, neither did it show any toxic effect at pH values down to 6·0 in continuous flow cultures in the steady state.

However, phenylacetate seemed to be strongly inhibitory to initial growth when the culture was inoculated at pH 6·7, but not at pH 7·0.

Of the two media, M18 and M22, used to produce 30 g mycelial dry wt/l., the simpler one, M18, seemed adequate. A feature of the use of M18, as Fig. 1 shows, was that the mycelial dry weight varied much more than with the more dilute media.

Growth rate

The maximum growth rate of the mould must be known so that one can predict the maximum possible dilution rate. For continuous flow culture the growth rate is most conveniently expressed as the specific growth rate $\mu = \dfrac{1}{x}\dfrac{dx}{dt}$, where x is the organism concentration and t the time. If growth occurs at a constant exponential rate then μ is constant and independent of x. There seems to be a consensus of opinion that moulds do not grow at a constant exponential rate as do bacteria but that the rate of increase becomes linear after a time (Brown, 1923; Hawker, 1950). Clearly, as judged by the rate of growth of a colony on a surface, a commonly used method (e.g. Brown, 1923), a constant exponential rate of increase is not achieved; but the results of Smith (1924), who studied the growth of a single hyphal system, that is, the increase in length of the parent hypha and all its branches, show that growth at a constant exponential rate does occur.

We found that under appropriate conditions, i.e. with an unlimited supply of nutrients, constant physical environment, and no entry of toxic substances into the system, growth followed the exponential law $\dfrac{dx}{dt} = \mu x$, where μ is constant. The necessary conditions were realized in batch cultures. This is shown by the straight line obtained on plotting the logarithm of the dry weight against time (Fig. 2). The slope of the line gives a maximum specific growth rate (μ_m) of 0·075 hr^{-1}. Therefore the doubling time (t_d) was 9·2 hr. This was the doubling time observed during the initial batchwise growth following inoculation.

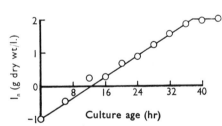

Fig. 2. Changes in the logarithm of the dry weight/l. with time after inoculation in a batch culture of strain Wis 47-1564. Medium M3 (without phenylacetate), with 20 g of glucose/l., pH 6·7.

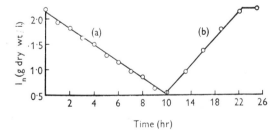

Fig. 3. Changes in the logarithm of the dry weight/l. with time in a continuous culture of strain Wis 47-1564 in medium M3, with 20 g of glucose/l., pH 7·4: (a) with dilution rate (0·296 hr^{-1}) greater than the maximum specific growth rate; (b) after stopping the medium flow.

Alternatively one can determine the maximum growth rate by a continuous flow method. If the growth in continuous culture is in accordance with the theory of Monod and of Novick & Szilard and the dilution rate (D) is increased until the organism is flowing out of the culture faster than it is being produced, that is $D > \mu_m$, an excess of nutrients should be available and the organisms should grow at the maximum rate. Under this condition the logarithm of the dry weight plotted against time should be a straight line with a negative slope equal to $\mu_m - D$, and from this by inserting D one obtains μ_m. A plot of log dry weight against time for a continuous flow culture with D about twice μ_m is shown in Fig. 3. This dilution method gave a value of 0·136 hr^{-1} for μ_m. When the flow was stopped and the culture allowed to grow batchwise the slope of the logarithmic plot shown in Fig. 3 gave the value of 0·139 for μ_m, which is in very good agreement with that obtained in the dilution method. The good agreement between the two methods was confirmed many times. The results given in Fig. 3 were obtained with a culture which had been kept for over 1000 hr in continuous culture. The much faster growth rate obtained then (doubling time 5·1 hr) compared with that in the initial batchwise growth shown in Fig. 2 (doubling time 9·2 hr) is attributed partly to the pH difference and partly to better adaptation to the environment.

Influence of pH and temperature. Maximum growth rates were generally determined by the dilution method and confirmed by subsequent batchwise growth. The influence of pH on growth rate is depicted in Fig. 4. The peak growth rate occurred between pH 7·0 and 7·4, the upper limit for growth being about 8·0. There was considerable variation in the maximum growth rate at any given pH value, as Fig. 4 shows. The variation in growth rate was well outside the experimental error, which was estimated as less than 10 %. The cause of this variation may be that after any environmental change, even an apparently small one like a change in pH from 6·5 to 7·0, the organism has to be grown for a long time in the new environment before it is fully adapted and able to show its true maximum growth rate. Investigation of this effect was beyond the scope of this work.

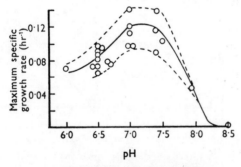

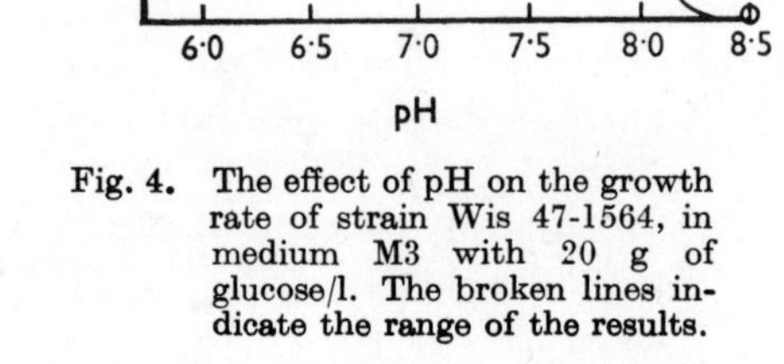

Fig. 4.   The effect of pH on the growth rate of strain Wis 47-1564, in medium M3 with 20 g of glucose/l. The broken lines indicate the range of the results.

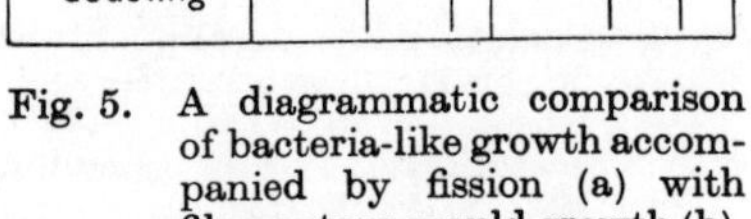

Fig. 5.   A diagrammatic comparison of bacteria-like growth accompanied by fission (a) with filamentous mould growth (b).

The effect of temperature over the range 23–32° on the growth rate at pH 7·4 was investigated. There were small but not significant decreases in the maximum specific growth rate at both ends of the range.

Both pH and temperature changes, it should be noted, were made gradually, generally 0·1 pH units or 1° at a time with 2 hr intervals, as a precaution against the shock to the organism which was noted in the continuous flow culture of bacteria (Pirt & Callow, 1958*b*).

Morphology

The growth of a filamentous mould such as *P. chrysogenum* is apical and not naturally accompanied by fission, as occurs with bacteria. The two types of growth are represented in Fig. 5. Whereas growth from a single bacterial organism should in 3 doubling times produce 8 organisms, filamentous growth should result in an eightfold increase in the length of the parent hypha and its branches. If culture of the filamentous organism were indefinitely prolonged, e.g. by continuous culture, it seems that the size of the mould fragment should increase indefinitely. However, it was found that in a continuous culture in the steady state the number of individual mould fragments was roughly constant (of the order of 10^6/ml with 1% dry weight at pH 7); hence fission of the mould must have occurred. Sporulation was not responsible for the increase in the number of fragments, and it was concluded that scission of the mould hyphae was caused by the shearing action of the stirrer. Elsewhere (Pirt & Callow, 1959*b*) we have shown that the morphology of the mould depends on the pH value and the nutrition during growth.

Yield of mycelium

The weight of organism produced per unit weight of carbon source used is of considerable importance in the quantitative formulation of growth processes. Monod (1942) showed that if a bacterial population were limited solely by the amount of carbon source available, then the weight of organism produced is directly proportional to the weight of carbon source utilized, a finding which he formulated in the important equation $x = Y\Delta s$, where x is the dry weight of organism produced, Δs is the amount of carbon source utilized and Y is a constant. Bacteriologists (e.g. Herbert, Elsworth & Telling, 1956) generally call Y the 'yield constant' and mycologists have called it the 'economic coefficient' (see Foster, 1949): the latter term seems to have considerable priority.

We found that the economic coefficient for *P. chrysogenum* in continuous flow cultures with excess of available oxygen was constant over a wide range of conditions. For strain Wis 47-1564 its average value (g dry wt/g of glucose utilized) was 0·443 (with standard deviation 0·043) over a wide range of conditions. This value is about the maximum reported for moulds. The coefficient was independent of the amount of glucose utilized over a wide range (up to 80 g/l.), of pH in the range 6·0–7·5, of temperature in the range 23–32°, of specific growth rate in the range 0·04–0·08 hr^{-1}, and of the strain used.

Oxygen demand

Calculations of the oxygen demands of cultures have been based on Q_{O_2} measurements (e.g. Rolinson, 1952) carried out on samples not under actual fermentation conditions and with no account being taken of growth rate. The calculated demands, therefore, may be very far removed from the true ones. Pirt (1957) showed that the oxygen demand, R_0 (m-moles O_2/l./hr) of a growing culture can be expressed as

$$R_0 = P\, \mu \Delta s$$

where P is the oxygen demand constant (moles O_2 consumed/mole substrate carbon used), μ is the specific growth rate and Δs the consumption of substrate carbon (m-mole/l.). P is characteristic of the organism and the conditions. However, it will be seen that P must lie between 0 and 1 if the carbon source has the empirical formula CH_2O, e.g. carbohydrate, and this enables one to make rough estimates of the oxygen demand and to set precise limits to it. For a continuous culture in the steady state one can substitute dilution rate for μ.

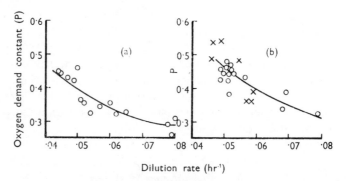

Fig. 6. The variation of the oxygen demand constant (P, moles of oxygen consumed/mole of glucose carbon) of *P. chrysogenum* with dilution rate when there was an excess of available oxygen, for (a) strain Wis 47-1564, (b) strain Wis 54-1255. Circles, at pH 7·4; crosses, at pH 6·7–7·0.

The oxygen uptake rates at various dilution rates were determined by analysis of oxygen in the effluent gas from the culture. From the values obtained, together with determination of the glucose utilized, the values of P were calculated. The variation of P with dilution rate when there was an excess of available oxygen is shown in Fig. 6. P decreased with increase in growth rate. In the middle of the growth rate range P was about 0·4, which is about the same as the values reported for bacterial and yeast species (Pirt, 1957). Variations of pH in the range 6·0–7·5 and temperature in the range 23–32° did not appreciably affect the value of P. There was a difference between the mean oxygen demands of strains Wis 47-1564 and Wis 54-1255, e.g. with D at 0·050–0·055, where most of the data were obtained, the mean value of P was 0·39 for Wis 47-1564 and 0·45 for Wis 54-1255. The maximum oxygen uptake rate observed was 76 m-mole/l./hr with a dry weight of 33 g/l.

The ratio, moles CO_2 produced:moles oxygen consumed was 1·1±0·1 under practically all the conditions used.

Carbon balances

Mycelium, carbon dioxide and penicillin were the only products determined. Table 1 shows some typical carbon balances obtained. The carbon in the mycelium and carbon dioxide accounted for virtually all the glucose carbon utilized at dilution rates up to 0·06 hr⁻¹, but at higher dilution rates, with strain Wis 47-1564, a significant part of the carbon utilized (15% with D at 0·08) was unaccounted for

Table 1. *Carbon balances in continuous flow cultures of* P. chrysogenum *with different dilution rates*

Dilution rate (hr⁻¹)	Glucose supplied (g/l.)	Glucose utilized (g/l.)	Carbon recovered* in		
			CO_2	Mycelium	Total
Strain Wis 47-1564†					
0·049	19·9	19·3	41·2	54·9	96·1
0·052	20·0	19·4	43·4	54·3	97·7
0·057	20·0	19·3	43·6	54·7	98·3
0·060	20·0	19·3	43·2	53·0	96·2
0·079	20·4	19 9	31·2	49·7	80·9
Strain D‡					
0·042	21·4	21·0	44·1	50·4	94·5
0·061	80·2	78·4	43·8	48·3	92·1

* As percentage of carbon in glucose utilized. † In medium M3, pH 7·4.
‡ In medium M3 with 21·4 g of glucose/l. and in medium M22 with 80·2 g of glucose/l., pH 6·7.

and may have been converted to unknown products. This behaviour resembles that of an *Aerobacter* species (Pirt, 1957), which at low growth rates and under fully aerobic conditions converts glucose entirely to CO_2 and cell carbon, but at growth rates near the maximum produces some organic acid. Yeast (Maxon & Johnson, 1953) shows analogous behaviour, but producing ethanol instead of an organic acid.

Penicillin production

The concentration of penicillin reached in a fermentation is proportional to: the mycelial dry weight concentration (x, mg/ml); the residence time (hr) of the mould in the culture, which in the case of a continuous flow culture is the reciprocal of the dilution rate (D); and to the metabolic quotient, q_{pen}, i.e. the number of units of penicillin produced/hr/g dry weight. Thus,

$$\text{Penicillin concn. (units/ml)} = \frac{x q_{pen}}{D}$$

Table 2 shows penicillin production by strain Wis 47-1564 in a continuous flow culture, and it can be seen that both the dilution rate and the dry weight may be varied considerably without significantly affecting q_{pen}. Since x and D can be varied at will, the factor determining penicillin concentration which is characteristic of the organism is q_{pen}, which is clearly an important parameter in the selection of an organism for the continuous process.

The fact that q_{pen} was stable throughout a culture which lasted nearly 1000 hr means that selection of lower yielding variants did not occur. This result is in marked contrast to that in batch cultures, in which q_{pen} falls practically to zero

Table 2. *The relationship between* q_{pen},* *culture age and dilution rate in a continuous flow culture of* P. chrysogenum†

Culture age (hr)	Dilution rate (hr⁻¹)	Dry weight (mg/ml)	Penicillin concn. (units/ml)	q_{pen}*
Medium M3; glucose supplied, 19 g/l.				
120	·080	2·71	27	0·80
312	·051	8·37	175	1·07
504	·050	8·63	166	0·96
815	·046	7·58	146	0·89
Medium M16; glucose supplied, 38 g/l.				
959	·048	14·7	280	0·92

* Rate of penicillin production in units/hr/mg dry weight of mycelium.
† Strain Wis 47-1564; at 25° and pH 7·4.

Table 3. *A comparison of penicillin production in batch and single-stage continuous flow cultures of* P. chrysogenum *strain Wis 47-1564*

Reference	Nitrogen and carbon sources	Duration of culture or residence time in culture (hr)	Maximum penicillin concn. (units/ml)	Maximum rate of penicillin accumulation (units/hr/ml)
Batch culture				
Pirt (unpublished)	NH₄⁺, glucose, lactose, acetate, lactate	100	300	3
Anderson *et al.* (1953)	Corn steep, lactose	90	1000	11
Continuous flow culture				
This paper	NH₄⁺, glucose	20	280	14

in 100–150 hr. A comparison of penicillin production in batch culture and in continuous flow culture is given in Table 3. The former had the highest rate of accumulation of penicillin, but the penicillin concentration was lowest in the continuous culture owing to the short residence time of the mould therein.

In several cultures in which the conditions differed from those given in Table 2 an appreciable fall in q_{pen}, which was attributed to selection of lower yielding variants of the mould, was observed.

DISCUSSION

This work shows that the Monod-Novick-Szilard continuous flow culture principle, first developed for bacteria, applies equally to a filamentous mould. The practical upper limit of the flow rate with a *P. chrysogenum* culture was 0·1 culture volumes/hr.

The great advantage of the continuous flow technique is that it permits much closer control of the environment of the organism; it should therefore greatly facilitate study of the effect of environment on the mould. An example of this is the elucidation by Pirt & Callow (1959*b*) of the effect of pH and nutrition on the morphology of *P. chrysogenum* in submerged culture. In the work described here the growth rate, oxygen demand, yield of mycelium and penicillin production with much improved environmental control are reported.

The stability of the penicillin production rate over a long period and its relatively high value in continuous flow culture are noteworthy; this is another advantage which the continuous flow process possesses over batch culture. The production of a high penicillin concentration will require means of increasing the residence time of the mould in the culture without causing either a fall in the rate of penicillin production/g of mycelium (q_{pen}) or selection of lower yielding strains of the mould, conditions which, we expect, will require a two-stage process (Pirt & Callow, 1959c). This aspect is now under investigation.

The authors acknowledge with pleasure the support and encouragement given by Dr. D. W. Henderson, the technical assistance of Mr. J. E. D. Stratton and the assistance of the team of process workers. Elementary analyses were carried out by the Chemical Defence Experimental Establishment and the statistical analyses by Mr. S. Peto and his group. The authors are grateful to Professor M. P. Backus for the gift of strain Wis 54-1255.

REFERENCES

ANDERSON, R. F., WHITMORE, L. M., BROWN, W. E., PETERSON, W. H., CHURCHILL, B. W., ROEGNER, F. R., CAMPBELL, T. H., BACKUS, M. P. & STAUFFER, J. F. (1953). Penicillin production by pigment-free molds. *Industr. Engng Chem. (Industr.)* **45**, 768.

BACKUS, M. P. & STAUFFER, J. F. (1955). The production and selection of a family of strains in *Penicillium chrysogenum. Mycologia* **47**, 429.

BARTLETT, M. C. & GERHARDT, P. (1958). Design and testing of a pilot plant for single-stage continuous antibiotic fermentation. *Abstracts, Int. Congr. Microbiol., Stockholm*, p. 408.

BROWN, W. (1923). Experiments on the growth of fungi on culture media. *Ann. Bot., Lond.,* **37**, 105.

CALLOW, D. S. & PIRT, S. J. (1956). Automatic control of pH in cultures of micro-organisms. *J. gen. Microbiol.* **14**, 661.

CHAIN, E. B., PALADINO, S., CALLOW, D. S., UGOLINI, F. & VAN DER SLUIS, J. (1952). Studies on aeration—I. *Bull. World Hlth Org.* **6**, 73.

ELSWORTH, R., MEAKIN, L. R. P., PIRT, S. J. & CAPELL, G. H. (1956). A two litre scale continuous culture apparatus for micro-organisms. *J. appl. Bact.* **19**, 264.

FOSTER, J. W. (1949). *Chemical Activities of Fungi.* p. 157. New York: Academic Press.

HAWKER, L. E. (1950). *Physiology of Fungi.* p. 28. London: University of London Press.

HERBERT, D., ELSWORTH, R. & TELLING, R. C. (1956). The continuous culture of bacteria: a theoretical and experimental study. *J. gen. Microbiol.* **14**, 601.

HERBERT, D. (1958). Some principles of continuous culture. In *Recent Progress in Microbiology,* p. 381. Stockholm: Almqvist & Wiksell.

HUMPHREY, J. H. & LIGHTBOWN, J. W. (1952). A general theory of plate assay of antibiotics with some practical applications. *J. gen. Microbiol.* **7**, 129.

JARVIS, F. G. & JOHNSON, M. J. (1947). The role of the constituents of synthetic media for penicillin production. *J. Amer. chem. Soc.* **69**, 3010.

KOLACHOV, P. J. & SCHNEIDER, W. C. (1952), Sept. 2. Continuous process for penicillin production. *U.S. pat.* no. 2,609,327.

MÁLEK, I. (1958). *Continuous Cultivation of Microorganisms: a Symposium.* Prague: Czech Acad. of Sciences.

MAXON, W. D. & JOHNSON, M. J. (1953). Aeration studies on propagation of bakers' yeast. *Industr. Engng Chem. (Industr.)* **45**, 2554.

MONOD, J. (1942). *Recherches sur la Croissance des Cultures Bactériennes.* Paris: Hermann et Cie.

PIRT, S. J. (1957). The oxygen requirement of growing cultures of an *Aerobacter* species determined by means of the continuous culture technique. *J. gen. Microbiol.* **16**, 59.

PIRT, S. J. & CALLOW, D. S. (1958a). Observations on foaming and its inhibition in a bacterial culture. *J. appl. Bact.* **21**, 211.

PIRT, S. J. & CALLOW, D. S. (1958*b*). Exocellular product formation by micro-organisms in continuous culture. I. Production of 2:3-butanediol by *Aerobacter aerogenes* in a single-stage process. *J. appl. Bact.* **21**, 188.

PIRT, S. J. & CALLOW, D. S. (1959*a*). The growth of *Penicillium chrysogenum* in continuous culture. *J. appl. Bact.* **22**, 11.

PIRT, S. J. & CALLOW, D. S. (1959*b*). The continuous flow culture of the filamentous mould *Penicillium chrysogenum* and the control of its morphology. *Nature, Lond.*, **184**, 307.

PIRT, S. J. & CALLOW, D. S. (1959*c*). Exocellular product formation by micro-organisms in continuous culture. II. A two-stage process and its application to the production of 2:3-butanediol. *Selected Scientific Papers*. **2**, (2), 292. Rome: Istituto Superiore di Sanità.

ROLINSON, G. N. (1952). Respiration of *Penicillium chrysogenum* in penicillin fermentations. *J. gen. Microbiol.* **6**, 336.

SMITH, J. H. (1924). On the early growth rate of the individual fungus hypha. *New Phytol.* **23**, 65.

STRANGE, R. E., DARK, F. A. & NESS, A. G. (1955). Interference by amino acids in the estimation of sugars by reductometric methods. *Biochem. J.* **59**, 172.

WHIFFEN. A. J. & SAVAGE, G. M. (1947). The relation of natural variation in *Penicillium notatum* to the yield of penicillin in surface culture. *J. Bact.* **53**, 231.

(*Received 27 August,* 1959)

Editor's Comments on Papers 14 and 15

14 Herbert: *The Chemical Composition of Micro-organisms as a Function of Their Environment*

15 Herbert: *A Theoretical Analysis of Continuous Culture Systems*

In Paper 14 Herbert awakened microbiologists to the implications of cellular dimensions and cellular aspects of microbial growth, to the "plasticity" of microbial metabolism and cell composition, and to the importance of the cultivation system for its influence in controlling the environment of the cells and, hence, of the growth taking place in a culture. This paper, which revealed some of the early results to be obtained from the new technique of continuous culture and the newer one of synchrony, is also notable for its wise and well-intentioned remarks on a number of matters, especially, perhaps, of the need for doing a job properly.

In the following article (Paper 15), although continuous culture systems are primarily under consideration, it is easy to see how other different systems may be placed in relation to them. Many workers have used and appreciated the advantages of multistage continuous cultures, but Herbert's contribution stands out from these others by its breadth and depth of focus. By comparing and distinguishing among the different systems and methods for growing cells, he not only indicated their potential for influencing the results obtained in growth studies, but also showed that the results obtained from such studies might reflect the system of growth rather than the object of growth.

Copyright © 1961 by the Society for General Microbiology

Reprinted from *Symp. Soc. Gen. Microbiol.*, **11**, 391–416 (1961)

THE CHEMICAL COMPOSITION OF MICRO-ORGANISMS AS A FUNCTION OF THEIR ENVIRONMENT

DENIS HERBERT

Microbiological Research Establishment, Porton, Wiltshire

There are few characteristics of micro-organisms which are so directly and so markedly affected by the environment as their chemical composition. So much is this the case that it is virtually meaningless to speak of the chemical composition of a micro-organism without at the same time specifying the environmental conditions that produced it. Such a statement as 'the ribonucleic acid content of *Bacillus cereus* is 16·2%' is by itself as incomplete and misleading as the statement 'the boiling-point of water is 70°'. While the latter statement may be true (on the summit of Mount Everest, for example), it is incomplete and misleading because it says nothing of the effect of pressure on boiling-point. The former statement is likewise true but misleading, since the RNA* content of *B. cereus* can vary from about 3 to 30%, depending on the environment in which the cells are grown.

The extent to which chemical composition is affected by the environment has come to be realized only fairly recently, and a great deal of early work on the chemical composition of micro-organisms (reviewed by Porter, 1946) is unfortunately invalidated through lack of adequate environmental control. No attempt will be made to review this early work, while even recent work is too extensive to be treated comprehensively in a single article. Attention will therefore be confined to a number of selected aspects of the subject, chosen because they illustrate some general principles or throw some light on problems of microbial growth or physiology. This aspect is emphasized, for there is little point in performing chemical analyses for their own sake; the important thing is what the results of the analyses tell us about microbiology in general.

The various factors comprising 'the environment' of micro-organisms may conveniently be grouped under the headings of: (1) physical or physico-chemical factors, and (2) chemical factors. So far as the former are concerned, there is little to relate; scarcely any studies have been made on the effects of pressure, temperature, etc., on the chemical

---

* Abbreviations used: RNA = ribonucleic acid; DNA = deoxyribonucleic acid; PHB = poly-β-hydroxybutyric acid; TPN = triphosphopyridine nucleotide.

composition of micro-organisms, but such data as exist indicate zero or negligible effects. As might be expected, the most important factor affecting the chemical composition of micro-organisms is the chemical composition of the environment. A significant fact here is that quantitative changes in the composition of the environment may be as important as qualitative ones. For example, in a culture medium containing only glucose, an ammonium salt and trace minerals, merely changing the ratio of the concentrations of glucose and ammonia may produce a more profound effect than replacing the ammonia by a mixture of amino acids.

The effects of such changes in the chemical environment on the chemical make-up of micro-organisms may be either qualitative (production in some environments of cell components completely absent in other environments), or quantitative (production of more or less of a cell component invariably present). The most striking examples of qualitative changes in cell composition are found in connexion with adaptive enzymes and certain types of antigen, but they will not be discussed here as they are the subjects of other contributions to this Symposium. This review therefore will be confined to the discussion of quantitative changes in major cell components, namely proteins, nucleic acids, polysaccharides (excluding antigens) and lipids.

PROTEINS AND NUCLEIC ACIDS

The nucleic acid content of bacteria can vary within very wide limits. In different bacterial species values for RNA content as low as 1·5 % and as high as 40 % have been recorded, while in a single strain a 16-fold variation in RNA content has been observed under different conditions of growth (Table 1). Changes in DNA content are less extreme, but still considerable.

The known association between nucleic acids and protein synthesis suggests that such changes would be intimately connected with the growth of the cell, and that environmental factors which affected growth would also affect the nucleic acid content. This is indeed the case, and in fact the effects of environment on the growth of the cell and on its chemical composition are here so intimately linked that it is virtually impossible to discuss one without the other. Some brief mention of the effect of environment on growth must therefore be made (without, it is hoped, trespassing too much on the subject-matter of another contribution to this Symposium).

When physico-chemical factors such as temperature and pH are favourable, the chemical composition of the environment determines not only the possibility of microbial growth but also its rate. The minimal requirements for growth vary with the synthetic ability of the organism; some, for example, can synthesize all the nitrogenous constituents of the cell from a single carbon compound and ammonia, while others are unable to synthesize one or more amino acids, which have to be present in the environment before the organism can grow. The rate of growth is affected by the extracellular concentrations of essential nutrients, the relationship between growth-rate and concentration being a type of 'saturation curve' (Monod, 1942) similar to the velocity against sub-strate–concentration curves found for enzyme action. Even when a cell is 'saturated' with all essential nutrients, it will grow still faster if additional (non-essential) nutrients are supplied, so that it no longer has to synthesize them. There are many such possibilities, so that cells may be made to grow at a number of different growth rates by supplying them with media of different chemical compositions. Alternatively, different growth rates may be induced by regulating the concentration of a single essential nutrient; the concentration may be held constant, in spite of utilization of the nutrient due to growth, by using a continuous culture apparatus of the 'chemostat' type (Monod, 1950). Both methods have been used to study the effect of growth rate on the chemical com-position of bacteria. In addition, changes in chemical composition have been studied during the course of the so-called 'growth cycle' that occurs when bacteria are inoculated in the traditional manner into a vessel con-taining a limited volume of culture medium. This type of experiment, which will be referred to as 'growth in a closed system '(Herbert, 1960), is historically the earlier and will be discussed first.

Changes in chemical composition during growth in closed systems

When bacteria are inoculated into a limited volume of culture medium, the following fairly standard sequence of events occurs. (1) A period of lag, followed by (2) a period of accelerating growth; the occurrence and duration of these periods depends on the previous history of the inoculum. (3) A period of exponential growth at a constant growth rate (the so-called 'logarithmic phase'), when growth follows the equation:

$$\frac{1}{x}\frac{dx}{dt} = \frac{d(\ln x)}{dt} = \mu = \frac{\ln 2}{t_d} \tag{1}$$

(x being the dry weight of cells/ml., μ the exponential growth rate and t_d the doubling time). During this period extracellular nutrients are being rapidly used up. (4) A period of declining growth rate, whose onset (in an adequately aerated and buffered medium) is due to the exhaustion of some nutrient (not necessarily an essential one) and whose duration depends greatly on the complexity of the culture medium. (5) A stationary phase, during which cells remain viable but no longer grow, initiated by the exhaustion of an essential nutrient.

(The above sequence of events is usually called 'the growth cycle'— a term which, I suggest, should speedily be abandoned, since it conveys a quite misleading impression that this sequence is a necessary and inevitable feature of bacterial growth, whereas it is in reality a sequence forced upon the organisms by sequential environmental changes which are inevitable when growth occurs in a closed system. Unfortunately, bacteriologists have been inoculating flasks of culture media for so long now that they have come to regard this as part of the natural order of events, instead of as a convenient but highly artificial experimental procedure.)

Table 1. *Cell mass and RNA content in 'resting' and 'log phase' cells of a number of bacterial species (from Wade & Morgan, 1961)*

| | | Cell mass (picograms)† | | RNA content (%) | |
| | | Resting cells | Log phase cells | Resting cells | Log phase cells |
Organism	Medium*				
Aerobacter aerogenes	CCY	0·11	0·40	4·4	26·6
Bacillus anthracis	TMB	—	—	1·5	24·0
Bacillus cereus	TMB	1·97	3·77	3·9	31·5
Chromobacterium prodigiosum	CCY	0·12	0·35	7·8	32·1
Chromobacterium violaceum	TMB	0·17	0·56	7·2	30·3
Clostridium welchii	TMB	0·91	2·19	32·2	42·2
Corynebacterium hofmannii	CCY	—	—	25·4	51·0
Salmonella typhi	CCY	0·19	0·34	10·5	35·9
Escherichia coli	CCY	0·12	0·41	15·5	37·0
Pasteurella pestis	TMB	0·13	0·15	5·9	20·1
Proteus vulgaris	TMB	0·18	0·36	12·6	35·0
Staphylococcus aureus	CCY	0·19	0·24	5·2	10·0

* CCY, casein-yeast extract medium; TMB, tryptic meat digest medium.
† One picogram = 10^{-12} g.

Bacteria grown in this way are most commonly studied during either phase (3) of the above sequence ('log phase cells'), or phase (5) ('stationary phase' or 'resting' cells). It has been known for some time that these two types of cell differ considerably both in size and in chemical composition. Log phase cells are large and have a high RNA content and low DNA content; resting cells (at least those produced by growth in the usual type of complex culture medium) are considerably smaller and have a much lower RNA content but a relatively higher DNA content. These generalizations are valid for a quite wide range of bacterial species, as illustrated by the data of Table 1.

More recently, changes in cell size, RNA content and DNA content have been followed for several different organisms throughout the growth of a culture. The type of result obtained is shown in Fig. 1, which is a composite (and slightly idealized) representation of the data of several workers (Hershey, 1938, 1939; Wade, 1952*a*, *b*; Mitchell &

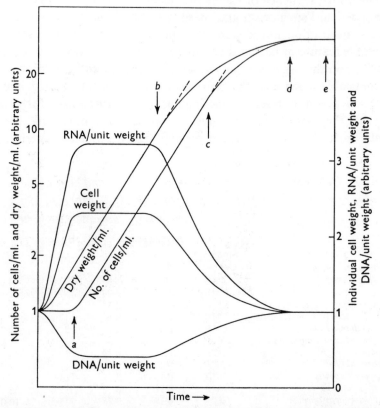

Fig. 1. Changes in cell size and chemical composition during growth of a bacterial culture. (Idealized curves after Wade and others.) The culture is inoculated at zero time with fairly young 'resting' cells. The increases in number of cells and dry weight/ml. are plotted on a logarithmic scale (left-hand ordinate). Values for individual cell weights (ratio of dry weight/ml. to total count/ml.), RNA content and DNA content are plotted on an arithmetic scale (right-hand ordinate). The units are arbitrarily chosen to make the initial values of all variables equal to 1·0.

Moyle, 1951; Gale & Folkes, 1953; Malmgren & Hedén, 1947). Bacterial dry weight/ml. and number of cells/ml. are plotted on a logarithmic scale, while RNA and DNA contents and cell mass (the ratio of dry weight to total count, i.e. the average dry weight of one cell) are plotted on arithmetic scales, the units being arbitrarily chosen to make the initial values of all variables equal to 1 (Maaløe, 1960). The culture is inoculated at zero time with resting cells taken from a previous culture at about the point marked by arrow *e*.

When such fairly 'young' resting cells are put into fresh medium, exponential growth (as measured by increase in dry weight) begins almost immediately, but there is no corresponding increase in cell numbers for a considerable period (up to point *a* on the curve). During this period of 'division lag' there is a rapid increase in cell size and even more rapid increase in RNA content until the values characteristic of log phase cells are reached at point *a*. Thereafter dry weight and cell numbers increase at the same rate while cell size and RNA content remain constant, until the end of the log phase at point *b*. Here the opposite effect occurs, growth rate decreasing more rapidly than division rate and the cell size consequently falling until, by the time growth has ceased at *d*, the cell mass has returned to the value characteristic of resting cells. Simultaneously the RNA content falls even faster until this too, at *d*, has returned to the original resting cell value. (This is due not to breakdown of RNA but to its slower synthesis during this phase.)

Changes in DNA content of the cell are the opposite of those found with RNA, the DNA content being lowest during the log phase. As the cell mass changes in the opposite direction to the DNA content, the quantity of DNA/cell remains at least approximately constant—there is some disagreement on this point, which it seems will be resolved by Maaløe's discovery (v.i.) that the amount of DNA/*nucleus* is much more nearly constant.

The foregoing account of changes in size and composition during the growth of a culture and of differences between log phase and resting cells has become fairly generally accepted in recent years, and is of some importance since a great deal of microbiological research is carried out with resting cells (which are the usual result of putting up an 'overnight' culture). As well as differing from log phase cells in size and chemical composition, resting cells are usually 'tougher' than log phase cells, being more resistant to physical disruption and osmotic shock and maintaining viability better on storage. However, work now in progress at M.R.E. (Fig. 2) suggests that all of the above applies only to growth in complex types of culture medium.

In Fig. 2, which is a previously unpublished experiment of Strange, Dark & Ness (1961), the growth of *Aerobacter aerogenes* is shown (*a*) in a complex (meat digest) medium, and (*b*) in a mannitol–ammonia–salts medium (with mannitol as limiting nutrient). The growth in complex medium is similar to the idealized curves of Fig. 1, the log phase being followed by an extended phase of deceleration before growth finally ceases. (As others have observed in such media (Monod, 1949), there are apparently two or perhaps three phases of exponential growth at successively decreasing rates, probably indicating successive exhaustion of non-essential but growth-accelerating nutrients.) Cells sampled in the log phase, deceleration phase and stationary phase show a progressive decline in RNA and increase in DNA content, as found by other workers. In the minimal medium, on the other hand, exponential growth is maintained almost to the end, ceasing abruptly and with a hardly detectable deceleration phase as the last traces of mannitol are exhausted. In this case the RNA and DNA contents of log phase cells and resting cells are virtually identical.

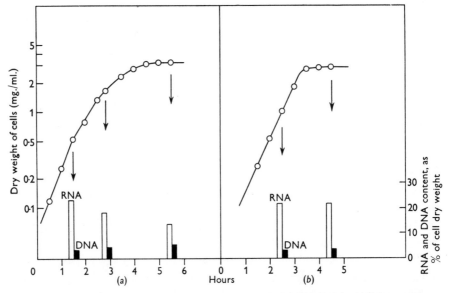

Fig. 2. Growth of *Aerobacter aerogenes* in (*a*) tryptic meat digest broth and (*b*) a minimal medium (NH₃—salts + 0·5 % mannitol as limiting factor) illustrating differences in chemical composition of 'resting' cells. From an unpublished experiment of Strange, Dark & Ness (1961); cultures grown in stirred fermenters with optimal aeration and continuous control of pH; samples for analysis taken at points indicated by arrows.

These results strongly suggest that the changes in RNA content at the end of growth in a complex medium are due to the extended period of growth at increasingly slow *rates*. This does not occur when growth terminates abruptly through exhaustion of a single essential nutrient, and the resting cells produced in such cases may be regarded as 'frozen' log phase cells. The effect of growth rate on the chemical composition of cells has been studied by both the methods described at the beginning of this section, with results described below.

Changes in chemical composition during steady-state growth at varying rates

As previously mentioned, the effect of growth rate on chemical composition has been studied by two methods: (1) unrestricted growth (i.e. all nutrients present in non-limiting concentrations) in media of different nutritional complexity, and (2) growth controlled by regulating the concentration of a single essential nutrient, using a flow-controlled type of continuous culture apparatus or 'chemostat'. Maaløe and co-workers at Copenhagen have used method (1), while at Porton we have used method (2); Magasanik, Magasanik & Neidhardt (1959) have used a combination of both.

Whichever method is used, it is essential that growth should continue in an essentially unchanging environment for long enough to allow steady-state* conditions to be established, since the regulatory mechanisms of the cell need some time to adjust to the medium; in practice this means at least three generations of exponential growth and preferably five or more. This can with care be achieved in the conventional closed-system culture, but very much to be preferred is Maaløe's technique of repeated dilution of the culture with fresh medium, the cell density always being kept well below the maximal attainable value. (An alternative method would be to use the type of continuous culture apparatus known as a 'Turbidostat' (Bryson & Szybalski, 1952) in which flow rate is regulated by cell density in such a way as to keep the latter at a level well below the maximum attainable in the medium.) With method (2), steady-state growth can be maintained for as long as is required at any desired growth rate, this being an inherent feature of the 'chemostat' type of continuous culture apparatus.

(1) *Chemical composition during unrestricted steady-state growth.* An extensive study of *Salmonella typhimurium* has been made by Schaechter, Maaløe & Kjeldgaard (1958), in which the organism was grown by the repeated dilution technique in twenty-two different growth media giving doubling times ranging from 22 to 97 min. The results were summarized by Maaløe (1960) in the graphs reproduced in Fig. 3, in which the following variables are plotted on a logarithmic scale against the growth rate:† in (A), the dry weight/cell, weights of RNA and DNA/cell

* Campbell (1957) has introduced the expression 'balanced growth': 'growth is *balanced* over a time interval if, during that interval, every extensive property of the growing system increases by the same factor'. This is an admirably precise definition, but I cannot see that it means anything more than the well-established term 'steady-state growth'.

† The measure of growth rate used by Maaløe is the reciprocal of the doubling time i.e. $1/t_d$. This is $1/\ln 2$ or 1.44 times greater than the exponential growth rate μ of equation (1).

and number of nuclei/cell: in (B) the dry weight/nucleus, RNA/nucleus and DNA/nucleus. The ordinate scale units are chosen to make the values of all of these equal to 1·0 at zero growth rate (cf. Fig. 1). In Fig. 4 the data of Fig. 3 are replotted so as to show the RNA and DNA contents expressed, not as quantities/cell, but as percentages of the total dry weight.

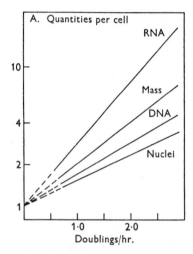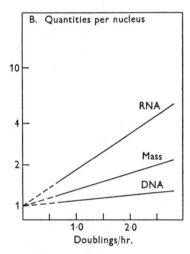

Fig. 3. Cell size and chemical composition of *Salmonella typhimurium* during unrestricted balanced growth (from Maaløe, 1960). The organism was grown in a series of culture media producing different rates of growth. The following values are plotted against growth rate: A, number of nuclei/cell, dry weight/cell, amount of RNA/cell and amount of DNA/cell; B, dry weight/nucleus, RNA/nucleus and DNA/nucleus. Dotted lines are extrapolations to zero growth rate from the regions covered by the experimental values. The units on the ordinate are as follows: 2·6 μg. DNA, 13 μg. RNA and 155 μg. dry weight/10⁹ cells.

Figure 3A shows that all the variables plotted increase exponentially with growth rate, though at different rates. The amount of RNA/cell increases more rapidly than the cell mass as the growth rate is increased, while the DNA/cell increases less rapidly. In other words, the RNA *content* of the cell, expressed as a percentage of the dry weight, increases with increasing growth rate, while the percentage of DNA decreases, as shown in Fig. 4. In this organism the average number of nuclei/cell also increases exponentially with growth rate, at nearly the same rate as the amount of DNA/cell, so that, as shown in Fig. 3B, the weight of DNA/ *nucleus* shows very little change with growth rate, although the RNA/ nucleus increases considerably.

The dotted lines in Fig. 3 indicate extrapolations back to zero growth rate from the regions covered by the experimental values. It is an interesting fact that the zero growth rate values for cell size, average

212

number of nuclei and chemical composition agree well with actual values obtained by the analysis of 'resting cells'.

(2) *Chemical composition during nutrient-limited steady-state growth.* Figures 5 and 6 show data obtained by Herbert *et al.* (1961) on cell mass and on RNA, DNA and protein contents of *Aerobacter aerogenes* and *Bacillus megaterium* grown over a wide range of growth rates in a continuous culture apparatus of the chemostat type. The theory of this type of apparatus has been discussed in detail by Monod (1950), Novick & Szilard (1950) and Herbert, Elsworth &

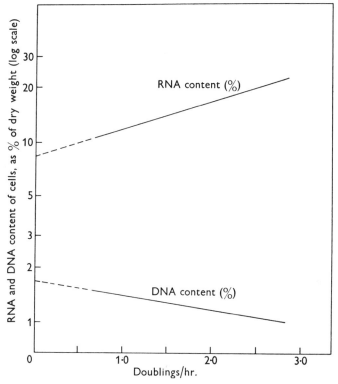

Fig. 4. Chemical composition of *Salmonella typhimurium* during balanced growth at different growth rates. The data of Maaløe *et al.* shown in Fig. 3 are plotted to show RNA and DNA contents as percentages of the dry weight (i.e. g. RNA or DNA/100 g. dry weight of cells).

Telling (1956). It will suffice here to say that if the dilution rate D (ratio of flow rate to culture volume) is held constant, the organisms grow up until the concentration of some essential nutrient is reduced to a level which makes the exponential growth rate μ equal to D. The system is now in a steady state, which is maintained indefinitely as long as D remains unaltered.

This means that, although medium of constant composition is fed to the growth vessel, the environment in which the cells are growing is different for each growth rate in respect of the concentration of limiting nutrient. In Fig. 5, for example, each experimental point corresponds

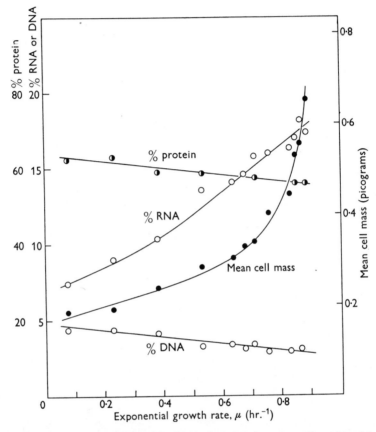

Fig. 5. Growth of *Aerobacter aerogenes* in continuous culture; protein and nucleic acid contents and mean cell mass as functions of the growth rate. The organism was grown in a continuous culture apparatus of the chemostat type at a number of different flow rates and the cells analysed after at least 2 days steady-state growth at each flow rate. Nucleic acid and protein contents expressed as percentage of cell dry weight; mean cell mass = dry weight/ml. divided by total count/ml. Culture medium: glycerol–NH$_3$–salts, with glycerol as limiting factor. (Data from Herbert *et al.* 1961.)

to the growth of the organism in a different concentration of glycerol, all other medium constituents being present in excess. The concentration of limiting nutrient can be made as low as desired simply by reducing the flow rate sufficiently; hence very much lower growth rates (doubling times up to 15 hr. or more) can be studied with this method than can be achieved by method (1).

26   MS XI

214

The results of such experiments with the chemostat are broadly similar to those of Maaløe and his colleagues in respect of changes in cell size, RNA content and DNA content (the number of nuclei was not studied). The protein content of the cells was also measured, but is seen to undergo no striking changes; it falls a little as the growth rate in-

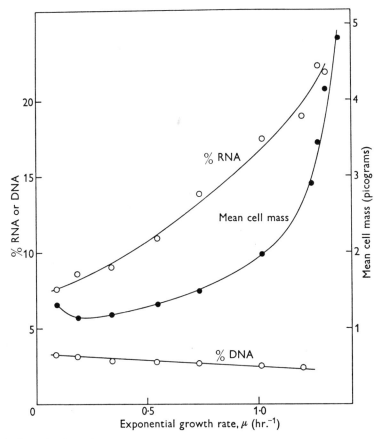

Fig. 6. Growth of *Bacillus megaterium* in continuous culture; nucleic acid content and cell mass as functions of growth rate. The organism was grown in a chemostat in a casein hydrolysate-mannitol medium and the cellular nucleic acid content and mean cell mass determined as in Fig. 5. (Herbert *et al.* 1961.)

creases, to an extent which can largely be accounted for by the increased proportion of RNA.

The data of Figs. 5 and 6 are plotted on an arithmetic scale; if logarithmic ordinates are used (cf. Figs. 3 and 4), the values for cell mass in both organisms, and for RNA content in the case of *Bacillus megaterium*, give approximately straight-line plots in the region of higher growth rates but fall off at lower growth rates; with *Aerobacter aerogenes* the change in RNA content with growth rate is closer to a linear than an exponential relationship. In this respect, therefore, there is not complete agreement between results obtained with the chemostat and by the unrestricted growth technique. In view of the different ranges of growth rate covered and the fact that species differences have already been observed in spite of the small number of organisms yet studied, this does not seem too serious; the general agreement in the over-all pattern of the results seems more important than the question of whether certain variables are exact exponential functions of growth rate, or only approximately so.

To the writer, at least, this general pattern of the results appears as one of the most striking examples of environmental control of chemical composition; it is remarkable that such profound changes in size, morphology and chemical make-up of the cell can be produced simply by altering the concentration of a single nutrient. Even more significant, however, is the fact that *the same pattern of changes seems to occur whatever the nature of the nutrient that is varied.* For example, the experiment of Fig. 5, with the nitrogen source in excess and glycerol as limiting nutrient, was repeated with a medium containing exactly the same chemical ingredients but with the carbon/nitrogen ratio altered so as to make the nitrogen source (NH_4^+ ion) the limiting nutrient. *The results were virtually identical;* not merely was the general pattern the same, but bacteria growing at the same rate in either experiment had the same size and chemical composition. Similarly, Schaechter *et al.* (1958) found that some of their growth media of quite different chemical composition gave identical growth rates during unrestricted growth; such media all produced cells of the same size and chemical composition.

Such results seem to show that the chemical composition of the cell is primarily dependent on the rate at which it is growing and is affected by the chemical composition of the growth medium only in so far as this affects the growth rate. The situation is not so simple as this, however, as was shown by the experiments of Schaechter *et al.* (1958) on the effect of temperature on chemical composition. On repeating the experiments of Fig. 3 at a lower temperature (25° instead of 37°), these workers found that for each different growth medium, cells of the same size and chemical composition were produced at both temperatures, although for all media the growth rate was halved at the lower temperature. Cell size and chemical composition would therefore seem to be determined by the pattern of metabolic activities imposed on the cell by the growth medium.

26-2

Chemical composition of cells during synchronous growth

All the data on cell size and chemical composition discussed so far were obtained from samples containing many billions of cells taken from un-synchronized cultures containing cells of all possible ages;* they there-fore represent cells of some sort of average age and it is worth considering what this may be. The age distribution in a steadily growing culture is related to the distribution of generation times (Powell, 1956) and has the interesting property, not generally known, that the youngest organisms are present in the greatest numbers.

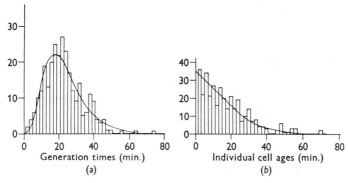

Fig. 7. Frequency distributions of (*a*) generation times, (*b*) individual cell ages, during unsynchronized growth of *Bacillus mycoides* (Powell, 1955, 1956).

Figure 7*a* shows the distribution of generation times in a culture of *Bacillus mycoides* during steady-state growth, from observations of Powell (1955); as is usually the case, the frequency distribution curve is unimodal and positively skew. Figure 7*b* shows the corresponding distribution of ages, i.e. the relative frequencies of occurrence of organisms of different ages in a sample taken from a culture at any instant. This has a J-shaped frequency distribution curve with a maxi-mum at zero age. The mean generation time of this organism was 28·7 min., but the median age was only 11 min.; i.e. half of the organisms in a sample from such a culture would be less than 11 min. old. Data obtained from large samples are therefore representative of cells in rather early stages of their individual life cycles.

It would be of the greatest interest to know whether the chemical composition of the individual cell undergoes any systematic changes

---

* The 'age' of an organism (to be distinguished from the age of a culture) is the time that has elapsed since its inception through fission of its parent; a 'young' organism is one whose age is small compared with the mean generation time, whatever the value of this may be.

during the course of its life. A direct approach to this problem is beyond the range of the most ultra-micro analytical techniques known at present; it is true that the dry weights of individual bacteria have been determined with the interference microscope (Ross, 1957) and their total nucleic acid contents by ultraviolet microspectrophotometry (Malmgren & Hedén, 1947), but only the former could be used for repeated measurements on the same cell throughout its life and even these would not be very accurate. In a culture undergoing synchronous growth, however (that is, a culture in which all the cells are growing and dividing 'in step'), samples taken at intervals throughout a division cycle should give the information needed, provided that all the cells in the culture are undergoing normal steady-state growth. There is the further possibility of identifying changes in composition occurring at certain times in the cycle with the process of cell division.

So exciting are these possibilities that great interest was aroused by the discovery that growing cultures could be artificially synchronized by subjecting them to treatments such as temperature changes, illumination changes (in certain photosynthetic organisms), and starvation of certain nutrients (see review by Campbell, 1957). Interest was intensified when it was found that changes in chemical composition did in fact occur in the course of synchronous division cycles, discontinuous or stepwise synthesis of DNA, RNA and protein being reported by various workers (see Campbell, 1957). However, more recent evidence (e.g. Schaechter, Benzton & Maaløe, 1959) suggests that these effects are artefacts and that the treatments used to induce such 'forced synchrony' produce unbalanced growth; the changes in chemical composition observed are probably akin to those found by Kjeldgaard, Maaløe & Schaechter (1958) when cultures in steady-state growth are abruptly shifted from one medium to another. It seems doubtful, therefore, that such artificially synchronized cultures can be said to reproduce the normal division cycle, at any rate during the first one or two generations. (After several generations a condition of steady-state growth will be regained, but by then, owing to the normal scatter in individual generation times (Fig. 7), most of the synchrony will have disappeared.)

A better method of obtaining synchronous cultures appears to be the filtration technique of Maruyama & Yamagita (1956) which *selects* from a normally growing population those small cells which have just been formed by division. Abbo & Pardee (1960) introduced refinements of technique which avoided temperature shifts or medium starvation during the filtration process and their results (Fig. 8) appear to give a true picture of the normal division cycle.

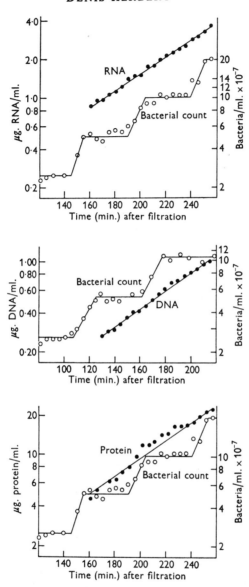

Fig. 8. Synthesis of RNA, DNA and protein during growth of synchronously dividing *Escherichia coli*. (From Abbo & Pardee, 1960.)

The results of Fig. 8 show that, contrary to what has been observed in artificially synchronized cultures, RNA, DNA and total protein all increase exponentially at the same constant rate over the whole division cycle; in other words, their *ratio* remains constant. So far as these components are concerned, therefore, we must conclude that the

219

chemical constitution of the individual bacterial cell remains constant throughout its life cycle; undoubtedly this must be a fact of fundamental importance in bacterial physiology.

POLYSACCHARIDES

It is now time to turn to another class of major cell components, namely the polysaccharides. A great deal of work has been done in this field on the effect of environmental factors on capsular polysaccharides and the polysaccharide components of surface antigens; these are discussed elsewhere in this Symposium. The present contribution will be confined to the intracellular polysaccharides, of which by far the most important is glycogen.

The glycogen content of bacteria and yeasts (moulds appear to have been little studied) can vary very greatly; values ranging from 2% to over 30% have been recorded in the same organism under different growth conditions. Little progress was made in elucidating the factors necessary for glycogen storage so long as micro-organisms were grown in complex culture media of unknown constitution. The use of simple chemically defined media has in the last few years clarified the situation considerably.

Holme & Palmstierna (1956) grew *Escherichia coli* in a simple glucose-ammonium salt medium with either glucose or ammonia as limiting nutrient. When glucose was limiting, the glycogen content of the cells remained low throughout the growth of the culture; when ammonia was limiting, the glycogen content was also low during the early stages of growth but increased enormously during and shortly after the final stages (i.e. when the ammonia concentration was approaching zero). The results strongly suggested that glycogen deposition occurs when the ammonia concentration is very low; this was confirmed by Holme (1957) using the continuous culture technique, with results shown in Fig. 9.

In this experiment, the organisms were grown at a number of different flow rates in a chemostat type of apparatus, with ammonia as the growth-limiting nutrient. Under these conditions the ammonia is nearly all consumed so that its steady-state concentration is very low, increasingly so as the growth rate is decreased. As Fig. 9 shows, the glycogen content of the cells increases as the growth rate and ammonia concentration decrease, from c. 3% of the dry weight at the highest growth rate tested to c. 23% at the lowest. As the glycogen content of the cells increases, the protein content (as approximately measured by total N) decreases; this is not due to less protein being synthesized, but to its being 'diluted'

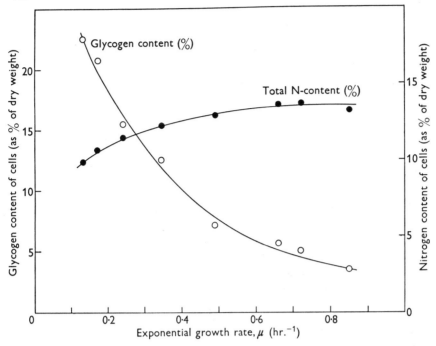

Fig. 9. Glycogen synthesis as a function of growth rate in *Escherichia coli* grown under conditions of nitrogen limitation. Plotted from data of Holme (1957, Table 2). The organism was grown in continuous culture at a number of different growth rates in a lactate–NH₃–salts medium with NH₃ as growth-limiting nutrient.

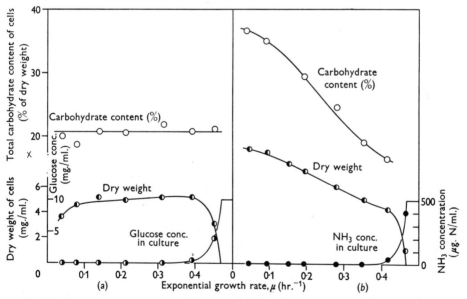

Fig. 10. Carbohydrate content of *Torula utilis* as a function of growth rate and limiting nutrient. (Unpublished data of Herbert & Tempest.) The organism was grown in continuous culture at a number of different growth rates in a glucose–NH₃–salts medium (*a*) with glucose as limiting nutrient, and (*b*) with NH₃ as limiting nutrient. Dry weight of cells in the culture and their total carbohydrate content (anthrone method), as well as steady-state levels of glucose and NH₃ in the culture, are plotted against growth rate.

with glycogen. Identical results were obtained when lactate replaced glucose as carbon source, the nitrogen content still being limiting.

Similar results to those of Holme have been obtained by the author and D. W. Tempest (unpublished) with a yeast (*Torula utilis*) and are shown in Fig. 10.

This figure shows the results of two continuous culture runs covering a wide range of flow rates (*a*) with glucose as limiting nutrient, and (*b*) with ammonia as limiting nutrient. With glucose limiting, the total carbohydrate content of the cells remains constant; while with ammonia limiting, the carbohydrate content increases with decreasing growth rate in the same manner as Holme observed with *Escherichia coli*. (It should be mentioned that the figures for 'total carbohydrate' were determined by the anthrone method and therefore include glucan and mannan as well as glycogen; separate analyses showed, however, that the changes were confined to the glycogen fraction.) The dry weight of cells obtained at each growth rate is also plotted and in Fig. 10*b* is seen to increase at low growth rates, the higher 'yield' being due to the higher glycogen content.

The results of Fig. 10*a* and *b* taken together suggest that high glycogen content is the result not of growth at a low rate but of growth in a low concentration of ammonia. The fact that synthesis of glycogen can occur at very low growth rates suggests that it will probably occur at zero growth rate, i.e. in resting cells; the results given in Fig. 11 show that this in fact occurs. In this experiment, washed yeast cells were shaken aerobically with glucose in a phosphate buffer, no nitrogen source being added so that growth, in the sense of protein or nucleic acid synthesis, could not occur. The cells doubled their dry weight in 2 hr., an increase which could entirely be accounted for in terms of glycogen synthesized (there was a barely significant increase in protein in the first half hour which we interpret as conversion to protein of the initial 'amino acid pool').

This experiment is by no means novel, numerous examples of 'carbon assimilation' in resting bacteria being known (Clifton, 1946), though the product of assimilation has by no means always been identified as glycogen. It has been included mainly to show the magnitude of the changes in chemical composition of the cell that can occur in such experiments; in this one, the carbohydrate content reached the very high level of 50% of the dry weight of the cell, the protein content falling to *c*. 35%. Such changes are sufficient to cause large changes in the refractive index of the cells, which have a characteristic appearance under the phase-contrast microscope.

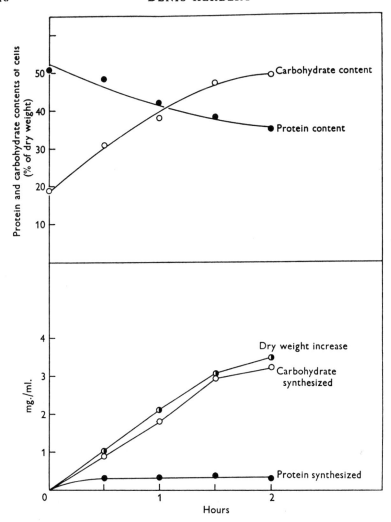

Fig. 11. Carbohydrate synthesis in *Torula utilis* in the absence of growth. Washed cells (4·3 mg./ml.) incubated aerobically in phosphate buffer (pH 5·5) containing 0·85 % glucose; dry weight, total carbohydrate and protein determined at intervals. (Unpublished data of Herbert & Tempest.)

LIPIDS

Work on bacterial lipids has virtually been confined to the *Myco-bacteria* and has been mainly concerned with problems of lipid chemistry. A little more is known about yeasts and moulds, although most pub-lished work is confined to a few species of high fat content which have mainly been studied in the fermentation industry with a view to indus-trial fat production.

With such organisms as *Rhodotorula gracilis* and *Endomyces vernalis* there is a fair amount of evidence (reviewed by Kleinzeller, 1948) that high lipid content, like high polysaccharide content, is associated with growth in nitrogen-deficient media. This is certainly the case with the lipid poly-β-hydroxybutyric acid (PHB).

This interesting substance, a long-chain polyester of β-hydroxybutyric acid discovered in *Bacillus megaterium* by Lemoigne in 1927, was for many years regarded as something of a biochemical curiosity. Interest in it has revived with the discovery that it is in fact a fairly common constituent of bacteria, having in recent years been found in many *Bacillus* species, *Azotobacter*, a wide range of Gram-negative organisms and certain photosynthetic bacteria such as *Rhodospirillum rubrum* (Doudoroff & Stanier, 1959). It may form up to 45 % of the dry weight of the cell, and is a major component of the sudanophilic 'lipid granules' found in several of the above species, although it is not itself sudanophilic. Recent work indicates that PHB should be considered along with the triglyceride fats and glycogen as a major reserve storage material in many bacteria.

Macrae & Wilkinson (1958*b*) studied the formation of PHB in *Bacillus megaterium* growing in a glucose-NH_4Cl medium: (*a*) with glucose limiting, (*b*) with NH_4Cl limiting. Their results bear a striking resemblance to those of Holme & Palmstierna (1956) mentioned above on the formation of glycogen in *Escherichia coli*. In carbon-limited cultures the PHB content remained fairly uniform and low throughout the growth of the culture, while when nitrogen was limiting there was a rapid synthesis of PHB up to very high levels in the period immediately preceding and following the cessation of growth due to exhaustion of the nitrogen source.

Like glycogen again, PHB may be formed in large amounts by resting cells supplied with glucose (and certain other substances such as pyruvate, acetate and butyrate) in the absence of any nitrogen source (Macrae & Wilkinson, 1958*a*; Doudoroff & Stanier, 1959); the latter workers showed that PHB can also be formed by photo-assimilation of acetate and butyrate in *Rhodospirillum rubrum*.

To summarize, it would appear that PHB (and perhaps other lipids) resemble glycogen in that their formation in the cell is largely controlled by the nitrogen content of the environment. Continuous culture studies should assist in investigating this point further; it would also be interesting to know whether glycogen and PHB are strictly alternative reserve storage materials or whether they can both be formed in the same organism.

DISCUSSION

Although this contribution has been deliberately restricted to a few selected topics, sufficient evidence has been brought forward to illustrate how greatly the chemical composition of micro-organisms may vary in response to the chemical composition of the environment in which they are grown, or to which they are exposed after growth. Of the major cell constituents, the protein content varies least but even so may undergo variations of 50–100 %; the DNA content may vary by two- to three-fold while the RNA, polysaccharide and lipid contents may vary by ten-fold or more and may amount to anything between a few per cent and nearly half the dry weight of the cell. These data reinforce the statement made at the beginning of this review that it is quite meaningless to write down a *single* set of analytical figures and state that 'this is *the* chemical composition of micro-organism X'. The chemical composition of a micro-organism can within wide limits be what we choose to make it.

The major chemical components of the cell may be broadly divided into two main groups:

> (i) storage materials (polysaccharides, lipids),
> (ii) basal materials (nucleic acids, proteins).

The distinction is a crude one, but is based on the important fact that the cell can, at a pinch, do without the former altogether, while the latter are absolutely essential. There appears to be a considerable difference between the way in which the concentrations of the two types of component within the cell are affected by the environment.

Storage materials appear to be laid down within the cell whenever the external environment (*a*) contains the necessary small molecules (glucose, acetate, etc.) from which they can be synthesized, and (*b*) does *not* contain nitrogenous materials necessary for growth, or at any rate contains them only in suboptimal concentration. Synthesis of such storage materials therefore can and does take place independently of growth.

It seems highly probable that the synthesis or breakdown of such storage materials depends on the concentrations within the cell of relatively few small molecules such as phosphate, coenzymes, etc., alterations in the concentration of which can shift enzymic equilibria in the direction of synthesis or breakdown. This is made probable by the work of Holzer (1959) on the effect of small external concentrations of (NH_4^+) ions on the concentrations of phosphate, α-ketoglutarate, TPN, etc., *within* the yeast cell. The internal concentrations of all of these are very markedly affected by small traces of (NH_4^+) in the external environment, and it is obvious how the synthesis of glycogen by the enzyme phosphorylase, for example, could be affected by the internal concentration of phosphate.

Changes in the internal concentrations of the 'basal components' are affected by the external environment in a much more complex way, and one intimately bound up with cell growth; these components are, relatively speaking, inert in cells that are not growing. It is not proposed to discuss here the complex relationships between the synthesis of RNA, DNA and protein, first because to do so would inevitably re-tread much of the ground covered at the 1960 Symposium of the Society, and secondly because the writer strongly agrees with Pirie (1960) that this is a field in which speculation has far outrun facts. Only the following broad generalizations will be made.

(1) In the light of present knowledge, it seems reasonable to regard the microbial cell as *essentially* a complex autosynthetic system; unlike metazoan cells, which spend most of their lives in a non-growing state, a microbial cell is only behaving naturally when it is growing exponentially.

(2) When such a complex self-replicating system is in the process of steady-state exponential growth, its basal elements must be in a state of dynamic equilibrium with the *internal* concentrations of the numerous small molecules—amino acids, purines, pyrimidines, etc.—from which they are synthesized. Changing the *external* environment, as by placing a cell in a different medium, will inevitably cause a complex shift in the concentrations of all components of the internal environment, which in turn will react upon the rates of synthesis of the various large molecules, decreasing some rates and increasing others, thus altering the ratios of their steady-state concentrations; in other words, changing the chemical composition of the cell. Some such concept seems necessary to explain the complex inter-relationships between growth rate, chemical composition of the cell and chemical composition of the growth medium, discussed in the first section of this review.

(3) The distinction between cell growth (defined as increase in mass) and cell division has long been made. These processes, though distinct, must obviously be related. What might not have been expected, however, is the relationship disclosed by recent work between growth rate, cell size (intimately connected with the division process) and the chemical composition of the cell.

These generalizations are not particularly original and, even if approved of, may be thought too imprecise to be useful. They do suggest two conclusions, however: (i) that in future work in this field, more attention should be paid to the effect of the external environment upon the concentrations of components of the internal environment, such observations preferably being made upon cells during steady-state growth; (ii) more work needs to be done on the relations between the external environment, the composition of the cell *and the cell size.*

It is customary to conclude a review of this sort with an improving moral, and for good measure two will be provided, whose implications are not confined to the field of the review. The first is, that when studying effects of environment on micro-organisms it is a good idea to know precisely what the environment is—in other words, micro-organisms should be grown wherever possible in chemically defined media. Anyone who thinks this a truism would do well to read some of the hundreds of papers on the chemical composition of bacteria in the older literature, in which countless man-hours of painstaking analytical work are rendered virtually valueless by accompanying statements such as 'the organisms were grown in Difco Bacto-peptone horse serum broth'—or a similar decoction. It is certainly striking, on surveying the literature of this field, to note how almost all the results of real value were obtained by workers using chemically defined growth media. It may occasionally still be necessary to grow micro-organisms on decoctions of unknown composition, but how often is the reason not simple laziness?

The second moral is that useful deductions can seldom be drawn from growth experiments unless they are conducted in such a manner as to obtain *steady-state growth.* Again it is striking to notice, in the present field, how often the most valuable results have been obtained by workers who have fully realized this principle. To which may be added the rider that a method guaranteed to make it difficult to achieve steady-state growth is the usual one of placing some medium in a glass container, inoculating it with micro-organisms and allowing events to run their course.

ACKNOWLEDGEMENTS

The writer wishes to express his gratitude to his colleagues R. E. Strange, D. W. Tempest and H. E. Wade of the Microbiological Research Establishment, Porton, for allowing him to quote their unpublished results.

REFERENCES

ABBO, F. E. & PARDEE, A. B. (1960). Synthesis of macromolecules in synchronously dividing bacteria. *Biochim. biophys. Acta*, **39**, 478.

BRYSON, V. & SZYBALSKI, W. (1952). Microbial selection. *Science*, **116**, 45.

CAMPBELL, A. (1957). Synchronization of cell division. *Bact. Rev.* **21**, 263.

CLIFTON, C. E. (1946). Microbial assimilations. *Advanc. Enzymol.* **6**, 269.

DOUDOROFF, M. & STANIER, R. Y. (1959). Role of poly-β-hydroxybutyric acid in the assimilation of organic carbon by bacteria. *Nature, Lond.* **183**, 1440.

GALE, E. F. & FOLKES, J. P. (1953). Nucleic acid and protein synthesis in *Staphylococcus aureus*. *Biochem. J.* **53**, 483.

HERBERT, D. (1960). A theoretical analysis of continuous culture systems. In *Continuous Cultivation of Micro-organisms*. Soc. Chem. Ind. Monograph (in the Press).

HERBERT, D., ELSWORTH, R. & TELLING, R. C. (1956). The continuous culture of bacteria: a theoretical and experimental study. *J. gen. Microbiol.* **14**, 601.

HERBERT, D., SPURR, E., GOULD, G. W. & PHIPPS, P. J. (1961). In preparation.

HERSHEY, A. D. (1938). Factors limiting bacterial growth. II. Growth without lag in *Bacterium coli* cultures. *Proc. Soc. exp. Biol., N.Y.* **38**, 127.

HERSHEY, A. D. (1939). Factors limiting bacterial growth. IV. The age of the parent culture and the rate of growth of transplants of *Escherichia coli*. *J. Bact.* **37**, 285.

HOLME, T. (1957). Continuous culture studies on glycogen synthesis in *Escherichia coli* B. *Acta chem. scand.* **11**, 763.

HOLME, T. & PALMSTIERNA, H. (1956). Changes in glycogen and N-containing compounds in *E. coli* B during growth in deficient media. I. Nitrogen and carbon starvation. *Acta chem. scand.* **10**, 578.

HOLZER, H. (1959). Enzymic regulation of fermentation in yeast cells. In *Ciba Symp. on Regulation of Cell Metabolism*. London: J. and A. Churchill Ltd.

KJELDGAARD, N. O., MAALØE, O. & SCHAECHTER, M. (1958). The transition between different physiological states during balanced growth of *Salmonella typhimurium*. *J. gen. Microbiol.* **19**, 607.

KLEINZELLER, A. (1948). Synthesis of lipides. *Advanc. Enzymol.* **8**, 299.

LEMOIGNE, M. (1927). Études sur l'autolyse microbienne. Origine de l'acide β-oxybutyrique formé par autolyse. *Ann. Inst. Pasteur*, **41**, 148.

MAALØE, O. (1960). The nucleic acids and the control of bacterial growth. *Symp. Soc. gen. Microbiol.* **10**, 272.

MACRAE, R. M. & WILKINSON, J. F. (1958a). Poly-β-hydroxybutyrate metabolism in washed suspensions of *Bacillus cereus* and *Bacillus megaterium*. *J. gen. Microbiol.* **19**, 210.

MACRAE, R. M. & WILKINSON, J. F. (1958b). The influence of cultural conditions on poly-β-hydroxybutyrate synthesis in *Bacillus megaterium*. *Proc. Roy. phys. Soc. Edinb.* **27**, 73.

MAGASANIK, B., MAGASANIK, A. K. & NEIDHARDT, F. C. (1959). Regulation of growth and composition of the bacterial cell. In *Ciba Symp. on Regulation of Cell Metabolism*. London: J. and A. Churchill Ltd.

MALMGREN, B. & HEDÉN, C.-J. (1947). Studies on the nucleotide metabolism of bacteria. III. The nucleotide metabolism of the Gram-negative bacteria. *Acta path. microbiol. scand.* **24**, 448.

MARUYAMA, Y. & YANAGITA, T. (1956). Physical methods for obtaining synchronous culture of *Escherichia coli*. *J. Bact.* **71**, 542.

MITCHELL, P. & MOYLE, J. (1951). Relationships between cell growth, surface properties and nucleic acid production in normal and penicillin-treated *Micrococcus pyogenes*. *J. gen. Microbiol.* **5**, 421.

MONOD, J. (1942). *Recherches sur la croissance des cultures bacteriennes.* Paris: Hermann et Cie.

MONOD, J. (1949). The growth of bacterial cultures. *Annu. Rev. Microbiol.* 3, 371.

MONOD, J. (1950). La technique de culture continue; théorie et applications. *Ann. Inst. Pasteur,* 79, 390.

NOVICK, A. & SZILARD, L. (1950). Experiments with the Chemostat on spontaneous mutations of bacteria. *Proc. nat. Acad. Sci., Wash.* 36, 708.

PIRIE, N. W. (1960). Biological replication considered in the general context of scientific illusion. *New Biology,* 31, 117.

PORTER, J. R. (1946). *Bacterial Chemistry and Physiology.* New York: John Wiley and Sons.

POWELL, E. O. (1955). Some features of the generation times of individual bacteria. *Biometrika,* 42, 16.

POWELL, E. O. (1956). Growth rate and generation time of bacteria, with special reference to continuous culture. *J. gen. Microbiol.* 15, 492.

ROSS, K. F. A. (1957). The size of living bacteria. *Quart. J. micr. Sci.* 98, 435.

SCHAECHTER, M., BENZTON, M. W. & MAALØE, O. (1959). Synthesis of desoxyribonucleic acid during the division cycle of bacteria. *Nature, Lond.* 183, 1207.

SCHAECHTER, M., MAALØE, O. & KJELDGAARD, N. O. (1958). Dependency on medium and growth temperature of cell size and chemical composition during balanced growth of *Salmonella typhimurium. J. gen. Microbiol.* 19, 592.

STRANGE, R. E., DARK, F. A. & NESS, A. G. (1961). Private communication.

WADE, H. E. (1952a). Observations on the growth phases of *Escherichia coli,* American type 'B'. *J. gen. Microbiol.* 7, 18.

WADE, H. E. (1952b). Variation in the phosphorus content of *Escherich₁ ₂ coli* during cultivation. *J. gen. Microbiol.* 7, 24.

WADE, H. E. & MORGAN, D. M. (1961). Private communication.

Reprinted from *Society of Chemical Industry Monograph 12*, Society of Chemical Industry, London, 1961, p. 21–53

A THEORETICAL ANALYSIS OF CONTINUOUS CULTURE SYSTEMS

By D. HERBERT

(*Microbiological Research Establishment, Porton, Wilts.*)

Criteria for the classification of continuous culture systems are discussed and a primary sub-division into 'closed' and 'open' systems is suggested; only the latter are capable of steady-state operation for indefinite periods. A preliminary classification of continuous culture systems is attempted and the different operating features of the various types are discussed. Several types are subjected to mathematical analysis which should make quantitative predictions of their behaviour possible. For the continuous operation of a given microbiological process, one particular type of system will usually give more efficient results than any other, and the factors involved in making a correct choice of system are discussed.

Introduction

NOWADAYS the term 'continuous culture' is applied to almost any process involving passage of fluids through some sort of reactor system in which they are exposed to microbiological action. Thus all the following processes, when operated continuously, have at various times been described as examples of 'continuous culture'.

(i) Production of bakers' yeast, food yeast, *Chlorella* cells, etc.
(ii) Production of ethanol, butanediol and other metabolic products by continuous deep fermentation
(iii) Production of acetic acid in 'vinegar towers'
(iv) Production of antibiotics and vitamins
(v) Transformation of steroids
(vi) Disposal of industrial wastes
(vii) Various sewage disposal processes, including trickling filters, the activated sludge process and anaerobic digestion.

The general application of the term 'continuous culture' to such diverse systems probably does no harm so long as it is not taken as implying that all these processes are basically much alike. Unfortunately there is a tendency to think that all continuous culture processes must have a great deal in common simply because they are continuous, as opposed to batch processes. Such resemblances as are due to the continuous nature of the processes are often, however, quite superficial.

There are indeed some underlying principles common to all

continuous-culture systems, but these have to do with the fundamental mechanisms of bacterial growth and apply with equal force
to batch culture. In most other respects, different types of continuous
culture systems may be as unlike each other as any of them is unlike
a batch process.

The industrial microbiologist who wishes to improve a traditional
process is therefore not only faced with a decision between batch
and continuous culture; he has also to choose what kind of continuous
culture system is best suited to the requirements of his microbiological process. There is little in the literature to help him in such
a decision.

Theoretical analysis of continuous culture systems has so far
been confined to one basic type, namely the single-stage stirred
fermentor. The operating principles of this type of continuous
culture system are now fairly well understood, and theoretical
predictions of its behaviour have been well borne out in practice.
While this is satisfactory so far as it goes, it must be admitted that
theoretical studies of all the other types of continuous culture
system (multi-stage, feed-back, pipe flow, etc.) are almost totally
lacking. Compared with the chemical industry, where every possible
type of continuous system has been analysed in great detail, our
theoretical knowledge of continuous culture is very deficient.

This paper is an attempt to remedy this situation and consists of:

(a) a preliminary essay in classification of continuous culture
systems according to their basic operating principles
(b) detailed analysis of some key examples.

Criteria for the classification of continuous culture systems

The classification of continuous culture systems may be
approached from two directions:

(1) the nature of the fermentation process (i.e., the microbiological
and biochemical approach)
(2) the type of operation (i.e., the chemical engineering approach).

The fusion of these two viewpoints will doubtless result in the
much-discussed new science of microbiological engineering. It will
always remain true, however, that the microbiological and biochemical aspects must be considered first.

(1) The fermentation process

All industrial microbiological processes, whether continuous or
batch, may be classified according to their purpose under two
headings:

(i) the production of micro-organisms

(ii) bringing about desired chemical transformations.

The latter may in turn be sub-divided into:

(*a*) formation of a desired end-product (e.g., production of industrial solvents, antibiotics, steroid transformations)

(*b*) decomposition of a given starting-material (e.g., sewage disposal, destruction of industrial wastes).

From the biochemical viewpoint, however, these sub-divisions are unimportant and all chemical transformation processes have one major feature in common which sharply differentiates them from processes where the aim is to produce microbial cells; namely, that chemical transformations need not necessarily be accompanied by growth.

This is a fact whose importance is often underrated by industrial microbiologists. It is well known that many chemical transformations can be effected by cells that are not growing or even by cells that are no longer alive, and that these transformations are brought about by the enzymes that the cells contain; and further that the most controllable way of effecting many transformations, and the one giving the highest yields, is to use *resting* cell suspensions, or enzymes prepared from them. Yet in spite of this, the most common way of effecting microbial transformations is to add the starting material to a *growing* culture. It might often be more efficient to effect a complete separation of the growth and transformation steps, growing the micro-organisms in one stage under conditions producing cells of maximum enzyme content, and subsequently adding the substrate to be transformed to the resting cell suspension. Two-stage continuous culture systems should be well adapted to effecting such transformations on a continuous basis.

Another important biochemical distinction is between:

(*a*) *catabolic processes*, in which complex compounds are broken down to simpler ones (e.g., glucose to ethanol in alcoholic fermentation)

(*b*) *biosynthetic processes*, in which simple molecules are built up into more complex ones (e.g., penicillin production).

The former are usually *exergonic* and can readily be brought about by resting cells or even enzyme preparations. The latter are usually *endergonic* and must be coupled with an exergonic reaction (e.g., glucose breakdown) to provide the energy for the biosynthesis; this may sometimes be difficult to bring about with resting cells and may necessitate (until more is known of the biochemistry of the process) the use of growing cultures.

It can easily be seen that to bring about these different types of chemical transformations in the most efficient manner may require quite different kinds of continuous culture process, which may in turn be different from the type of process required for the continuous production of microbial cells.

(2) *Operation types*

The chemical engineering approach to continuous culture processes also affords several different criteria for classification. One important distinction is between *homogeneous* and *heterogeneous* systems.

A homogeneous system is one whose composition is uniform throughout, while a heterogeneous system exhibits concentration gradients of cells and of substrates within the system. Heterogeneous systems may exist in single-phase or multi-phase types, while homogeneous systems are necessarily single-phase. The operational kinetics of homogeneous and heterogeneous systems may obviously differ markedly, but there are equally important biological differences. During steady-state continuous operation, all the micro-organisms in a homogeneous system are growing under the same environmental conditions which do not change with time; all the cells within the system at all times will therefore be in the same physiological state. In a heterogeneous system on the other hand, micro-organisms in different parts of the system are exposed to different environments, will have had different histories and will be in different physiological states. Even during continuous steady-state operation (where this is possible), micro-organisms passing through a heterogeneous system will undergo something akin to the 'growth cycle' of a batch culture. This constitutes a very important difference between the two systems.

All continuous culture is conducted in some form of reactor, and one might consider classification criteria based on *reactor types*. For single-phase systems this classification runs parallel to the previous one, for most reactors lie somewhere between the extremes of the completely-mixed type (e.g., the stirred fermentor) which is a homogeneous system, and the ideal tubular type with piston flow and no mixing, which is heterogeneous. Multi-phase reactors (e.g., packed towers) are necessarily heterogeneous.

The distinction between single-phase and multi-phase systems is liable to become increasingly important, because of the new forms of multi-phase systems that have developed lately. Until recently, the only known multiphase systems were of the solid–liquid type in which the micro-organisms formed a separate solid phase, either by growing as a pellicle or through adhesion to some solid support as in packed towers, trickling filters, etc. (A different type of solid–liquid system would be one in which the cells were freely suspended and the substrate a solid.) Recent research in the petroleum industry has led to the development of liquid–liquid[1] and liquid–gas[2] systems in which the substrates are water-immiscible liquid or gaseous hydrocarbons.

In addition to the above criteria, another is now introduced which appears to be particularly useful for classification purposes; viz., the distinction between *'closed'* and *'open'* systems. 'Closed' systems are those in which the microbial cells are wholly retained within the system, either by a semi-permeable membrane, by being adherent to some part of it or because they are continuously returned to it by re-cycling. 'Open' systems, on the other hand, are those from which cells continuously emerge in the effluent; i.e., those in which the effluent is a culture rather than a cell-free fluid. (It is emphasised that the terms 'closed' and 'open' are not used here with the specialised thermodynamic meanings.)

In all closed systems, nutrients are continually flowing into or through the system while no cells leave it. The number of cells in the system will therefore continue to increase indefinitely, until in the end some factor other than nutrient supply (lack of oxygen, for example) arises to put a check on further growth; the majority of the cells will then eventually die. In other words, closed systems cannot achieve a steady state. In open systems, on the other hand, cells continually leave the system at the same rate at which new cells are generated within it, so that steady-state operation is possible.

The distinction between closed and open systems appears to be more important than any of those made previously and is the basis of the classification scheme which follows.

A classification of continuous culture systems

A preliminary classification of continuous culture systems is given in Figs. 1 and 2; these are believed to include all types of continuous culture systems yet described (although others might be imagined).

Although the importance of the fermentation process was stressed in the preceding section, the classification is actually made entirely on the basis of operation types. This is because (as explained more fully later) a system of given operation type will behave in much the same way whatever the nature of the micro-organism or fermentation process. Continuous culture systems are therefore divided into the two main groups of open systems (Fig. 1) and closed systems (Fig. 2); each group may be further sub-divided into homogeneous and heterogeneous (and mixed) systems, and each sub-group again divided into single-stage and multi-stage systems.

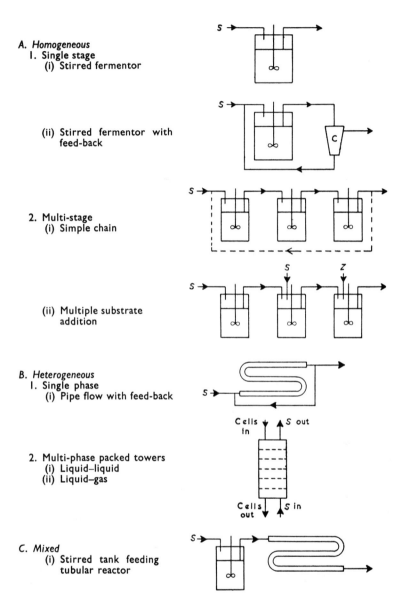

A. *Homogeneous*
  1. Single stage
      (i) Stirred fermentor

      (ii) Stirred fermentor with
          feed-back

  2. Multi-stage
      (i) Simple chain

      (ii) Multiple substrate
          addition

B. *Heterogeneous*
  1. Single phase
      (i) Pipe flow with feed-back

  2. Multi-phase packed towers
      (i) Liquid–liquid
      (ii) Liquid–gas

C. *Mixed*
      (i) Stirred tank feeding
          tubular reactor

FIG. 1.   *Types of 'open' continuous culture systems*
S indicates substrate addition, C indicates continuous centrifuge or
settling tank

A. *Homogeneous*
   (i) "Cellophane bag"
      cultures

   (ii) Stirred fermentor with
       100% feed-back of cells

B. *Heterogeneous*
  1. Single phase
    (i) Pipe flow with 100%
       feed-back of cells

    (ii) Partitioned tank with
       100% feed-back of cells

  2. Two-phase
    (i) Pellicle growth

    (ii) Packed towers

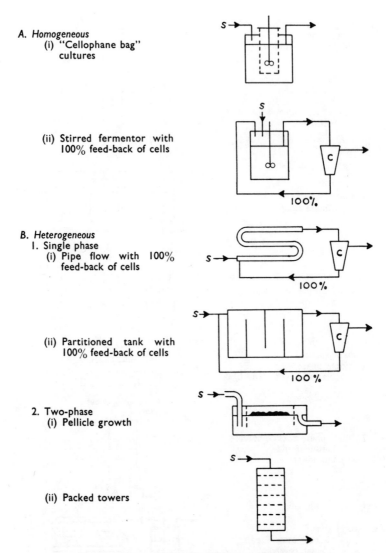

FIG. 2 *Types of 'closed' continuous culture systems*
S indicates substrate addition, C indicates continuous centrifuge or
settling tank

I. *Open systems*

A. *Homogeneous*

Homogeneous open systems are all based on the completely-mixed tank or 'stirred fermentor', used in one or more stages with or without feed-back.

(1) *Single-stage systems*

(i) *The single stirred fermentor.*—This is the simplest possible homogeneous open system, consisting of a stirred tank into which growth medium is fed and from which culture emerges at the same rate, the volume being maintained constant by some form of constant-level take-off.

Such systems may be controlled:

(*a*) by controlling the flow-rate; this type of system was the first to be subjected to mathematical analysis by Monod[3] and independently by Novick and Szilard,[4] who named it the 'Chemostat';

(*b*) by controlling the cell density (e.g., Bryson[5]); this arrangement is often called a 'Turbidostat'.

Much has been written on the differences between these two types of apparatus, but they are essentially equivalent and lead to the same end-result.[6]

(ii) *Stirred fermentor with feed-back.*—The single stirred fermentor may be modified by the introduction of feed-back, in which a fraction of the cells issuing from the fermentor is continuously returned to it by means of a continuous centrifuge, settling-tank or other means. The concentration of cells in the fermentor is thus higher than in the effluent, to an extent determined by the amount of feed-back. (This must always be less than 100% or the system would be closed.) Re-cycling systems of this type have been used in the manufacture of food yeast and fermentation alcohol.

(2) *Multi-stage systems*

These consist of chains of two or more stirred fermentors; three is the largest number that has yet been used in industrial practice. Medium may be fed to the first stage only, or to subsequent stages also, and the system may be further modified by feed-back. Such systems have been used for a variety of industrial processes, from yeast manufacture to industrial waste disposal. Their behaviour is complex and no adequate analysis of it has been given so far, so that most studies have been conducted on a purely trial-and-error basis. Many variants of such systems could be devised; the following are typical:

(i) *The simple chain.*—By this is meant a chain of stirred fermentors each one of which feeds the next, culture medium being fed to the first stage only. The system may be modified by introducing feed-back from later to earlier stages (Fig. 1).

(ii) *Chain with multiple substrate addition.*—In this type of system, substrate is introduced not only into the first fermentor of the series, but also into subsequent stages. It is, perhaps, not sufficiently realised that this modification is sufficient to change completely the overall behaviour of the system (this is discussed in detail later). A further modification is to introduce a different substrate at a later stage in the system (Fig. 1). This may be a substance which can undergo transformation by the micro-organism, without supporting growth. This should be an excellent system for bringing about chemical transformations, and is analysed in detail below.

B. *Heterogenous open systems*

(1) *Single phase systems*

(i) *Pipe flow reactors.*—The pipe flow or tubular reactor is the type system, but it may be mentioned that the partitioned tank type of reactor (Fig. 2) is essentially the same in principle. They may be regarded as moving batch cultures, i.e., each element of fluid moves along the pipe without much mixing with the fluid behind or ahead of it, and the cells in it grow essentially as in a batch culture. Thus there is an increasing concentration of cells from inlet to outlet and moreover, the cells at different points along the tube will be in different physiological states. In these systems some degree of feed-back (but less than 100%) is essential for steady-state operation.

(2) *Multi-phase systems*

Two types have been described:

(i) *Liquid–liquid and* (ii) *Liquid–gas.*—In both cases, the bacterial culture in the aqueous phase trickles down a bubble-cap column while the organic substrate, consisting of a water-immiscible liquid or gas, passes upwards. The method has been applied to liquid[7] and gaseous[8] petroleum fractions. Obviously many variations could be devised, including the use of feed-back, but the most interesting feature of such systems is their counter-current operation, which is possible only with systems containing two mobile immiscible phases.

C. *Mixed systems*

For the sake of completeness, one can include certain systems, necessarily multi-stage, containing both homogeneous and heterogeneous elements. These have not been much used, but the example

B

illustrated in Fig. 1 of a stirred fermentor feeding a pipe-flow reactor, is one that might have industrial applications and is discussed in more detail below.

II. *Closed systems*

The second main class of continuous culture systems, illustrated in Fig. 2, comprises all those in which the micro-organisms are wholly retained within the system. As mentioned above, several types of open system can be converted to closed systems by effecting 100% re-cycling of cells to the reactor; in addition there are a number of systems which are essentially 'closed' in nature.

A. *Homogeneous closed systems*

(i) *Cells enclosed by semi-permeable membrane.*—The best known example, illustrated in Fig. 2, is the 'Cellophane tube culture', in which cells grow inside a Cellophane tube dipping into a vessel through which fresh medium continually flows. This has only been used in the laboratory and is unlikely to be of industrial importance, but is included to show its resemblance in principle to other types of closed system.

(ii) *Stirred fermentor with* 100% *feed-back of cells.*—This system is identical in design with the open system I.A.1.(ii), the difference in operation being in the degree of re-cycling of cells, which in this case amounts to 100%. This difference, however, is enough to alter completely the nature of the system, which now becomes exactly equivalent to the 'Cellophane bag' culture mentioned above. Systems essentially of this type have been used for the production of fermentation alcohol.

B. *Heterogeneous closed systems*

(1) *Single-phase systems*

(i) *Pipe-flow reactor, or* (ii) *partitioned tank, with* 100% *cell feed-back.*—As mentioned in the discussion of open systems, the tubular reactor and the partitioned tank are essentially equivalent, and both become closed systems when operated with 100% feed-back of cells to the reactor. Type (i) has not been much used in industry but many anaerobic digestion tanks used in sewage works are essentially similar to type (ii).

(2) *Two-phase systems*

These are all of the solid–liquid type, their characteristic feature being that the micro-organisms are retained within the reactor either by adhesion to some solid support or by themselves forming the 'solid' phase.

(i) *Pellicle growth.*—This type of system is possible with any bacterium or mould which grows as a self-adherent surface pellicle or mycelial mat in unstirred or weakly stirred liquid culture, and is operated by continuously flowing fresh medium beneath the stationary pellicle. It has been used mainly in laboratory applications (e.g., Kuska[7]) but some industrial examples have been described. These have often been run in an intermittent or semi-continuous manner, the culture fluid being drained off and replaced at intervals without disturbing the pellicle.

(ii) *Packed towers.*—In this type of system, which is widely used, the micro-organisms grow as a film on a solid support of open structure (varying from birch twigs to coke) with which the tower is packed and over which the growth medium continually trickles. The best known examples, which are of considerable economic importance, are the 'vinegar towers' of the 'quick' vinegar process, and sewage works trickling filters. On the laboratory scale, the soil perfusion apparatus of Lees[8] is an example of this type of system with the addition of feed-back.

Comments on the classification scheme

This classification scheme is preliminary and tentative, and undoubtedly is not the only scheme that could be devised. Nevertheless, it is claimed that the criteria used for classification correspond to real and important differences between the various types of continuous culture systems. Admittedly some of the distinctions drawn are quantitative rather than qualitative; an open system with cell feed-back, for example, will tend increasingly to resemble a closed system as the amount of feed-back approaches 100%. In practice, however, the different categories remain clearly distinguishable. It is hoped that the scheme will serve a practical purpose in promoting ready recognition of the category to which a given continuous culture system belongs, and hence of its salient operating features.

One question that may arise is whether the entire group of 'closed' systems can accurately be described as true examples of 'continuous culture'. A genuinely 'continuous' process should be capable, at least theoretically, of operating for an infinite period, which is not true of any closed system but is true (in theory) for open systems which can attain true steady-state conditions. It could be argued that closed systems ought really to be regarded as batch cultures with continued replenishment of nutrients. This has much justification, but for now the designation of 'continuous culture' for these systems will be retained, if only because most of them are almost universally so described in the literature. The present classification should at least serve to emphasise the differences between these 'quasi-continuous' systems and 'true' continuous culture.

Mathematical analysis of continuous culture systems

Analysis of any continuous culture system must take into account (i) the flow characteristics of the reactor system employed, (ii) the kinetics of the microbial growth process. These are best considered separately.

Continuous flow reactors

Danckwerts[9] has shown that the flow characteristics of a reactor are best described by the distribution of residence-times of minute particles or molecules passing through the system. If df equals the fraction of the material having residence-times between t and $(t + dt)$, we may write in general:

$$df = \varphi(t)\, dt \qquad . \qquad . \qquad . \qquad . \quad (1)$$

where $\varphi(t)$ is the residence time frequency function. For an ideal tubular reactor, with piston flow and no mixing, all particles have the same residence-time, equal to the mean residence-time, $\bar{t}$, which in this case is given by:

$$\bar{t} = v/f = 1/D \qquad . \qquad . \qquad . \quad (2)$$

where v is the volume of the reactor, f the flow-rate, and D is the ratio f/v or flow per unit volume—the dilution rate. For any real tubular reactor, there will be a certain amount of mixing or axial diffusion in a direction parallel to the flow, which will cause a spread of residence-times about the mean; the resulting distribution of residence-times will be a single-peaked curve, the frequency function $\varphi(t)$ having approximately the form of the Gaussian function (Fig. 3). The mean residence-time, however, still remains equal to v/f.

The mean residence-time for the completely-mixed stirred reactor is also v/f, but the distribution of residence-times is completely different, and can easily be shown to be an *exponential* distribution with a frequency function

$$\varphi(t) = De^{-Dt} \qquad . \qquad . \qquad . \quad (3)$$

as plotted in Fig. 3.

The 'wash-out' curves of the two types of reactor are also very different. Suppose the reactor to be filled with, for example, a bacterial suspension of uniform concentration x_0, the total quantity being $Q = vx_0$; the cells are supposed not to be growing. At zero time this begins to be washed out by a steady inflow of sterile growth medium at a dilution rate D, and we record the changes with time of (i) the concentration of the effluent, x, and (ii) the total quantity Q remaining in the fermentor. For ideal piston flow, x will remain equal to x_0 until $t = \bar{t}$, when it will drop abruptly to zero, while Q will decrease linearly with time, becoming zero at $\bar{t}$. When some axial diffusion is present, the curves will be modified as in Fig. 3 (lower). For a completely-mixed reactor the curves for x/x_0 and Q/Q_0 are identical (Fig. 3) since the concentration of the effluent is always equal to the mean concentration; the wash-out curves have the same form (negative exponential) as the residence-time distribution curve, being given by:

$$-\frac{dx}{dt} = -\frac{dQ}{dt} = Dx \qquad . \qquad . \qquad . \quad (4)$$

or:

$$\frac{x}{x_0} = \frac{Q}{Q_0} = e^{-Dt} \qquad . \qquad . \qquad . \quad (5)$$

(It will be seen that we are really plotting here the integrated residence-time distribution or distribution functions, $f = \int_t^\infty \varphi(t)dt$; the exponential distribution is unique among frequency-distributions in having a distribution function identical in form with the frequency function.)

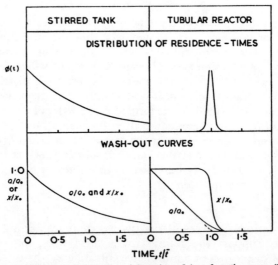

STIRRED TANK   TUBULAR REACTOR

DISTRIBUTION OF RESIDENCE – TIMES

WASH-OUT CURVES

TIME, $t/\bar{t}$

FIG. 3. *Residence-time distribution* (above) *and 'wash-out' curves* (below) *for complete-mixed reactors* (left) *and tubular reactors* (right) Ordinates: see text. Abcissae: time in dimensionless units of $t/\bar{t}$ ($\bar{t}$ = mean residence-time)

The wash-out curves play an important part in determining the behaviour of continuous culture systems employing reactors of these types.

Bacterial growth; kinetics and stoicheiometry

Theoretical discussions of bacterial growth usually start from the familiar 'exponential growth' equation:

$$dx/dt = \mu x \qquad . \qquad . \qquad . \qquad . \quad (6)$$

where x is the concentration of organisms (dry weight of cells per unit volume) at time t and μ is the exponential growth-rate. The value of μ is not constant, but was found by Monod[10] to depend on the concentration of growth substrates in the culture medium.

Consider a medium in which all essential growth substrates are in excess except one—the limiting substrate. Then the growth-rate is found to vary with the concentration of this substrate (s) in the manner shown in Fig. 4. The relationship found is a 'saturation function', μ being approximately proportional to the substrate concentration when this is low but tending to a maximum value, μ_m known as the growth-rate constant, as the substrate concentration increases. Monod's data[10] are quite well fitted by the equation:

$$\mu = \mu_m \left(\frac{s}{K_s + s} \right) . \qquad . \qquad . \qquad . \quad (7)$$

where K_s is a 'saturation constant' numerically equal to the substrate concentration at which the growth-rate is one-half the maximum value (i.e., $\mu = \mu_m/2$). Figure 4 is a plot of this equation. The value of K_s is usually very low (of the order of mg./l. for carbohydrate substrates[10] and μg./l. for amino-acids[11]) and this fact is of great importance in continuous culture applications.

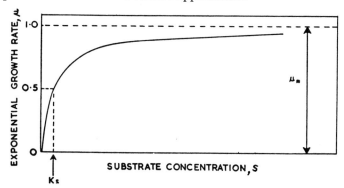

FIG. 4. *Relation between exponential growth-rate (μ) and concentration of limiting substrate* (s)
K_s = saturation constant; μ_m = growth-rate constant

From equations (6) and (7) we therefore have the following equation, expressing the growth-rate as a function of substrate-concentration:

$$\frac{dx}{dt} = \mu_m x \left(\frac{s}{K_s + s} \right) \qquad \cdot \qquad \cdot \qquad \cdot \qquad (8)$$

For a complete theory it is necessary to know the relationship between growth of bacteria and utilisation of substrate, usually expressed as the 'yield', Y, defined as:

$$Y = -\frac{dx}{ds} = \frac{\text{weight of bacteria formed}}{\text{weight of substrate used}} \qquad \cdot \qquad \cdot \qquad (9)$$

The simplest assumption is that Y is constant, which was found to hold by Monod[10] and others in batch cultures. Hence from equations (8) and (9) we have for the rate of substrate utilization:

$$-\frac{ds}{dt} = \mu_m \frac{x}{Y} \left(\frac{s}{K_s + s} \right) \qquad \cdot \qquad \cdot \qquad \cdot \qquad (10)$$

Equations (8) and (10) are equations for the rates of growth of bacteria and disappearance of substrate which, it is emphasised, apply equally to batch or continuous culture. Together with the 'wash-out' equations given above, they allow mathematical treatment of any open single-phase continuous culture system.

While discussion has so far been confined to the growth of bacteria, it is well known that the growth of yeasts also can be expressed by equations (8) and (10); some mycologists have been reluctant to admit, however, that the concept of exponential growth is applicable to moulds. It is of some importance, therefore, that Pirt and Callow[12] have found the exponential growth and yield equations above to be applicable to the filamentous mould *Penicillium chrysogenum*, which grew in continuous culture in much the same manner as bacteria and yeasts.

Continuous culture applications: the single-stage stirred fermentor

This, the simplest type of open homogeneous system, was the first to be treated mathematically by Monod[3] and Novick and Szilard[4] and subsequently in more detail by Herbert *et al.*[13] It will be discussed here only briefly, in order to illustrate general principles.

All continuous culture problems involving stirred fermentors can be treated by setting up balance equations for cells, substrates, products, etc., in which the overall rate of change in concentration of any component is given as the resultant of the rates of all the processes tending to increase or decrease it. In general these will be: (i) increase due to inflow to the fermentor—this is equal to the inflow rate multiplied by the concentration of component in the inflow; (ii) decrease due to outflow—this is given by the 'wash-out' equation (4) discussed above; (iii) increase (of cells) due to growth—this is given by equation (8) above, and (iv) decrease (of substrate) due to utilisation—this is given by equation (10).

In the present example, let the single stirred fermentor contain the constant volume of culture v, with cell concentration x and substrate concentration s; fresh culture medium of substrate concentration s_R is fed in at a steady flow-rate f. The dilution rate or flow per unit volume is $D = f/v$, and is also equal to $1/\bar{t}$, the reciprocal of mean residence time.

In this case there is no entry of cells to the fermentor and the balance equation for cells (individual terms referring to rates in each case) is:

$$\text{Increase} = \text{growth} - \text{output}$$

or:

$$\frac{dx}{dt} = \mu_m x \left(\frac{s}{K_s + s} \right) - Dx \ . \qquad . \qquad . \quad (11)$$

while the balance equation for substrate is:

$$\text{Increase} = \text{input} - \text{output} - \text{consumption}$$

or:

$$\frac{ds}{dt} = D s_R - D_s - \mu_m \frac{x}{Y} \left(\frac{s}{K_s + s} \right) \qquad . \qquad . \quad (12)$$

Here as in other cases we arrive at two simultaneous differential equations in x and s, whose solutions will give a complete description of the behaviour of the system, in transient as well as in steady states. Unfortunately these equations can be integrated only in the simplest case, and usually no simple treatment of transient states is available. However, solutions for the steady state can always be obtained by equating dx/dt and ds/dt to zero; this gives two simultaneous algebraic equations in x and s which in principle can always be solved.

In the present case, writing $dx/dt = 0$ in equation (11) and rearranging gives:

$$\mu_m \left(\frac{s}{K_s + s} \right) = D \qquad . \qquad . \qquad . \quad (13)$$

In other words, the growth-rate in the steady state is equal to the dilution rate; this is a most important characteristic of the single-stage continuous fermentor. Writing $ds/dt = 0$ in equation (12) gives a second equation in x and s which with equation (13) allows expressions for the steady-state concentrations of both cells and substrate to be obtained; they are

$$s = K_s \left(\frac{D}{\mu_m - D} \right) \qquad . \qquad . \qquad . \qquad (14)$$

$$x = Y(s_R - s) \qquad . \qquad . \qquad . \qquad (15)$$

From these equations the steady-state concentrations of cells and substrate in the fermentor can be predicted for any experimental values of the dilution rate D and concentration of inflowing substrate s_R, provided the values of the microbial 'growth constants' μ_m, K_s and Y are known; these can always be determined from batch culture experiments. By plotting the predicted values of x and s (from equations 14 and 15) against D, a graphical 'picture' is obtained of the way in which the system operates when s_R is kept constant and the dilution rate D is varied; this has been done in Fig. 5.

It will be evident from equation (14) that the value of the steady-state substrate concentration in the fermentor is independent of the concentration of inflowing substrate s_R and the cell concentration x; in fact, for a given organism and growth medium, the value of s is determined solely by the dilution rate, D. This explains the stable and self-regulating properties of the system; the dilution rate

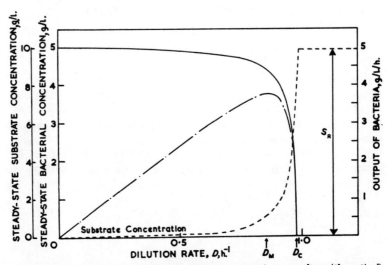

FIG. 5. *Steady-state relationships in a single-stage continuous culture (theoretical)*
The steady-state values of substrate concentration, cell concentration and output at different dilution rates are calculated from equations (14) and (15) for an organism with the following growth constants: $\mu_m = 1 \cdot 0$ h.$^{-1}$, $Y = 0 \cdot 5$ and $K_s = 0 \cdot 2$ g./l.; and a substrate concentration in the inflowing medium of $s_R = 10$ g./l.

———————— bacterial concentration
— — — — output of bacteria
— — — substrate concentration

determines the substrate concentration which in turn fixes the growth rate at a value equal to D, which is a condition for steady-state operation.

Equation 14 also indicates, as shown in Fig. 5, that over most of the range of possible dilution rates the steady-state value of s is very low (i.e., the added substrate is almost all consumed); only at dilution rates close to the 'wash-out point' does unused substrate appear in the culture. Correspondingly, the cell concentration remains at a high level over most of the range, dropping fairly abruptly near the wash-out point. Figure 5 also shows the output of cells (i.e., the product of cell concentration and dilution rate, Dx) as a function of dilution rate; this curve goes through a maximum at the dilution rate D_M which will be the optimum dilution rate for cell production. Graphical presentation as in Fig. 5 is recommended both for ready appreciation of the overall behaviour of the system and for easy detection of departures from 'theoretical' behaviour.

It is evident that the foregoing is about the simplest mathematical account of continuous culture that could be devised, particularly in respect of the few and simple assumptions made about the kinetics and stoichiometry of microbial growth, and it may be questioned whether it does not present a greatly over-simplified picture of the facts. The experimental data obtained by the writer on *Aerobacter*

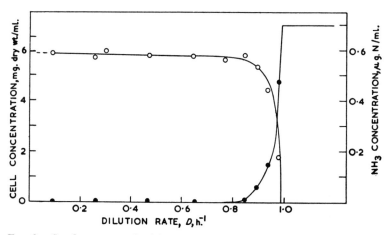

FIG. 6.   *Steady-state growth of* Aerobacter aerogenes *in single-stage continuous culture (for comparison with Fig. 5), with NH₃ as growth-limiting substrate* Organisms were grown with excess aeration at a number of different flow-rates; dry weight of cells and concentration of substrate determined after at least 2–3 days' steady-state growth a teach flow-rate. Medium contained 0·05 M-NH₃, 2% glycerol, 0·05 M phosphate and trace minerals; pH controlled at 7·4; temp. 37°. Glycerol was in excess at all flow-rates.
      o————o cell concentration   ●————● NH₃ concentration

aerogenes and plotted in Fig. 6 for direct comparison with the 'theoretical' curves of Fig. 5, show that in some cases at least the theory fits the facts very well. In other cases deviations have been observed[6], [13] due either to imperfect mixing (so that the wash-out rate is no longer given exactly by equation (4)) or to variation of the yield Y with growth-rate, instead of the constancy assumed in equation (9). Such deviations, however, are quantitative rather than qualitative and the overall picture does not greatly depart from Fig. 5; the theory could easily be elaborated to accommodate them, but this does not seem necessary at present.

Single-stage continuous culture with feed-back

An elaboration of the single-stage system which has often been practised is to feed back continuously a fraction of the emergent cells to the fermentor as shown in Fig. 7.

In this system growth medium of substrate concentration s_R enters a fermentor containing culture of volume v at a steady flow-rate f; the concentrations of cells and substrate in the fermentor are x_1 and s. The culture leaving the fermentor enters a continuous centrifuge which returns to the fermentor a concentrated cell suspension (concentration Cx_1) at a flow-rate αf, C being termed the concentration factor (C always > 1) and α the volumetric feed-back ratio, i.e., the ratio of feed-back flow to medium flow. Through its other outlet the centrifuge discharges at flow-rate f the final effluent in which the cell concentration x_2 is lower than that in the fermentor.

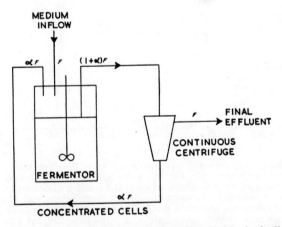

FIG. 7. *Single-stage stirred fermentor with feed-back of cells*
f = medium inflow rate (l./h.); $(1 + \alpha f)$ = flow-rate of culture from fermentor; αf = feed-back flow-rate of concentrated cells from centrifuge to fermentor.

The total flow-rate through the fermentor is therefore $(1 + a)f$ and the cell feed-back fraction is given by:

$$\frac{\text{cells fed back to fermentor}}{\text{cells issuing from fermentor}} = \frac{fCx_1}{(1 + a)fx_1} = \frac{C}{1 + a} \quad (16)$$

The dilution rate for the fermentor is

$$D_1 = (1 + a) f/v \quad . \qquad . \qquad . \qquad . \quad (17)$$

but this is not identical with the overall dilution rate D defined as:

$$D = \frac{\text{total flow through system}}{\text{total fermentor volume}} = f/v \quad . \qquad . \quad (18)$$

From a consideration of the cell concentrations and flow-rates into and out of the centrifuge, we may write:

$$\frac{\text{cell concentration in effluent}}{\text{cell concentration in fermenter}} = \frac{x_2}{x_1} = (1 + a - aC) \quad (19)$$

The cell balance equation for the fermentor is:

$$\text{increase} = \text{feed-back} - \text{outflow} + \text{growth}$$

or:

$$\frac{dx}{dt} = DCx_1 - (1 + a)Dx_1 + \mu_m x_1\left(\frac{s}{K_s + s}\right) \qquad . \quad (20)$$

Hence in the steady state:

$$\mu_m\left(\frac{s}{K_s + s}\right) = (1 + a - aC)D \qquad . \qquad . \quad (21)$$

In this case, unlike the single fermentor without feed-back, the steady-state growth-rate is less than the dilution rate D_1 (cf. equations 21 and 13).

The balance equation for substrate is:

$$\text{increase} = \text{input} + \text{feed-back} - \text{outflow} - \text{consumption}$$

$$\frac{ds}{dt} = Ds_R + Ds - (1 + a)Ds - \frac{\mu_m x_1}{Y}\left(\frac{s}{K_s + s}\right) \quad . \quad (22)$$

From equations (19), (20) and (22) the steady-state concentrations of substrate and cells can be shown to be:

$$s = \frac{K_s AD}{\mu_m - AD} \quad . \qquad . \qquad . \qquad . \quad (23)$$

$$x_1 = \frac{Y}{A}(s_R - s) \quad . \qquad . \qquad . \qquad . \quad (24)$$

$$x_2 = Y(s_R - s) \quad . \qquad . \qquad . \qquad . \quad (25)$$

where for convenience A is written for $(1 + a - aC)$.

These equations are plotted in Fig. 8, together with the output of cells, Dx_2; for comparison, curves for cell concentration and output

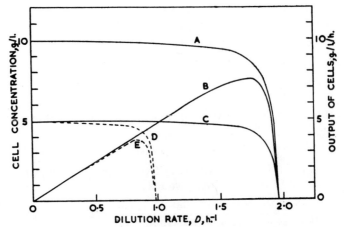

FIG. 8. *Steady-state relationships in a single-stage continuous culture with feed-back (theoretical)*

Curves are plotted from equations (23)–(25) for an organism with the following growth-constants: $\mu_m = 1 \cdot 0$ h.$^{-1}$; $Y = 0 \cdot 5$; $K_s = 0 \cdot 2$ g./l.; and a substrate concentration in the inflowing medium of $s_R = 10$ g./l.

Continuous curves: with feed-back (volumetric feed-back ratio $a = 0 \cdot 5$; cell concentration factor $C = 2 \cdot 0$). *Dotted curves:* without feed-back

curve A   cell concentration in fermentor
curve B   output of cells
curve C   cell concentraion in outflow
curve D   cell concentration without feed-back
curve E   output without feed-back

of cells from the same system without feed-back are also plotted (dotted lines).

By comparison of the behaviour of the two systems it will be seen that the addition of feed-back (*a*) allows the system to be run at much faster flow-rates, and (*b*) greatly increases the output of cells. These effects of feed-back are simply explained by the higher concentration of cells maintained in the fermentor, whose contents can therefore utilise substrate faster and so accommodate higher medium inflows. Figure 8 has been calculated for values of *a* and *C* giving a cell concentration in the fermentor twice as high as that obtained without feed-back, which has the effect of doubling the output of cells. In this example, therefore, the application of feed-back to the process would allow the same output to be obtained from a fermentor of half the size.

There are of course limits to the application of feed-back in aerobic processes, for as the cell feed-back fraction is increased a cell concentration will eventually be reached where it becomes impossible to keep the cells supplied with oxygen. This does not apply to anaerobic processes, in which the application of feed-back should usually be highly advantageous. Andreev[14] describes a feed-back system used in continuous alcoholic fermentation of soft wood hydrolysates; ingenious use was made of a small settling-tank, instead of a continuous centrifuge, for concentrating the cells prior to their return to the fermentor, using specially selected branching strains of yeast which sedimented rapidly.

Tubular (pipe flow) continuous fermentors

Tubular reactors are commonly used for continuous processes in the chemical industry, but have been little used by microbiologists either in industry or in the laboratory. This is partly due to the practical difficulties of temperature control, pH control and aeration, which are much easier to carry out in stirred tanks; it may also be partly due to the lack of any theoretical treatment of the subject in its microbiological applications.

An exact mathematical treatment of continuous culture in 'real' tubular reactors, taking into account the hydrodynamic factors affecting axial diffusion (cf. Danckwerts[15]) as well as the microbial growth kinetics, would indeed be difficult. As a first approximation, growth is considered in 'ideal' tubular reactors, with pure piston flow and no mixing.

The conditions of microbial growth in such reactors differ pro-foundly from those in stirred fermentors. In the absence of mixing, each element of fluid may be considered to retain its individuality as it travels along the tube, exchanging only negligible amounts of material with neighbouring elements; the same bacteria remain within it throughout and therefore grow under batch culture condi-tions, so that each fluid element may be regarded as a small moving batch culture. It will be evident that continuous culture can only be accomplished by inoculating every fluid element as it enters the tube; the use of continuous inoculation or feed-back is therefore essential for steady-state operation of this type of system. (The operation of this type of reactor may be visualised from the following example: a bacteriologist is continually inoculating flasks of sterile culture medium at one end of a long, slowly-moving conveyer belt; the system will continue to produce bacteria only as long as the supply of inoculum lasts. If this is maintained by occasionally sending back to him a grown flask of culture from the far end of the conveyer, the system can operate indefinitely. The growth in each flask depends only on the initial inoculum and the time for which it has been travelling along the conveyer; each point in space along the conveyer will correspond to a given point in time of the progress of a batch culture.)

The occurrence of some longitudinal mixing blurs, but does not essentially alter, the general picture. The system will then operate without feed-back, each fluid element entering the tube being inoculated from those ahead of it (a type of internal feed-back); but if feed-back is applied it will operate more efficiently, particularly at high flow-rates. When feed-back is applied, the changes in concentration along the length of the tube will be much the same for an ideal piston-flow system as for one with some longitudinal mixing; i.e., the feed-back tends to mask the effect of longitudinal mixing.

For a simple quantitative treatment, consider a long pipe of total volume v, with a steady inflow of culture medium containing substrate at concentration s_R, at a flow-rate f. A feed-back pipe of negligible volume conveys culture at a flow-rate af from the outlet back to the inlet, where it is mixed throughly with the inflowing culture medium before entering the pipe; no further mixing occurs thereafter. For the whole system the overall dilution rate D and mean residence time $\bar{t}$ are given by:

$$\bar{t} = v/f = 1/D \quad \cdot \quad \cdot \quad \cdot \quad \cdot \quad (26)$$

The total flow in the pipe is $(1 + af)$; hence the internal dilution rate D_i and mean residence time $\bar{t}_i$ are given by:

$$\bar{t}_i = \frac{v}{(1 + a)f} = \frac{1}{D_i} \quad \cdot \quad \cdot \quad \cdot \quad (27)$$

(cf. equations 17 and 18 for the stirred fermentor with feed-back).

Consider the mixing of medium and feed-back streams at the inlet, and let x_0 and s_0 be the concentrations at the inlet, and x_ω and s_ω the concentrations at the outlet of the pipe. It is easy to show that:

$$x_\omega = x_0(1 + a)/a \quad \cdot \quad \cdot \quad \cdot \quad (28)$$
$$s_\omega = s_0(1 + a)/a - s_R/a \quad \cdot \quad \cdot \quad (29)$$

i.e., the ratios of the concentrations at the inlet and outlet are determined solely by the values of the volumetric feed-back ratio a, and of s_R. Hence, a steady state can exist only when the microbial growth during the passage along the pipe results in an increase from the initial value x_0 to the value $((1 + a)/a)x_0$ in a time $\bar{t}_i$, and also a decrease in substrate concentration given by equation (29).

An exact steady-state solution involves the integration of the growth equations (8) and (9) and leads to complex expressions. An approximate solution may be obtained simply, as follows. The maximum multiplication rate of the cells is:

$$dx/dt = \mu_m x \quad \text{or} \quad x/x_0 = \exp(\mu_m t) \quad \cdot \quad \cdot \quad (30)$$

Hence, steady-state conditions can be obtained only when:

$$\exp(\mu_m \bar{t}_i) \geqslant (1 + a)/a \quad \text{or:} \quad \frac{\mu_m}{(1 + a)D} \geqslant \ln\left(\frac{1 + a}{a}\right) \quad (31)$$

For greater values of D (i.e., smaller values of $\bar{t}_i$) the maximum growth-rate of which the cells are capable will not enable them to increase by the factor $(1 + a)/a$ in the available time $\bar{t}_i$, and the culture will wash out.

Writing D_c for the wash-out or critical value of D, we have:

$$D_c = \frac{\mu_m}{(1 + a) \ln ((1 + a)/a)} \qquad . \qquad . \qquad . \qquad (32)$$

For a single-stage stirred fermentor the wash-out value is:[13]

$$D_c = \mu_m \left(\frac{s_R}{K_s + s_R} \right) \qquad . \qquad . \qquad . \qquad (33)$$

which is always greater than the value in equation (32).

To summarise the properties of tubular fermentors:

(i) feed-back is essential for steady-state operation in the case of ideal piston flow, and improves the efficiency of operation even when there is considerable longitudinal mixing,

(ii) the outputs of cells from a tubular fermentor and a stirred fermentor run at the same overall dilution rate D will be virtually the same, almost complete conversion of substrate to cells being achieved over most of the working range in both cases,

(iii) the maximum dilution rate obtainable with a tubular fermentor is lower than with a stirred fermentor, and the maximum output is therefore lower;

(iv) tubular fermentors can be run at faster rates by increasing the feed-back ratio a; the effect of doing so, however, is to make them resemble more and more closely a completely-mixed fermentor and so to destroy their characteristic features.

Taking into account the practical difficulties associated with tubular fermentors—difficulties of pH or temperature control, tendency of cells to sediment, etc.—they would seem to have few advantages compared with stirred fermentors, at any rate for cell production. For other purposes, however, e.g., chemical transformations, they may be equally efficient (v.i.), and there may be applications (e.g., in certain waste disposal problems) where the tendency of cells to settle is actually an advantage.

Multi-stage stirred fermentors

The complexity of multi-stage systems increases greatly with the number of stages. In practice, however, it has seldom been necessary to use more than three stages, while the essential principles of multi-stage operation may be illustrated by a discussion of two stages only.

There are several operational types (Fig. 1) having superficial similarities in construction but differing greatly in operating principles and applications. In discussing them the following notation is used: for the rth fermentor in the chain, x_r and s_r are the concentrations, v_r the culture volume, f_r the total flow-rate leaving the fermentor, D_r the individual fermentor dilution rate ($= f_r/v_r$ in the absence of feed-back and $\bar{t}_r = 1/D_r$ the mean retention time. For the whole system, the overall dilution rate, D, and mean retention time, $\bar{t} = 1/D$, are defined (cf. equation (18)) as the total flow through the system divided by the total volume of all the fermentors, and for n fermentors is usually

$$D = \frac{f_n}{v_1 + v_2 + \ldots \ldots v_n} = \frac{f_n}{v_{tot}} \qquad . \qquad . \quad (34)$$

This is useful for comparing the output of the system with that of a single-stage system of the same total volume.

Simple chain of fermentors (Type I. A.2 (i), Fig. 1)

The fermentors, which may be of different sizes, are arranged in cascade with medium fed to the first of the series only. The first fermentor may be treated in exactly the same way as the single-stage fermentor and the equations for steady-state substrate concentration, cell concentration and wash-out rate D_c are given by equations (14), (15) and (33).

If $D_1 > D_c$, wash-out will occur in the first fermentor but not necessarily in the second; there are two possibilities: (i) if v_2 is sufficiently greater than v_1, D_2 can be less than D_c and the second fermentor will operate as a single-stage, receiving unchanged cell-free culture medium from fermentor 1, which now functions only as a medium holding tank; (ii) if, however, $v_1 \geqslant v_2$, then $D_2 > D_c$ and wash-out will occur in the second fermentor also.

The second stage differs from the first and resembles the single-stage system with feed-back, in that (excluding wash-out conditions) there is a constant input of cells to the fermentor. The balance equation for cells is:

increase = input − output + growth

$$\frac{dx_2}{dt} = \frac{v_1}{v_2} \cdot D_1\,x_1 - D_2\,x_2 - \mu_m\,x_2 \left(\frac{s_2}{K_s + s_2}\right) \qquad . \quad (35)$$

while the balance equation for substrate is:

increase = input − output − consumption

$$\frac{ds_2}{dt} = \frac{v_1}{v_2} D_1 x_1 - D_2 x_2 - \frac{\mu_m x_2}{Y} \left(\frac{s_2}{K_s + s_2}\right) \qquad . \quad (36)$$

The steady-state solutions of these equations for x_2 and s_2 are rather complicated expressions and will not be given here, but one may readily deduce the quite simple expression:

$$x_2 - \left(\frac{v_1 D_1}{v_2 D_2}\right) x_1 = Y \left(\frac{v_1 D_1}{v_2 D_2}\right) s_1 - s_2 \quad . \quad . \quad (37)$$

Since $v_1 D_1 / v_2 = D_2$ this reduces to:

$$x_2 - x_1 = Y (s_1 - s_2) \quad . \quad . \quad . \quad (38)$$

The difference $(x_2 - x_1)$ is the increase in cell concentration due to growth in the second fermentor, which is proportional to the difference in substrate concentrations between the first and second fermentor $(s_1 - s_2)$.

Now, over most of its working range of dilution rates, the limiting substrate in fermentor 1 will be almost completely consumed (i.e., s_1 will be very low) as shown in Figs. 5 and 6. Consequently, very little or no growth will take place in fermentor 2, whose function will be reduced more or less to that of a holding tank. If much growth is to occur in the second fermentor, the first must be run close to its wash-out point so that most of the substrate remains unconsumed and passes into the second stage. But in this case there will be so little growth in the first fermentor that it performs no useful purpose, becoming virtually a growth medium holding tank. It would be possible, by careful adjustment of flow-rates, to arrange that approximately equal amounts of growth occurred in each fermentor, but it can be shown that for all the above possibilities, the maximum output of cells obtainable from the system is never greater than from a single fermentor of the same total volume.

In many cases, therefore, there is little to be gained by the use of two (or more) fermentors, a single fermentor of equal capacity being more efficient. This is certainly the case where the object is to produce the maximum output of cell material, using a simple type of growth medium, e.g., the production of food or fodder yeast.

Multi-stage systems may show an advantage (a) when a complex type of growth medium is being used, (b) when it is necessary to produce cells of a particular quality (i.e., cells in a particular physiological state).

With complex growth media, there may be two or more substrates present which support growth at different rates. For a particular (fairly high) value of D_1, the more readily assimilated substrate would be used in the first fermentor, leaving the less readily assimilated substrate to pass over and support growth in the second fermentor. There may be cases where this is desirable,[16, 17] although the same result will often be attainable with greater output by running at a lower flow-rate with a single larger fermentor.

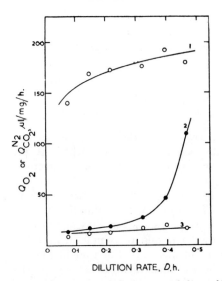

FIG. 9. *Effect of growth-rate on carbohydrate metabolism of* Torula utilis
The organism was grown in continuous culture at a number of different flow-rates on a glucose–NH_3–salts medium (glucose limiting) and samples removed for determination of rate of oxidation and anaerobic fermentation of glucose by Warburg manometric method

Q_{O_2} (glucose)$_2^2$ $Q_{CO_2}^{N_2}$ (glucose)   3 Q_{O_2} (endogenous)

The physiological state of micro-organisms is known to be affected by the conditions of cultivation (cf., Malek[18]). When bacteria are grown in continuous culture at different growth-rates, cells grown at a high growth-rate are large and have high RNA and low DNA contents;[6, 19] those grown at low growth-rates are small, with a low RNA and a higher DNA content, and intermediate forms are obtained at medium growth-rates. Figure 9 shows how the fermentative ability of *Torula utilis* is affected by the growth rate.

In this experiment, *T. utilis* was grown in a single-stage continuous fermentor at a number of different flow-rates, samples being removed from the fermentor at each flow-rate, washed by centrifuging and transferred to Warburg manometers for measurement of their respiration and anaerobic fermentation with glucose as substrate. It will be seen that the cells grown at low growth-rates had virtually no ability to ferment glucose, while fast-grown cells had high fermentative ability. Many such examples will undoubtedly emerge as our experience of continuous culture enlarges.

Now, with a single-stage process, there is one dilution-rate D_M which gives maximum output (Fig. 5), but it may happen that this produces cells in a different physiological state from that desired.

Multi-stage culture allows cells to be grown at maximum output at the first stage, followed by subsequent stages at different dilution-rates, to produce cells of the desired properties. This might well be necessary in baker's yeast production, for example, in order to produce yeast of the correct 'quality'.

Two-stage processes with multiple substrate addition

An obvious modification of the 'simple chain' of fermentors is to feed a separate supply of nutrient to the second and later stages. This trivial change in operational set-up has a profound effect on the behaviour of the system. The second substrate may be the same as the first, or a different one.

(i) *One substrate fed to all stages.*—Here the first stage is again identical with the ordinary single-stage system and equations (14), (15) and (33) apply. The second stage receives cells and substrate from the first fermentor, and a second supply of substrate from an independent feed system. The cell balance equation will be the same as for the two-stage simple chain, and is given by (35); the substrate balance equation is different, being given by:

$$\text{increase} = \text{input 1} + \text{input 2} - \text{output} - \text{consumption}$$

$$\frac{ds_2}{dt} = \frac{v_1}{v_2} \cdot D_1 s_1 + D_2 s_{R2} - D_2 s_2 - \frac{\mu_m x_2}{Y}\left(\frac{s_2}{K_s + s_2}\right) \quad . \quad (39)$$

where s_{R_1} and s_{R_2} are the respective substrate concentrations in the medium feeds to the first and second stages.

The characteristic operational features of the second-stage fermentor (which differ considerably from those of the 'simple chain' type) are as follows:

(a) The independent addition of substrate at this stage ensures that cell growth will always occur in the second fermentor, even if the culture entering it from the first stage is virtually exhausted of substrate.

(b) Wash-out can never occur in the second fermentor however large the value of D_2, so long as a finite concentration of cells enters it from the first stage.

In the steady state we have from the cell balance equation (35):

$$\frac{x_2}{x_1} = \frac{v_1 D_1/v_2}{D_2 - \mu_m \left(s_2/K_s + s_2\right)} \quad . \quad . \quad . \quad (40)$$

Thus, when $D_2 < \mu_m$, the fraction $s_2/(K_s + s_2)$ will fall to a low value, i.e., nearly all the substrate in the second fermentor will be consumed. The system is still stable when $D_2 > \mu_m$; in this case, when D_2 is large enough, s_2 will approach s_{R_2} and the concentration of cells x_2 will fall to a level less than x_1 although growth is occurring at a virtually maximum rate. As Novick[20] has pointed out, this is a feature of the system which is very attractive to those using continuous culture as a research tool, as it enables micro-organisms to be maintained indefinitely growing at their maximum rate in saturation levels of substrate.

Multi-stage systems of this type are therefore much more versatile than the simple chain system. It is still true, however, that the maximum output of cells from such a system cannot be greater (and will usually be less) than that of a single-stage system of the same total volume.

(ii) *Different substrates fed at later stages.*—This type of system is illustrated in Fig. 10; substrate s is fed to the first stage at concentration s_R, and a different substrate z is fed to the second stage at concentration z_R.

There are two variants of this system. In the first case, z is an alternative growth source to s (e.g., s might be glucose and z lactose); the analysis here follows essentially the same lines as the previous section. In the second case, z is a different type of substance from s and will not support growth but can be metabolised. This is a most important system industrially since it is applicable to many processes involving chemical transformations, and will therefore be discussed in some detail.

In the first stage the usual single-fermentor equations (14), (15) and (33) apply. Only the case (which will nearly always be adopted in practice) is considered when the first fermentor is run at a rate ensuring virtually complete utilisation of substrate; i.e., s_1 is negligible and from equation (15)

$$x_1 = Y(s_R - s_1) \sim Y s_R \qquad . \qquad . \qquad . \qquad (41)$$

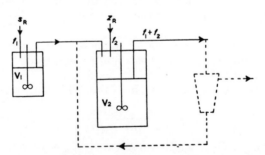

FIG. 10. *Two-stage continuous culture, with addition of second substrate to second-stage fermentor*
f_1 and f_2 are medium feed rates; s_R and z_R concentrations of substrates in feeds; v_1 and v_2 volumes of culture in fermentors. Possible application of feed-back indicated by dotted lines.

Then virtually no growth will occur in the second stage, but the cells will attack the second substrate z converting it to a desired product, p. The kinetics of such transformations in resting cells have been widely studied[21] and are found to resemble the action of an enzyme on its substrate; i.e., they obey the Michaelis–Menten equation:

$$- \frac{dz}{dt} = \frac{dp}{dt} = kx \left(\frac{z}{K_z + z} \right) \qquad . \qquad . \quad (42)$$

where k is a 'metabolic coefficient' having the dimensions of a velocity constant (t^{-1}) and K_z is the 'Michaelis constant' or half-speed concentration. The value of K_z is small (usually 10^{-3} M or less) and in practice will be negligible compared with z_R; equation (42) therefore reduces to:

$$- dz/dt = dp/dt = kx \qquad . \qquad . \qquad . \quad (43)$$

i.e., the reaction follows zero-order kinetics (at constant x) until virtually the whole of the substrate has been converted to product.

Let f_1 and f_2 be the rates of inflow of s and z at concentrations s_R and z_R respectively; i.e., $D_1 = f_1/v_1$ and $D_2 = (f_1 + f_2)/v_2$. In the second stage no growth occurs and the cells are simply diluted in the ratio D_1/D_2, so that

$$x_2 = \left(\frac{f_1}{f_1 + f_2} \right) x_1 \sim \left(\frac{f_1}{f_1 + f_2} \right) Y s_R \qquad . \qquad . \quad (44)$$

The balance equation for z is:

$$\text{decrease} = \text{conversion} + \text{output} - \text{input}$$

$$- \frac{dz_2}{dt} = kx_2 + \left(\frac{f_1 + f_2}{v_2} \right) z_2 - \left(\frac{f_2}{v_2} \right) z_R \qquad . \qquad . \quad (45)$$

Hence in the steady state:

$$z_2 = \left(\frac{f_2}{f_1 + f_2} \right) z_R - \left(\frac{v_2}{f_1 + f_2} \right) kx_2 \qquad . \qquad . \quad (46)$$

and similarly:

$$p_2 = \left(\frac{v_2}{f_1 + f_2} \right) kx_2 \qquad . \qquad . \qquad . \quad (47)$$

Now any appreciable output of z in the effluent represents a wastage. To reduce this to zero, the concentrations and feed-rates s_R, z_R and f_1, f_2 must be adjusted so as to make the two terms on the right-hand side of equation (46) equal. In other words, conversion will be complete and z_2 will be zero when

$$f_2 z_R = v_2 k x_2 = \left(\frac{f_1}{f_1 + f_2}\right) v_2 k Y s_R \text{ (from equation (44))} \quad (48)$$

or:

$$k Y \left(\frac{f_1 s_R}{f_2 z_R}\right) = \left(\frac{f_1 + f_2}{v_2}\right) \qquad . \qquad . \qquad . \qquad (49)$$

This equation enables feed concentrations and flow-rates to be calculated for fermentor systems of different sizes, and vice versa.

An interesting situation arises when the second stage is not a stirred fermentor but a tubular reactor, as in Fig. 1, type I.C. (i).

The solution here is simply obtained by integrating equation (43) which gives:

$$z_0 - z_\omega = k x_2 \bar{t} \qquad . \qquad . \qquad . \qquad . \qquad (50)$$

where z_0 and z_w are the concentrations at inlet and outlet respectively, and $\bar{t}$ the mean residence time.

Now $z_0 = f_2 z_R/(f_1 + f_2)$, and $\bar{t} = v_2/(f_1 + f_2)$; substituting these in equation (50) gives:

$$z_\omega = \left(\frac{f_2}{f_1 + f_2}\right) z_R - \left(\frac{v_2}{f_1 + f_2}\right) k x_2 \qquad . \qquad . \qquad (51)$$

identical with equation (46) for the stirred tank.

Danckwerts[9] has shown that for reactions with kinetics of first or higher order, the tubular reactor is more efficient than the stirred tank, while for reactions of the autocatalytic type (such as bacterial growth) the stirred tank is more efficient. It does not appear to have been pointed out that for zero-order reactions the efficiencies of the two types of reactor can be the same, as in the present case.

Returning to equation (49), it will be seen that the conversion of a given amount of substrate z requires the expenditure of a proportional amount of substrate s in order to produce cells; if the cells are not a useful by-product, this is wasteful. This wastage can be greatly reduced by the introduction of feed-back of cells to the second fermentor, as indicated by the dotted lines in Fig. 10. This amounts to using the same cells many times over and obviously allows the size of the first stage to be greatly reduced.

It is suggested that the two-stage system with feed-back, as shown in Fig. 10, will prove to be the most generally useful type for processes involving chemical transformations. It is extremely versatile, as the pH, degree of aeration etc., can be controlled at different levels in the two stages, so as to be optimal for growth in the first fermentor and for chemical transformation in the second fermentor.[22]

Discussion

The foregoing survey has covered only a fraction of the types of continuous culture systems which might be discussed with profit.

The two-phase counter-current systems, for example, involve several new principles of great theoretical interest, and it is unfortunate that our knowledge of them is limited to a few guarded references in the patent literature.[1,2]

The whole group of 'closed' continuous culture systems has also not been discussed, for the very different reason that discussion of general principles is here hardly possible. The nature of closed systems inevitably leads after any lengthy period of operation first to cessation of growth, then to death of the majority of cells in the system, followed by decay of the enzymes on which their metabolic activities depend. Such systems have complicated life-histories, different in each individual case.

It may be surmised that the author does not consider closed continuous culture systems to be particularly efficient. It is certain that nobody knows whether any such system, as used at present, is operating at even 1% of its theoretically possible efficiency.

One may also speculate whether it is anything but tradition that determines the use of closed systems of continuous culture for certain industrial processes. As will be explained in another paper in this Symposium, there is good reason to suppose that the most traditional of batch processes, namely brewing, can be conducted more efficiently as a continuous process. Is there any real reason why acetic acid manufacture, for example, should not be operated continuously in stirred fermentors?

It is hoped that the analysis in the first part of this paper will have served to emphasise that there are many different operational types of continuous culture system, with very different properties. For any given process, there will usually be one particular type of system more appropriate than any other. In order to choose the best system, theoretical analysis is necessary. It is hoped that the second part of the paper shows (a) that such analysis can be applied as successfully to complex systems as to simple ones, and (b) that it does not, fortunately, involve any difficult mathematics.

Some microbiologists may question the value of this type of theoretical approach, for the reason that they have been able to operate continuous culture processes without it. The purely empirical approach will often result in a process that works after a fashion, but seldom with maximum efficiency.

Continuous processes are more rewarding, but also more exacting than batch processes; to get the best out of them we need a much deeper knowledge of underlying principles. The situation in microbiological industry today is surely analogous to that existing formerly in the chemical industry when the change from batch to continuous processes was beginning. It was then found that more was needed to be known about fundamentals, and much work was stimulated on the kinetics and thermodynamics of process reactions, chemical engineering problems relating to the flow of materials through processing units, etc. The same type of research into basic problems is essential if equal success is to attend the application of continuous culture methods in microbiological industry.

References
[1] Zobell, C. E., U.S.P. 2,641,564
[2] Taggart, M. S., U.S.P. 2,396,900
[3] Monod, J., *Ann. Inst. Pasteur*, 1950, **79**, 390
[4] Novick, A., & Szilard, L., *Proc. nat. Acad. Sci., Wash.*, 1950, **36**, 708
[5] Bryson, V., *Int. Congr. Microbiol. No. 7* (Stockholm), 1958, p. 372
[6] Herbert, D., as reference 5, p. 381
[7] Kuska, J., *Continuous Cultivation of Micro-organisms, a Symposium*, 1958, 114 (Prague: Academy of Sciences)
[8] Lees, H., *Plant & Soil*, 1949, **1**, 221
[9] Danckwerts, P. V., *Industr. Chemist*, 1954, **30**, 102
[10] Monod, J., *La Croissance des Cultures Bacteriennes*, 1942 (Paris: Herman et Cie)
[11] Novick, A., as reference 7, p. 29
[12] Pirt, S. J., & Callow, D. S., *J. appl. Bact.*, 1960, (in press).
[13] Herbert, D., Elsworth, R., and Telling, R. C., *J. gen. Microbiol.*, 1956, **14**, 601
[14] Andreev, K. P., as reference 7, p. 186
[15] Danckwerts, P. V., *Chem. Engng Sci.*, 1953, **2**, 1
[16] Beran, K., as reference 7, p. 122
[17] Jerusalimskii, N. D., as reference 7, p. 62.
[18] Malek, I., as reference 7, p. 11
[19] Herbert, D., as reference 7, p. 45
[20] Novick, A., *Folia Microbiol.*, 1959, **4**, 395
[21] Stephenson, Marjory, 'Bacterial Metabolism', 1948 (London: Longmans, Green & Co.)
[22] Pirt, S. J., & Callow, D. S., *Selected sci. Pap. Inst. sup. Sanita*, 1959, **2**, 239

Editor's Comments on Paper 16

16 Myers and Clark: *Culture Conditions and the Development of the Photosynthetic Mechanism: II. An Apparatus for the Continuous Culture of Chlorella*

This paper, although devoted to algae, introduced some new ideas, theoretical and experimental, for the study of microbial growth. The ideas, which dealt primarily with attaining steady unchanging conditions of culture by continuous dilution of the cells in quite sophisticated equipment which used much instrumentation for that time, were important advances that foreshadowed the steady state and the turbidostat of a few years later.

16

Reprinted from *J. Gen. Physiol.*, **28**(2), 103–112 (1944)

CULTURE CONDITIONS AND THE DEVELOPMENT OF THE PHOTOSYNTHETIC MECHANISM

II. An Apparatus for the Continuous Culture of Chlorella

By JACK MYERS and L. B. CLARK

(*From the Department of Zoology and Physiology, University of Texas, Austin, and the Division of Radiation and Organisms, Smithsonian Institution, Washington, D. C.*)*

(Received for publication, July 24, 1944)

It has been a common experience that the photosynthetic behavior of algal cells may vary within rather wide limits depending upon previous conditions of culture. However, no attempt has been made to relate culture conditions to the subsequent type of behavior of the cells. A very great handicap to any such investigation is the internal variation which occurs with time within any one culture. In the first paper of this series (Myers, 1944) it has been pointed out that as a culture of Chlorella matures, the H^+ and NO_3^- ion concentrations of the medium, the conditions of carbon dioxide supply, and the effective light intensity and quality show marked variations. There are probably other variables as well; *e.g.*, the "inhibitors" of Pratt (1942). Consideration of the nature of these internal variables shows that all of them are functions of the population.

It has been a common practice in studies on photosynthesis to use algal cells from cultures harvested at some fixed period after inoculation. By using cells taken always at the same point on the growth curve a fair degree of reproducibility is to be expected, since all conditions which depend upon the population will be reproduced with some uniformity. Similar results may be obtained by periodically harvesting a part of the culture suspension and replacing it with fresh medium. Techniques of this kind have been reported by Felton and Dougherty (1924) for the culture of pneumococcus and by Ketchum and Redfield (1938) for the culture of marine diatoms. It would seem that if a culture could be continuously diluted so as to be maintained always at one point on its growth curve, then the effects of changing internal conditions might be eliminated entirely. This would

* The work of this paper was begun when one of us (J. M.) was supported by a National Research Council Fellowship at the Smithsonian Institution. Improvement in design of the apparatus and the accumulation of experimental data were done at The University of Texas under a grant from the University Research Institute. The authors are indebted to these institutions for support and assistance. Acknowledgement is made of the technical assistance of Mary Benjamin Smith.

afford at once (1) a source of experimental material of high uniformity and/or (2) a means of stablilizing internal variables so that relation of culture conditions to photosynthetic behavior might be systematically explored.

Following these considerations there has been developed an apparatus which maintains an algal[1] culture at a given density of population by automatically diluting the growing culture with fresh medium. This paper describes the apparatus and presents data illustrating its operation.

Description of the Apparatus

Fig. 1 is a diagrammatic, cross-sectional view of the apparatus. A number of parts have been distorted in position in order to place all of them in the same plane. The culture chamber[2] is made of three concentric glass tubes, affording an outer annulus, J, for circulation of constant temperature water and an inner annulus, A, in which the algal suspension is contained. Outside dimensions of the chamber are approximately 6.0 cm. diameter by 66.0 cm. long; the only critical dimension is the thickness of annulus A, which in all chambers used is 5 to 6 mm.

Any desired gas mixture is provided through the bubbler tube, B (shown in part) which simultaneously provides carbon dioxide and agitates the algal suspension. (The gas mixture is first humidified by bubbling through a column of liquid medium held at the same temperature as the chamber.)

Because of the requirement of a rather rapid air flow (2 to 6 cc. per sec.) the use of compressed gas mixtures has proved uneconomical. It has been necessary to devise a' means of obtaining constant air-carbon dioxide mixtures. Outside air is delivered *via* a diaphragm pump at a constant pressure obtained by allowing the excess air to escape against a head of 8 feet of water. Carbon dioxide is delivered from a cylinder at about 4 pounds pressure by suitable mechanical reducing valves. The two gases pass through orifices so chosen that the resulting gas mixture has a composition of about 4.4 per cent carbon dioxide. The actual rates of flow of the two gases are indicated by calibrated Venturi flow gauges and occasional Haldane analyses have been made as overall checks on the operation. The gas mixture is delivered at constant pressure by allowing the excess gas to escape against a head of about 6 feet of water.

Samples of the suspension are harvested as needed by opening a screw clamp and allowing the suspension to run out through the *withdrawal tube*. Fresh medium is added by activation of a solenoid valve, SV, which opens a rubber tube and allows the medium to run from a large aspirator bottle into the annulus A of the culture chamber.

By suitable precautions pure culture conditions can be maintained. The

[1] In principle it is applicable to other types of microorganisms as well.

[2] The first chamber was constructed by one of us (L. B. C.). Additional chambers have been obtained from E. Machlett and Son, New York City.

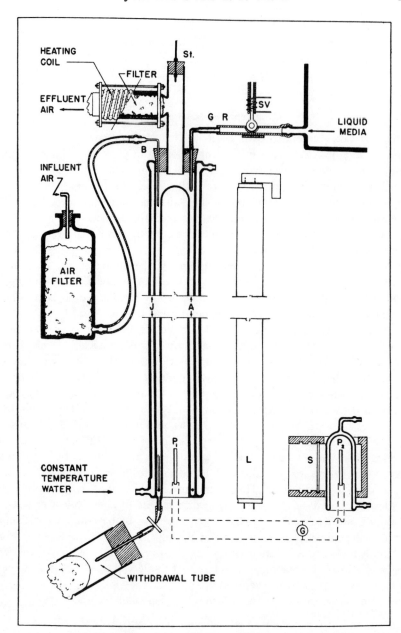

FIG. 1

withdrawal tube is protected by a glass skirt and cotton plug. After collection of a sample in a sterile flask a fresh cotton plug, previously autoclaved in a large test tube, is inserted in the glass skirt. Influent air is filtered through a 500 ml. aspirator bottle packed with cotton. Effluent air passes out through a glass tube packed with cotton. The cotton filters must be kept warm to insure against condensation of moisture. A heating coil around the effluent filter is shown partly cut away. A heater (not shown) for the influent filter is provided by a simple tin-can and light bulb arrangement. Before setting up, the assembly consisting of the chamber, withdrawal tube, influent and effluent filters is sterilized by autoclaving. During this procedure a cotton plug is used in place of the rubber stopper, St, and the glass inlet tube for the medium is not connected to the aspirator bottle but is protected by a vial and cotton wrapping at G. The aspirator bottle is autoclaved separately, the end of the rubber tube, R, being protected by a vial and cotton wrapping.

After autoclaving, the chamber is placed in position, the clamp to the withdrawal tube closed, and a current of air passed through. An inoculating suspension of cells from an agar slant is introduced through the uppermost opening and the cotton plug replaced by a sterile rubber stopper, St, fitted with a glass rod for ease in handling. The rubber tube, R, from the aspirator bottle is attached to the glass tube, G, with aseptic precautions.

The whole procedure of sterilization is indeed cumbersome but it is required only at infrequent intervals of a month or more. New aspirator bottles of medium can be inserted by carefully removing the rubber tube of the old bottle, flaming the glass tube, G, and slipping on the sterile rubber connection of a new bottle. In operation a sample (about 1 ml.) is withdrawn daily into a flask of glucose-peptone broth as a check against bacterial or mold contamination. Many cultures have been run for a month or more without contamination.

Illumination is provided by four tubular fluorescent or tungsten filament lumiline bulbs spaced symmetrically around the chamber. (Only one lamp is shown in the diagram.) Short period variations in light intensity are minimized by use of a voltage stabilizer which delivers 115 ± 1.0 volts to the lamps. Light intensity may be varied by changing the distance of the lamps from the chamber or by surrounding the lamps with sleeves made of wire screen.

Constant density of population is maintained by adding fresh medium as the culture grows. The solenoid valve, SV, is operated by a photometric device. Two photocells,[3] P_1 and P_2, are introduced into the current-balancing circuit recommended by Wood (1934) by connecting terminals of opposite polarity. The meter,[4] G, is a spotlight galvanometer which has a No. 922 RCA phototube mounted in place of the usual glass scale. The phototube

[3] General Electric rectangular, barrier-type photocells $\frac{29}{32}$ inch $\times$ $1\frac{45}{64}$ inch.
[4] Rubicon No. 3402; period 3.7 seconds; R = 335 ohms; CDRX = 3900 ohms.

feeds into an electronic relay employing a No. 2051 vacuum tube. The arrangement is such that an increase in light on the phototube energizes the relay. The relay operates the solenoid valve, SV.

The two photocells look at the same light bulb. P_1, the inside photocell, is screened by the annulus of algal suspension. P_2 is screened as desired by insertion of one or more uniformly exposed lantern slides, S. The outside photocell, P_2, is jacketed by a glass condenser through which is circulated the same constant temperature water used for the chamber; photocell and jacket are partially enclosed within a box arranged so that light enters only from the direction of the one light source. When low population densities (1 to 2 c.mm. cells per cc.) are desired, the inside photocell, P_1, is also screened with a piece of white paper so that its maximum output is about 50 microamperes. This decreases the sensitivity but greatly improves the stability of the photocell circuit.

In order to minimize ambient light effects on the photocells, the apparatus is partially enclosed in a box 30 x 30 x 36 inches high, painted white inside. The chamber just barely projects through a hole in the top of the box; and the air filters, solenoid valve, and aspirator bottle of medium are all mounted on top. The front of the box is a sheet of plywood with holes provided for ventilation and is easily removed when a sample is to be withdrawn.

In practice the original inoculum is allowed to grow until a desired population density is reached and the photocells are then balanced by inserting screens in front of the outside photocell. Thereafter a sample is withdrawn at a fixed time each day. The air flow is temporarily stopped and a small sample ($\sim$1 ml.) collected in a flask of sterile broth as a check against contamination. The rest of the culture is then withdrawn down to a mark placed a centimeter or so above the top of the inside photocell. This leaves an inoculum (50 to 100 ml. of suspension) for the next day's growth. As the algae multiply illumination on the inside photocell, P_1, is reduced. An off-balance current flows through the primary photocell circuit causing the galvanometer light spot to move across the phototube. The electronic relay then actuates the solenoid valve and allows new culture medium to flow in, diluting the algae and increasing the illumination on P_1 until a zero current again obtains through the galvanometer. In this way the algal suspension "grows" up the chamber. The design of the apparatus is such that conditions of illumination are independent of the total amount of the culture.

If samples are harvested from a culture at equal intervals, then the amount of the sample is an index of the rate of growth. It is also possible to harvest samples at any other desired time, provided that the culture is not allowed to run over. Daily samples of 200 to 300 ml. containing 200 to 600 c.mm. of algal cells are easily obtained.

Operation of the Apparatus

Satisfactory operation of the entire apparatus depends upon the stability and sensitivity of the primary photocell circuit. By maintaining the two photocells at constant temperature and keeping their illumination at a low level good stability may be obtained over a period of weeks. The over-all sensitivity is such that the solenoid never allows more than 1 per cent of the volume of the culture to flow in during one relay cycle.

One unit of the apparatus has been in operation for over a year and a second unit for about 6 months. Numerous difficulties have arisen. The present procedure of maintaining pure culture conditions is the result of a gradual improvement in technique. Development of an apparatus for producing a reliable gas mixture gave considerable trouble. In general the mechanical problem has been one of maintaining stability over long periods of time.

Several modifications in the preparation of culture media have been necessary. Originally we used a Knop's solution from which calcium was omitted (0.010 M $MgSO_4$, 0.012 M KNO_3, 0.009 M KH_2PO_4, 1.0 $\times$ 10^{-5} M ferric ion). Micro elements were provided by the addition of 1.0 ml. per liter each of the A5 and B6 solutions of Arnon (1938). These provide in the final medium 0.5 parts per million B, 0.5 ppm. Mn, 0.05 ppm. Zn, 0.02 ppm. Cu, and 0.01 ppm. each of Mo, V, Cr, Ni, Co, W, Ti. In this medium contained in a pyrex aspirator bottle there would develop in time a fine white precipitate. Concurrently, a culture provided with the medium would show a decreased rate of growth and lowered capacity for photosynthesis. Subsequently, we have purified the major salts by the adsorption procedure of Stout and Arnon (1939), increased the iron concentration to 13.3 $\times$ 10^{-5} M, and added sodium citrate to give 0.00056 M citrate as used by Hopkins and Wann (1927). With these modifications the culture media will remain clear indefinitely and rate of growth and capacity for photosynthesis are as great as obtained by any other method of preparation of medium.

It has also been found that after insertion of a fresh bottle of medium the rate of growth and capacity for photosynthesis may be lowered for a day or so. It has been possible to demonstrate that the Knop's solution in contact with the rubber tubing leaches some toxic materials out of the rubber during autoclaving. This difficulty has been minimized by (1) using rubber tubing previously boiled in dilute alkali and leached out in distilled water and (2) running several hundred milliliters of medium from a fresh bottle out into a sterile flask before attaching the rubber tubing to the glass inlet tube of the apparatus.

Presented in Table I are typical data obtained on a culture of *Chlorella pyrenoidosa* (Emerson's strain) over a period of 3 weeks. The scanty data on the first eleven samples are omitted since rate of growth and capacity for photosynthesis were apparently limited by iron concentration. After the ninth sample a new bottle of medium containing 9.0 $\times$ 10^{-5} M iron was inserted.

From the twelfth sample on all data are presented, though in several cases data are believed invalid due to serious failure of the temperature control

TABLE I

Typical Data Obtained on Chlorella pyrenoidosa Grown in the Continuous Culture Apparatus

Temperature 25.05°C. Light intensity $\sim$160 foot-candles as provided by four 20 watt "Daylight" fluorescent bulbs mounted 27 cm. from the chamber.

Sample No.	Sample size	Population		Maximum apparent rate of photosynthesis	
		Cells	Cell volume	Buffer 9	Buffer 11
	ml.	*10^9/c.mm.*	*c.mm./cc.*	*c.mm. O_2/min./ c.mm. cells*	*c.mm. O_2/min. /c.mm. cells*
12	216	—	1.33	—	—
13	—	0.0118	1.33	0.65	0.70
14	—	0.0120	1.33	0.62	0.66
15	—	0.0118	1.33	0.62	0.66
16	212	0.0121	1.33	0.64	0.68
17	216	—	—	—	—
(Added additional A₅ and B₆ to medium to give 1.5 ml./liter of each)					
18	224	—	1.33	0.63	0.69
(Added iron to give total concentration of 19. $\times 10^{-5}$ M)					
19*	211	—	—	—	—
20	226	0.0169	1.39	0.64	0.70
21	226	0.0168	1.33	0.66	0.71
(Inserted new bottle of culture medium)					
22*	123	—	—	—	—
23*	148	—	—	—	—
24	208	0.0141	1.33	0.64	0.68
25	198	—	—	—	—
26	213	—	—	—	—
27	200	—	1.31	0.63	0.67
28	205	0.0150	1.36	0.62	0.68
29	220	—	1.33	—	—
30	198	0.0180	1.36	—	—
31	202	—	1.39	0.62	0.68
32	210	—	1.39	—	—
33	198	0.0200	1.36	0.63	0.71
(Culture discontinued)					
Mean..................	210.5	—	1.346	0.633	0.687
Standard deviation......	9.5	—	0.025	0.0125	0.017
Maximum variation.....	28.0	—	0.08	0.04	0.05

* Temperature failure; data for this sample not valid.

mechanism. Mean values, maximum variation, and the standard deviation are given at the foot of each column.

Samples were removed from the chamber at 24 hour intervals and 79 ml. of suspension left for inoculum each time. The mean sample size (omitting samples 19, 22, and 23) of 210.5 ml. indicates a multiplication rate of 3.66 per 24 hours under the conditions employed. There is a slight downward trend in the data on sample size, probably reflecting a decay in light output of the lamps with time.

Densities of population of the samples were determined by means of hemo-cytometer counts (giving cells per c.mm.) and by centrifuging to give packed cell volumes (c.mm. cells per cc.). Hemocytometer counts were quite variable, probably reflecting to some extent the variations in culture media used. Each count represents the mean value from a sample of about 1000 cells so that the experimental error is about 5 per cent. Of all the data, these show the greatest variation.

Packed cell volumes were determined by centrifuging 15.0 ml. of suspension in a 15 ml. centrifuge tube. The packed cells were resuspended in a little of the same medium and transferred to a Bauer and Schenk tube of 3 ml. capacity and graduated in 0.004 cc. divisions. 1.33 c.mm of cells per cc. corresponds to 5.0 divisions which can be estimated to about 0.1 division. Constant values for the packed cell volume are obtained between 15 and 25 minutes centrifuging at a relative centrifugal force of 2150. It will be seen that the packed cell volume (c.mm. cells per cc.) is quite uniform with a max-imum variation of about 6 per cent and a standard deviation of about 2 per cent.

Capacity for photosynthesis of the cells was determined by the Warburg technique, using the Warburg buffers 9 and 11 to provide approximately saturating concentration of carbon dioxide, a temperature of $25 \pm 0.05°C.$, and a saturating light intensity. The light sources used were a grid of white fluorescent tubing ($\sim$1000 foot-candles) or two 60 watt lumiline bulbs ($\sim$450 foot-candles) immersed in the water bath just below the vessels. No differences in photosynthesis rates produced by the two sources can be detected when they are used at full intensity. For the manometric measurements a 5.0 ml. aliquot of the sample was pipetted into a 15 ml. graduated centrifuge tube. The cells were centrifuged out, suspended in distilled water, centrifuged out, suspended in Warburg buffer 9 (0.015 M K_2CO_3 + 0.085 M $KHCO_3$), centrifuged out, and suspended in buffer 9 to give 15.0 ml. 5 ml. of this suspen-sion were pipetted into each of two rectangular Warburg vessels of about 10 ml. volume. By a similar procedure 5.0 ml. of a suspension of cells in buffer 11 (0.005 M K_2CO_3 + 0.095 M $KHCO_3$) were delivered to each of two other vessels. Rate of photosynthesis was determined graphically by plotting 5 minute readings taken over a period of about an hour. The rate in terms of Δmm. pressure per minute was multiplied by the vessel constant and divided by the volume of cells used to obtain c.mm O_2 per min. per c.mm. cells. Dupli-cate rates generally agree within a few per cent; the averages of the duplicates for each buffer mixture are the values listed in the fifth and sixth columns of Table I. The data represent only the *apparent* rates of photosynthesis uncorrected for respiration. Respiratory rates were not measured.

Values of the apparent rate of photosynthesis per cubic millimeter of cells (fifth and sixth columns) show, for each carbon dioxide concentration, a maximum variation of 6 to 7 per cent and a standard deviation of about 2 per cent. The experimental errors in these values are contributed both by the errors of the manometric measurement and the error in determination of cell volume.

Additional data (not shown in the table) were obtained by occasional measurement of the pH of the suspension immediately after harvesting. Measurements with a Coleman glass electrode yielded pH values lying between 6.05 and 6.10. The fresh Knop's solution used had a pH of 5.0.

DISCUSSION

The data of Table I on maximum apparent rate of photosynthesis describe one arbitrarily chosen physiological characteristic of the cells. It is our experience that for a culture allowed to mature along the growth curve, the maximum rate of photosynthesis may decrease to a value as low as 0.05 c.mm. O_2 per min. per c.mm. of cells. Similar results were obtained by Sargent (1940, Table IV). Sargent's data (Table III) also illustrate the considerable variation in maximum rate of photosynthesis to be expected for cells of different cultures harvested at approximately the same age. It is evident that the maximum rate of photosynthesis is a characteristic which may vary widely between different batches of algal cells cultured by the usual procedures. The data (columns 5 and 6) of Table I illustrate the uniformity in maximum rate of photosynthesis to be expected of cells obtained from the continuous culture apparatus. Rate of growth also shows fair uniformity. Other physiological characteristics have not been examined. However, it is reasonable to expect that cells will be equally uniform in other characteristics since they are grown under highly uniform conditions.

In only one respect has the apparatus fallen short of theoretical expectations. A basic assumption is that the sample removed and the inoculum left in the chamber are identical as to concentrations of all components (*i.e.*, cells, inorganic ions, metabolites). This condition is not entirely attained. Because of the bubbling action, surface-active materials tend to accumulate at the upper liquid-gas interface and therefore do not distribute equally between the sample and inoculum. Foaming occurs, though a permanent foam is not formed until after several weeks of operation. However, no detrimental effects of the surface-active materials have yet been observed.

The data of Table I describe the operation of the apparatus as applied to one of the two purposes for which it was designed; *i.e.*, the production of uniform experimental material day after day. Application of the apparatus in studying the relation of culture conditions to photosynthetic behavior will be considered in later papers of this series.

SUMMARY

1. An apparatus has been developed which maintains a constant density of population of *Chlorella* by automatic dilution of the growing culture with fresh medium.

2. Cells harvested from the apparatus in daily samples are highly uniform in rate of growth and rate of photosynthesis measured under arbitrarily chosen conditions.

BIBLIOGRAPHY

Arnon, D. I., *Am. J. Bot.*, 1938, **25**, 322.
Felton, L. D., and Dougherty, K. M., *J. Exp. Med.*, 1924, **39**, 137.
Hopkins, F. B., and Wann, E. F., *Bot. Gaz.*, 1927, **83**, 194.
Ketchum, B. H., and Redfield, A. C., *Biol. Bull.*, 1938, **75**, 165.
Myers, J., *Plant Physiol.*, 1944, in press.
Pratt, R., *Am. J. Bot.*, 1942, **29**, 142.
Sargent, M. C., *Plant Physiol.*, 1940, **15**, 275.
Stout, P. R., and Arnon, D. I., *Am. J. Bot.*, 1939, **26**, 144.
Wood, L. A., *Rev. Scient. Instr.*, 1934, **5**, 295.

Editor's Comments on Papers 17 and 18

17 Málek: *The Physiological State of Microorganisms During Continuous Culture*

18 Řičica et al.: *Properties of Microorganisms Grown in Excess of Substrate at Different Dilution Rates in Continuous Multistream Culture Systems*

Málek and his school in Prague have set and held a lead in physiological aspects of growth and continuous culture that has complemented the work of the Porton group in theoretical matters. In the introductory section of his article (Paper 17), Málek gives a useful summary of the empirical development of continuous growth techniques. By referring to batch and continuous cultures as static and dynamic systems, Málek used terms that could be misapplied in relation to the cellular activities; consequently, the alternative "closed" and "open" designations of Herbert are preferred.

In his paper Málek makes the important point that the Monod and related theories disregard the possibility of qualitative changes taking place in the cells and account for such changes that occur by quantitative alterations of an ideal condition instead. It is unfortunate that in presenting his concept of the physiological state, Málek did not define it more explicitly; perhaps this was intentional because he was then uncertain how it might be done. The problem of the physiological state is still unsolved, but from this paper multistage continuous cultures are already seen as the experimental means for attempting its solution.

The paper by Řičica and colleagues describes a most versatile technique, the theoretical and experimental potential of which is considerable. The results presented in this paper demonstrate both the enormous range and variability of growth activity and the possibility of being able to examine this behavior systematically in both restricted and unrestricted conditions of growth.

Reprinted from *Continuous Cultivation of Microorganisms: A Symposium,* Academia: Publishing House of the Czechoslovak Academy of Sciences, Prague, 1958, pp. 11–28

THE PHYSIOLOGICAL STATE OF MICROORGANISMS DURING CONTINUOUS CULTURE

Ivan Málek

In the course of the last 10 years extraordinary attention has been paid to continuous culture. We can mark an increase in the number of papers which show that it is about to become a new method enabling us to deepen our knowledge about the multiplication of microorganisms and at the same time gives us a stable living material for experimental studies of variability and mutability and for biochemical analyses. It also points the way towards a substantial increase of productivity for a number of fermentations.

Up to 1945 there existed but a few papers describing the use of this method. Moreover these papers usually did not consider this method as a new and basic one and one which corresponds to the reproductive capacity of microorganisms better than the commonly used static methods. This is true for the work of Rogers and Whittier (1930) who tried to draw an analogy between a bacterial culture and a multicellular organism, and cultivated bacteria (*Streptococcus lactis*) at a rather slow rate of flow of the nutrient medium on the one hand, and using a microbial filter under which the nutrient medium was repeatedly renewed on the other hand. Their most striking results deal with the lactic acid production in continuous culture. Cleary, Beard and Clifton (1935) used the continuous method to study the basis of the stationary phase of common batch culture methods. The papers of Jordan and Jacobs (1944, 1947, 1948) also deal with an analysis of the stationary phase. Moyer (1929) used the continuous method only to increase the bacterial mass for chemical analysis. Among the theoretical papers from this period, only those of Utenkov (1929, 1942) and Málek (1943) considered continuous culture as a new experimental method from the very beginning. Utenkov, who was probably the first to take up the continuous flow method systematically, proceeded in 16 years of experimental work (from 1922) from the assumption that the batch culture method does not sufficiently reveal the true characteristics of microorganisms, and does not permit satisfactory regulation of the development of cultures. It is

11

only to be regretted that Utenkov's paper remained unknown to the world microbiological literature. It was from a similar of point view that Málek studied the continuous flow method (1943) and stressed its advantages in studying multiplication, variability and pathogenicity of microorganisms.

Practical exploitation of the method was only slightly developed. Lebedev (1936) developed the method of continuous alcoholic fermentation. In Germany, fermentations have been patented which are based on continuous flow fermentation (Lupinit, Norddeutsche Hefeindustrie, 1934); later on, a group of American research workers — Unger, Stark, Scalf, Kolachov (1942) and others — worked out procedures for continuous production of yeast and alcohol, and pointed out their advantages and greater productivity as compared with the static "batch" methods.

Later papers (after 1945), however, almost without exception, consider continuous flow culture as a new method. It has been studied technically both from the experimental-laboratory viewpoint (Castor (1947), Bactogen-Monod (1950), Chemostat-Novick and Szilard (1950), Anderson (1953), Málek (1943, 1952), Northrop (1954), Kubitschek (1954), Graziosi (1956, 1957), Davies (1956), Karush, Iacocca, Harris (1956), Formal, Baron, Spilman (1956), Perret (1956)) and from the laboratory-production viewpoint (Malmgren and Hédén (1952), Málek (1955), Elsworth et al. (1956), Pirt (1957)) as well as applied to microcultures (Pirfilev — personal communication, Rosenberg (1956)). A methematical treatment has been given for the theory of reactions in continuous flow systems in general (Denbigh (1944), Pasynskij (1957)) and in particular to the multiplication of microorganisms (Monod (1950), Northrop (1954), Herbert et al. (1956), Maxon (1955), Jerusalimskij (1958). The method has been considered from the point of view of experimental possibilities and perspectives (Monod (1950), Málek (1943, 1952), Maxon (1955), Novick (1955), Powell (1956)). The following theoretical problems have been studied by this method: induction of antibiotic properties (Sevage and Florey (1950)), the formation of mutants and other problems of genetics (Novick and Szilard (1950), Bryson (1953), Zelle (1955), Lee (1953), Moser (1954)) microbial adaptation (Verbina (1955), Graziosi (1956, 1957)), growth of fungi (Duché and Neu (1953), Hofsten et al. (1953)), bacterial multiplication (*Brucella*: Gerhardt (1946); *Aerobacter*: Herbert et al. (1956), Pirt (1957); *Escherichia coli*: Málek (1943, 1950); *Streptococcus haem. A.*: Karush et al. (1956); *Salmonella typhi*: Formal et al. (1956); *Mycobacterium tb.*: Švachulová and Kuška (1956)), and development of bacterial cultures (Málek et al. (1952, 1953, 1953b, 1955), Macura and Kotková (1953), Ševčík (1952), Jerusalimskij (1956, 1958)).

Other papers deal with the industrial application of the method (Harris et al. (1948), Victorero (1948), Adams and Hungate (1950), Šarkov and his school (1950), Elsworth et al. (1956), Málek et al. (1955, 1957)).

Experience has been acquired not only with the continuous culture of bacte-

12

ria, yeasts and molds but also of some algae (Ketchum, Bostwick, and Red-field (1938), Myers et al. (1944) cf. Novick, Tamiya (1957)) and protozoa (Browning and Lockinger (1953), Vávra (1958)).

In this review we intend to consider only theoretical papers and results particularly with regard to the physiological state of microorganisms in continuous flow cultures. Therefore its mathematical aspect and the kinetics of growth of microorganisms will not be covered, since this question has been sufficiently studied in earlier papers (Monod, Northrop, Maxon, Herbert et al.).

Let us begin with a few remarks on terminology. When referring to "continuous cultures" we shall usually have in mind continuous flow cultures, as compared with continuous culture in the broader sense of the word, where culture with periodic renewal of nutrient medium is included (for this type of culture sometimes the word "semi-continuous" according to Maxon is used). Continuous flow culture is thus the extreme case of periodic continuous culture; the technical difference between these two is usually only in the length of the interval between additions of medium, because even in a continuous flow culture the nutrient medium is usually added intermittently. This difference is at times of great importance for the process. In contrast to continuous culture we shall consider the static single-run culture i. e. the common batch culture. Sometimes the term "dynamic culture" is used even for a batch culture when stirred. It is felt that this usage is incorrect because this type of culture remains basically static in character. It is typical of a true dynamic culture that a dynamic steady state is established.

What do we actually mean by the physiological state of microorganisms in continuous culture?

A lot of experience obtained in studying common static cultures have shown that the number of microorganisms follows the typical growth curve. However, only the exponential part of this curve is significant for multiplication of microorganisms in common continuous cultures, because only then is uniform multiplication taking place. Therefore, this part of the curve is usually treated as a whole quite uniformly, as no striking change of culture can be observed there. A number of papers (Malmgren and Hédén (1951), Hinshelwood (1947), Valyi-Nagy (1955) and many others) have shown that microorganisms are undergoing changes even in this exponential part of the curve. Thus, for instance, the ribonucleic acid content of cells does not remain constant during the exponential period, i. e. the period of uniform cell division, but usually drops very rapidly at the very beginning of the curve. On the other hand there are a number of reasons for the assumption that the beginning of multiplication coincides with the end of the lag-phase, i. e. with the late lag-phase, because then a great production of microbial material can be observed as a consequence of regular microbial proteosynthesis, as shown by an increase in enzymic activity, by a rise in the ribonucleic acid content, and by an increased sensitivity towards external conditions (Malmgren and Hédén (1951), Winslow, Walker et al. (1939) and other authors). All this indicates that the physiological state of microorganisms, as manifested by enzymic activity, sensitivity, proteo-

13

synthesis, RNA content etc., is undergoing changes in static cultures even during the phase of full multiplication. The conditions, however, which lead towards this change have not been sufficiently elucidated.

Now, to which part of the curve which has been thus divided can we compare the physiological state of microorganisms under conditions of continuous multiplication? A logical consideration leads to the conclusion that under conditions fully ensuring the multiplication of microorganisms, the physiological state will most likely correspond to the state of a culture in the late lag-phase or at the beginning of the log-phase. But what is the case when microorganisms are grown under conditions which are below this level, and when slower multiplication is taking place; does the physiological state of cultures then change in a similar sense as can be observed in static cultures?

This question appears to be rather important, since the physiological state can considerably influence multiplication itself and the kinetics thereof. Only when we have fully answered this question can we purposefully regulate continuous processes and exploit them for theoretical or practical ends. It is particularly important in cases when we intend to grow microorganisms over a long period with a maximum activity of multiplication, or when we wish to obtain products which in static cultures are associated with changes in the state of the culture, and appear therefore either at the end of the exponential, or during the stationary phase. The question arises, however, whether we can draw any parallel between continuous and static cultures in terms of the individual phases of development.

It is most probable that laws governing the multiplication of microorganisms in static cultures during the phase when the nutrient medium contains sufficient amounts of all necessary components, and when it is not affected by metabolites produced, are similar to those of continuous cultures. But even then there is an important difference between these two types of cultures: In static cultures the concentration of nutrients is usually excessive at the moment of beginning growth, when it is intended to last for several generations; in continuous cultures, on the contrary, the concentration of added nutrients is diluted into the whole volume of nutrient medium with the microbial culture, which can thus assimilate it immediately. When the basic nutrient factors or the source of energy are optimally balanced with respect to the requirements of a multiplying culture, we reach the stage where nutrients are assimilated on addition and cannot be determined in the culture fluid although constantly replenished. This is another aspect of the steady state of continuous cultures, and very important from the point of view of optimal exploitation of nutrients —a definite requirement in practical application. Is this difference in any way reflected in the physiological state of microorganisms in culture, as to the manner of absorbing nutrients and in the manner of their utilisation? Such a possibility cannot be excluded. What is the case when microorganisms are

14

cultivated in such a way that the rate of continuous feed is slower than would correspond to the growth rate as required by Monod for cultures in which self-regulation is operating? Is the answer given by merely slowing down multiplication, or is there a concurrent change of physiological state as is known in the latter stages of the exponential curve of a static culture? It cannot be excluded that some change could take place together with a different response of cultivated microorganisms to external conditions. Evidence on this would be of great experimental importance.

These are some of the questions that should help us to show the relation between the basic problem of the physiological state of cultures under conditions of continuous multiplication, and the application of this method in theory and practice. These questions have thus far been accorded little experimental attention, despite the importance for evaluation of obtained and obtainable results.

The attitude toward this question is connected with another problem of particular importance during the first period of study of continuous cultures, and of general biological importance: whether conditions of static or dynamic cultures better meet the physiological requirements of microorganisms. By the term "requirement" we have in mind the relation of microorganisms to their environment as developed in the course of their exposure to physiological conditions in nature.

Older papers — with the exception of those of Utenkov and Málek — assumed that the growth-curve characteristics derived from static cultures have an absolute validity, and that they reflect the natural development of the physiological state of microorganisms and their requirements. Cleary et al. (1935) therefore did not study continuous multiplication from the point of view of microbial multiplication, but rather in order to learn more about the stationary phase of static cultures. This attitude was apparently an expression of the fact that in that period a diagnostic raison d'être still prevailed in microbiology, for which static cultures are most convenient. Particular attention has been given to the study of the so called "maximum concentration" according to Bail (1929). There were however more profound biological reasons for this view: under natural conditions (e. g. in the soil etc.) microorganisms do not as a rule multiply in a homogeneous solution with a constant afflux of nutrients, but rather on structures where they behave similarly as in artificial cultures on solid media — they form colonies which are analogous to the well-known growth-curves of liquid static cultures (e. g. Vinogradskij (cf. Volodin 1952), Novogrudskij (1950)). From this the conclusion is usually drawn that microorganisms have developed a fixed form of multiplication which corresponds to the conditions of static cultures, and that even when they can multiply without limitation, cultures undergo developmental changes that correspond to the typical growth-curve of static cultures. The practical

15

consequence of such assumptions was that microorganisms were cultivated in continuous cultures with an afflux of nutrients at a rate considerably lower than would correspond to the growth rate. Another consequence of this thinking was that it was not generally believed possible to cultivate microorganisms in continuous cultures ad infinitum without some degeneration or change that would correspond to the stationary and decrease phases of static cultures. This opinion constituted an obstacle in accepting continuous culture methods and developing them in fermentation. This opinion has been shown to be wrong by work describing continuous cultures of microorganisms over long periods of time without signs of degeneration (Herbert et al. (1956), Málek (1955) and others).

A logical conclusion from the above findings would appear to be that static culture and the developmental changes observed with it are artefacts. The lag-phase appears as an artefact because, in the course of it, a culture changed by previous static cultivation must adapt itself to regular multiplication; the decrease of biosynthesis in the latter part of the exponential phase also appears as an artefact, as well as the whole of the stationary phase. The natural conditions which most fully correspond to the dynamics of microbial multiplication exist only in continuous culture. Therefore, only by using this method can we produce cultures which are really physiological, and correspond to the real physiological and biochemical characteristics of microorganisms. This view is supported by the results of continuous cultures which have shown that microorganisms can be grown under such conditions for indefinite periods in a vegetative form, with optimal results as to the production of living matter and optimal enzymic activity, etc. From the above the conclusion can be drawn that in static cultures only the initial, the late lag, and the exponential phases, can be considered as physiological, because only then can the synthesis of living matter, and all phenomena connected with it, proceed in an uninhibited manner. Everything else in the static growth curve appears to be an artefact caused by conditions inadequate for the multiplication of microorganisms.

This conclusion does not take into account the fact that some of the developmental changes in static cultures, at least in some microorganisms, have the character of fixed traits: above all the sporulation of bacilli (and, apparently, also of actinomycetes) and probably also the formation of resting forms of bacteria in the stationary phase. These traits are a biological fact which shows that in the course of development bacilli had to adapt to conditions somewhat analogous to those of static cultures. But the same microorganisms can be kept constantly in a vegetative state under suitable conditions. Furthermore a number of important products, which are certainly not a laboratory artefact, and which are of great practical importance, are formed only during the latter phase of static culture (antibiotics etc.). It cannot be affirmed

16

therefore that continuous flow cultures in their commonly used simple form correspond to all the physiological variables and requirements of microorganisms as which have developed. They meet the optimal requirements of fast vegetative multiplication of microorganisms; therefore, all the phenomena which are associated with the multiplication of microorganisms, be it the production of living matter, some basic enzymic processes, the influence of the environmental conditions on actively multiplying cells, or spontaneous mutability, can be conveniently studied only in continuous cultures. Only these cultures, when well set up technically and under constant conditions, can produce stable material for such investigation. On the other hand, in order to study phenomena caused by changes in environmental conditions due to the activity of microorganisms themselves, it is necessary to resort to a static culture, or to modify suitably a continuous culture method.

The majority of research workers using the continuous culture method base their views on the second assumption and do not take into account the possibility of changes of the physiological state under changed conditions. Theoretical papers devoted to the mathematical basis of multiplication (Monod, Herbert, Maxon and others) presume the existence of an ideal state in which the rate of growth and the activity of metabolic processes represent values dependent only very simply on the flow rate (dilution). This ideal state is hypothetically reached by assuming a system in equilibrium when the dilution rate (i. e. the ratio of emptied volume per unit of time to the volume of the culture) is equal to the multiplication rate (i. e. the number of divisions per unit of time) multiplied by 0,69 ($= \log_e 2$). Should some change in the physiological state of the culture occur, a new equilibrium would be formed, on the basis of which the experiment proceeds. Therefore no qualitative change in the physiological state of microorganisms is considered; it is taken to be constant in the course of a given experiment. The steady state which exists at different rates of flow is usually considered to differ only quantitatively. This abstraction is necessary as a basis for experimentation and it can be neglected within certain limits of multiplication rate. This assumption served as the basis for methods of the type of Monod's Bactogen or Novick's and Szilard's Chemostat, which choose an afflux of nutrients which remains below the level ensuring a maximum growth rate, and limit one important nutrition component. Other methods proceeding along the same line are those of the Turbidostat type, as used by Northrop, Bryson, Anderson and others, when a photo-cell helps to keep the density constant and the afflux changes in intensity. By means of both of the above-mentioned methods, the rate of growth can be automatically controlled and kept constant — in the first case by means of the self-regulating ability of the continuous flow system kept below the level of maximum growth rate, in the second case by means of an external mechanism. Even workers using these methods to study important physiological processes (Monod, Duché and Neu, Karush et al.) or genetic processes (Novick and Szilard and others) work under conditions that disregard the physiological state and its changes. Thus they have the possibility of studying a number of important biochemical, physiological and genetic problems considerably more exactly, and with a more constant and more physiological microbial material than is possible in static cultures, but they do not fully exhaust the advantages of continuous methods for studying more thoroughly the biological factor itself and applying the results in practical fermentations. It follows from the conclusion of Monod's paper that he is aware of this simplification. Northrop stresses in the introduction of his paper that "any change in the concentration of any substance indicates a change in organism" and Maxon in his paper points out that a rigorous mathematical treatment of all factors is not even possible, and so far not useful, because a fermentation is a highly complex living system.

It seems particularly important to investigate to what extent we are justified in drawing a parallel between the slow growth achieved in the Chemostat by limiting one of the important nutrient factors (cf. e. g. Novick and Szilard 1950), and the faster growth in the physiological state. It remains a question what rôle a similar limitation might play in a continuous culture. This question is discussed by Powell (1956) from the mathematical point of view and in relation to the generation time of individual microorganisms. He proves that a continuous flow culture "discriminates heavily against organisms of unusually long generation time". Such long generation time certainly reflects a definite physiological state. Powell assumes in his conclusion that a continuous culture will stabilize itself within a certain range of physiological activity of microorganisms. But the question appears to be even more complicated: it is assumed, on the basis of a statistical treatment of the whole population, that the culture is homogeneous, but such is not the case even in a well stirred continuous culture. Let us therefore consider the effect of limiting one of the important nutrient factors: will the result of this procedure be that all the microorganisms present in the population consume an equal minimum amount of the limited factor and therefore multiply themselves more slowly (and in turn influence their physiological state), or that microorganisms metabolically more active consume the limited factor, depriving less active bacteria? In the latter case the culture can be divided into a portion with a normal generation time corresponding to temperature etc., and into a portion with an abnormally long generation time, or not dividing at all. The resulting generation time would reflect the statistical difference of these cases, but would not give a true picture of the physiological state of the microorganisms present. From this point of view it is interesting to note the remark of Karush et al. (1956) that

18

282

"the efficiency of utilisation of glucose increases with increasing growth rate". The above is intended to point out the importance of a more thorough investigation of growth limitation in continuous cultures achieved by limiting one important source. It is particularly important because under the conditions of limited growth, mutability has been studied. The relation of this phenomen to the physiological state deserves attention.

The assumption that only continuous culture is physiological is a starting point for practical applications in simple fermentative productions, as for instance, the production of yeast. Even with these simple processes, the question of physiological state can play an important role. For instance, in yeast cultivation the question arises of the behaviour of glycolytic activity when aeration is prolonged, or what will happen with maltase activity during prolonged culture on molasses. Similar questions arise as to physiological state when growth takes place on complex media, e. g. wood hydrolysates or sulphite liquors, and the capacity to utilise a mixture of carbohydrates, the role of diauxia, etc.

We have now compiled sufficient proof to show that the question of the physiological state of microorganisms is important for the knowledge of conditions and possibilities of continuous culture. But so far little attention has been given to this question. Utenkov proceeds from it rather systematically by his method, which he has called "microgeneration". He worked out in detail the possibilities of combination of the continuous flow method in various applications (e. g. aerobic culture, the possibility of combining media without interrupting the experiment, and the like) with static methods. His broad experimental work is available only in a brief summary of his doctorate thesis from 1942, where the principles of this method are presented together with a summary of results from 16 years of work (from 1922) concerning the development and variability of 40 different species of microorganisms under continuous flow conditions as compared with static conditions. As mentioned above, all his work is motivated by the endeavour to influence the physiological state of microorganisms through a suitable manipulation of cultivation. Utenkov is convinced that through a suitable choice of cultivation method it would be possible to keep microorganisms constantly at the different stages of their development. The continuous culture method represents for him the method for maintaining constant the stage of active vegetative multiplication and its practical exploitation. He studied with particular attention the first stage of continuous growth immediately after inoculation, as well as the occurrence and significance of atypical bacterial forms: He noticed that growth rate is higher in a continuous culture than in a static one, that the cultures reach a steady state, that under the conditions of continuous cultivation it is possible to preserve certain characteristics of microorganisms (e. g. virulence) which is of importance for the preparation of vaccines and for the possibility of producing long-term (permanent) modifications under changed conditions. He furthermore treated theoretically, and partly experimentally, the question of mutants. In conclusion he stressed the advantage of the method for studying microbial associations, and for practical tasks, because it forms a basis for the application of automatisation of microbial growth etc... Even if these experiments are mostly descriptive and not sufficiently analytical, they represent the basis of continuous cultures, and what is even more important, they always concentrate upon the microorganism itself and on its physiological state and its relation to the environment.

Powell (1957) also touches on the physiological state of microorganisms, and considers the growth rate and generation time of bacteria in relation to the conditions of continuous culture.

The physiological state of yeasts under continuous culture conditions have been taken up by Plevako, who studied the dynamics of subcultures from individual yeast cells (the material of this symposium (1958)). Jerusalimskij and his coworkers are studying it in detail with cultivation of *Cl. acetobutylicum* (1958).

The question of the physiological state of microorganism under continuous flow conditions has formed also the basis of our studies (Málek et al. 1943 und other papers). We assumed that static batch cultures are in antagonism with the dynamics of a multiplying culture, and do not comply with that part of physiology of microorganisms which is manifested by multiplication. From this antagonism arise those complicated population and physiological changes of microorganisms as are known from common batch cultures.

What are the possibilities of studying the physiological state of microorganisms? For this purpose we must bear in mind the analogy with experiences obtained in work with static batch cultures, because only here have those parameters been studied which reflect the physiological state, although care must be taken in making such comparisons. The physiological state manifests itself above all in the qualitative aspect of multiplication, and in the proteosynthesis associated with it. If we permit an analogy with static cultures it is necessary to determine whether the culture and its characteristics correspond to the culture from the first part of the exponential curve, or from a later stage. It appears very convenient therefore to estimate the ribonucleic acid content, because this reflects the most marked changes in this period. The physiological state is also reflected in the qualitative changes in enzymic systems, cell resistance, etc.

As far as the first, purely quantitative, aspect is concerned, it is important to study multiplication and its utilization of living sources. By comparing the data obtained with mathematically predicted results, we can estimate

20

whether the system corresponds to the optimal capacity of vegetative reproduction. The mathematical basis has been studied in a number of papers, and can well be used for the treatment of simple continuous systems.

That is of course possible only in a perfectly homogeneous stirred culture. But the conditions of such a culture need not necessarily meet all the aspects of a continuous culture. Thus, for instance, a common problem that had to be solved by some authors is that even in fully stirred cultures, some bacteria settle preferently on the glass of the culture flask, and grow there. This phenomenon is particularly striking when cultures are grown in small flasks where the only movement is provided for by the nutrient medium dropping into the flask, and by its removal. Under such conditions the culture separates into two distinct parts even at a very fast rate of flow, as we have seen in a long-term culture of *Escherichia coli*; bacteria multiply in the nutrient liquid on the one hand — there they resemble the actively multiplying organisms as to shape and size — and grow in a thick film along the glass of the flask on the other hand — where they have the shape of small rods, multiplying apparently at a slower rate. The former are like those from the first part of the exponential curve, the latter like those from the stationary phase. But of course such a simple type of cultivation cannot provide us with any more exact data than those presented by a mere description.

Therefore we have developed an **experimental method** based on multistage cultivation, where several flasks are connected, one to the other, and the microorganisms pass with the flow of the nutrient medium from one into the other. In a system thus arranged it is possible to study the influence of different degrees of nutrient exhaustion in the medium, as well as the influence of particular metabolites on the change of the physiological state of cultures. The culture then contains several steady states at different levels. By changing the number and size of flasks, and the rate of flow, their ratio can be affected as desired, this makes it possible to study the conditions of origin of some stages of development of microorganism, e. g. sporulation, etc. It also gives us the possibility to study the influence of the physiological state of the culture on the production of various metabolites, such as antibiotics, etc. Finally, by using this method, we can study in detail the influence of various factors on the physiological state of microorganisms by adding them to the second, or some other flask where they are in contact with a fully multiplying culture. This opens new approaches to the knowledge of the adaptive ability to various factors. With actively multiplying cultures we may investigate their capacity to respond more actively to environmental influences, and mutability caused and influenced by external factors. Jerusalimskij presents a mathematical treatment of such a multi-stage system.

To study the physiological state of microorganisms during continuous culture in a multi-stage system we first chose *Azotobacter chroococcum* (Málek (1952),

21

Macura and Kotková (1953)). This microorganism undergoes — when cultivated statically — a complicated form of development from rods to chroococcus bundles, and from these to cystoid forms. In a number of experiments in a glass apparatus it could be observed that with a suitable and sufficient rate of flow of the medium, ensuring fast multiplication, it was possible to maintain it permanently in the form of motile rods with undifferentiated plasma, while at a slower rate of flow the rods were shortened, encysted, and chroococcus forms appeared. Particularly striking was the difference between microorganisms in individual stages. Even if the steady state at the various stages did not differ very much as to the number of microorganisms there were striking differences in shape: in the first stage there is a predominance of actively multiplying rods with undifferentiated plasma, in the second and third stage the chroococcus forms predominate. This difference could be shifted by changing the rate of flow.

Our second model was *Escherichia coli* and *Salmonella enteritidis* and we took the sensitivity towards temperature and towards formaldehyde (Málek et al. 1953) as the criteria of physiological state.

We proceeded from the assumption that microorganisms during the stage of "physiological youth" (Walker, Winslow et al. 1933, 1939) exhibit an increased sensitivity to temperature, H^+ ions, phenol etc. (Sherman and Albus 1923) together with an increased metabolic activity. This sensitivity enables us to distinguish clearly between microorganisms from the beginning of the exponential phase and those from the latter part of this phase, as well as from the stationary phase. It has been shown experimentally that microorganisms cultivated in a continuous culture approach in sensitivity those from the beginning of the exponential phase. They are very much more sensitive than microorganisms obtained by means of a common static cultivation, and their sensitivity decreases in further stages even if it remains well above the level of static cultures. At the same time it was interesting to note that the culture always contained a certain number of resistant microorganisms. These experiments again supported the view that with continuous flow techniques such cultures are developed which in their physiological characteristics resemble microorganisms from the beginning of the exponential phase.

Further studies were carried out using an aerobic sporulating bacillus from the group *B. subtilis* (Ševčík (1952), Málek et al. (1953)). These experiments showed above all, that under the conditions of an aerated continuous culture the microorganisms studied formed more robust rods or fibers that reproduced vegetatively without sporulation. Ševčík observed that the capacity to form these robust rods was preserved by the bacilli even in the first static subculture. When there was a sufficient source of nitrogen present, together with a limited source of carbon (glucose), so that all carbon was utilised during the first stage, during the next stages sporulation, as a rule took place. Aside from

22

sporulation, we have studied the content of RNA and DNA, and the metabolic activity. We observed that during the stage when these robust fibers started to be formed, the content of RNA increased nearly to values at the beginning of the exponential phase in a static culture, and remained there for several days. During the second and third stage, after an analogous initial phase, a decrease could be observed to values lower than in the first stage, and an increase started only when sporulation set in. The metabolic activity, as measured by the comsumption of O_2 per mg of dry weight per hour, did not produce unambiguous results, probably because a non-uniform cellular development takes place. This inequality of individual cells in the population was a new, important finding; even cells in individual fibers behaved differently. There were fully live cells directly adjacent to dying ones. The same could be observed in the setting-in of sporulation in individual cells of chains and fibers. We are of the opinion that this inequality of individual cells in the population during a continuous culture constitutes a fact that must be taken into account much more seriously.

Further work concerning the physiology of microorganisms in continuous fermentation was carried out using the yeast *Saccharomyces cerevisiae*, baker's type. Because these experiments are to be discussed in another report of this symposium (Beran), I shall limit myself only to the conclusions. As is well-known, the growth of this yeast is a manifestation of mixed metabolism. During cultivation a certain amount of alcohol is always formed, qualitatively dependent on the rate of flow. If a high yield of yeast is desired it is necessary to prevent as much as possible the formation of alcohol, which can be done only by limiting the carbohydrate source. Furthermore, the fermentation is connected with certain characteristics which are of importance for the quality of the yeast, e. g. raising power in dough, and durability. We are dealing then with a complicated system, in which the physiological state of the culture is of importance for practical applications.

We carried out our experiments — for practical reasons using molasses — at different rates of flow. We worked most often at a rate corresponding either to a three-hour generation time — which was found to be optimal for our experimental purposes — or a four-hour generation time which is most often used in yeast manufacturing plants.

We tried to find an answer to the following questions: In which physiological phase is a continuous flow culture in respect to its development in a batch culture? To what extent do aerobic systems undergo changes caused by utilization of sugars, ethanol and acetic acid, as well as anaerobic systems responsible for anaerobic fermentation of glucose and maltose? For practical reasons we also studied the autolytic rate of these cultures and the relation of data obtained to the quality of produced yeast in the raising of dough.

On the basis of the RNA content, $Q_{O_2}^{air}$ values on ethanol and acetic acid,

23

and $Q_{CO_2}^{N_2}$ values on maltose, we reached the conclusion that at a flow rate corresponding to a 3—4 hour generation time, we obtain a continuous flow culture corresponding to the terminal part of the exponential phase of the growth curve.

The values of $Q_{O_2}^{air}$ on maltose, glucose, ethanol and acetic acid did not change for the whole duration of the experiment (120 hours) and did not depend on the rate of flow. On the other hand $Q_{CO_2}^{N_2}$ values on glucose were dependent on the rate of flow. At a rate corresponding to the calculated 2,6-hour generation time, the $Q_{CO_2}^{N_2}$ value did not change, at a rate corresponding to a three-hour generation time it fell, but remained at a rather high value. At a four-hour generation time, however, a sharp drop could be observed during the 96 hours of cultivation. We consider it important that for the entire duration of experiments the CO_2 from maltose retains an adaptive character under anaerobic conditions, which fact is in keeping with the physiological state toward the end of the exponential phase of the growth curve in static fermentations. We feel that this phenomenon can be of practical value. The autolytic rate did not exhibit any characteristic changes during incubation. On the other hand we observed a change in shape of the yeast cells, manifested by an elongation without changes in volume, in the course of a short-term continuous cultivation.

From these experiments it can be concluded that yeasts grown in continuous cultures do not change in the course of the cultivation as to their aerobic systems, and are in good physiological condition, corresponding to the terminal stage of the exponential phase of the growth curve.

In orientation experiments we have shown that by a suitable modification of the multi-stage construction, it is possible to reach a continuous formation of such evolutionarily complex metabolites as penicilin (Řičica, Málek 1955).

In conclusion to these experiments, to a certain extent orientational in character, it can be said that the question of physiological state of microorganisms in continuous cultures is a very important one, which deserves considerable attention. With bacteria which undergo, in static cultures, a morphologically demonstrable development (*Azotobacter*, sporulating bacilli) it was shown that even in a continuous culture they can undergo an analogous development when suitable conditions are preserved, particularly at slower rates of flow. The multi-stage continuous culture has proved to be a convenient method not only for the study of the physiological state of continuous flow cultures, but also for other questions. It will be necessary to develop it both technically and experimentally to such perfection as has been reached by the one-stage batch method of the chemostat or turbidostat type.

24

SUMMARY

1. A review of results obtained in developing the continuous culture method is given.

2. From an analysis of the world literature it has been concluded that not enough attention is being given to the question of the physiological state of microorganisms cultivated under the conditions of continuous flow culture, although it is an important question both theoretically and practically.

3. The following important questions arise:

a) Does the continuous flow culture optimally ensure all the necessary conditions of growth and development of microbial cultures?

b) Is it possible to draw a parallel between the physiological state of microorganism in a continuous culture and their development in static cultures?

c) In what range of flow rates does the physiological state of the culture remain constant, and corresponds to the optimal state existing, as a rule, at the beginning of static culture development?

4. We have mentioned possibilities that exist for the study of physiological state in a multi-stage continuous flow culture, and have mentioned results obtained by its application to the study of the development of *Azotobacter* cultures, to the determination of sensitivity toward temperature and disinfectants with *Escherichia coli* and *Salmonella enteritidis*, to the study of aerobic bacillus sporulation, and to the production of penicillin. We have reached the conclusion that the physiological state of continuous flow cultures of bacteria corresponds, under our conditions, to static cultures at the beginning of the exponential phase. We have called attention to the practical importance of the study of the physiological state of baker's yeast grown in continuous culture; in this case cultures have been compared rather with the terminal part of the static exponential curve on the basis of their RNA content and of the analysis of aerobic as well as anaerobic fermentation.

5. It is concluded that it is necessary to pay more attention to the physiological state of continuous flow cultures, particularly from the methodological and experimental point of view.

REFERENCES

ANDERSON P. A., 1953: Automatic Recording of the Growth Rates of Continuously Cultured Microorganisms. J. of Gen. Phys., 36: 733.

BRYSON V., 1952: The Turbidostatic Selector — a Device for Automatic Isolation of Bacterial Variants. Science 116: 48.

BRYSON V., 1953: Applications of the Turbidistat to Microbiological Problems, I: 396.

CASTOR J. G. B., 1947: Apparatus for Continuous Yeast Culture, Science, 166: 23.

CLEARY I. P., BEARD P. J., CLIFTON C. E., 1935: Studies of Certain Factors Influencing the Size of Bacterial Populations. J. of Bact., 29: 205.

25

DAVIES A., 1956: Invertase Formation in *Saccharomyces fragilis*. J. gen. Micr., 14: 109.

DENBIGH K. G., 1944: Velocity and Yield in continuous Reaction Systems. Trans. Faraday Soc., 40: 352.

DUCHÉ I., NEU J., 1950: Cultures des microorganismes en milieu continu. Acad. des Sciences.

ELSWORTH R., MEAKIN L. R. P., PIRT S. J., CAPELL G. H., 1956: A Two Litre Scale Continuous Culture Apparatus for Microorganisms. J. of Applied Bact., 19: 264.

Feinstbelüftungsverfahren für die Gärungsindustrie. Lupinges. Mannsheim, Pat. L. 10, No 10.

FORMAL S. B., BARON L. S., SPILMAN W., 1956: The Virulence and Immunogenicity of *Salm. typhosa* Grown in Continuous Culture. J. of Bact., 72: 168.

GERHARDT P., 1946: *Brucella suis* in Aerated Broth Cultures. Continuous Culture Studies. J. of Bact., 52: 283.

GRAZIOSI F., 1956: Metodo per la coltura continua dei batteri mediante un apparato turbidistatico. Giorn. di Microb., 1: 491.

GRAZIOSI F., 1957: Studie quantitative dell'adattamento di *Micrococcus pyogenes* e *Proteus vulgaris alla novobiocina*. Rend. dell'Ac. Naz dei Lincei ser. VIII, vol. XXII: 3.

HARRIS E. E., SAEMAN I. P., MARQUARETT R. R., HANNAN M. L., ROGERS S. C., 1948: Fermentation of Wood Hydrolysates by *Torula utilis*. Ind. Eng. Chem., 40: 1216.

HARRIS T. N., 1956: Growth of Group a *Hemolytic Streptococcus* in the Steady State, J. of Bact., 72: 283.

HÉDÉN C. G., HOLME T., MALMGREN B., 1955: An Improved Method for the Cultivation of the Microorganisms by the Continuous Technique. Acta path. et micr. Scand., 37: 42.

HERBERT D., ELSWORTH R., TELLING R. C., 1956: The Continuous Culture of Bacteria; a Theoretical and Experimental Study. J. gen. Micr., 14: 601.

HOFSTEN B., HOFSTEN A., FRIES N., 1953: The Technique of Continuous Culture as Applied to a Fungus *Ophiostoma multiannulatum*. I: 406.

JERUSALIMSKIJ N. D., 1952: Fiziologija razvitija čistych bakterialnych kultur (Autoreferate) Moskva.

JERUSALIMSKIJ N. D., RUKINA E. A., 1956: Issledovanije uslovij sporoobrazovanija u masljanokislych bakterij s pomoščiu kolloidnych gilľz. Mikrob. (Russian) 25: 649.

JORDAN R. C., JACOBS S. E., 1944: The Growth of Bacteria with a Constant Supply. I. Preliminary Observations of Bact. coli. J. of Bact., 48: 579. II. The Effect of Temperature .... J. Gen. Micr., 1: 121, 1947. III. The Effect of pH at Different Temperatures ... J. Gen. Micr., 2: 15, 1948.

KARUSH F., IACOCCA V. F., HARRIS T. N., 1956: Growth of Group *A-hemolytic Streptococcus* in the Steady State. J. of Bact., 72: 283.

KUBITSCHEK H. E., 1954: Modifications of the Chemostat. J. of Bact., 67: 254.

LEE H. H., 1953: The Mutation of *E. coli* to Resistance to Bacteriophage Tb. Arch. Bioch. et Biophys., 47: 438., cit. dle Novicka 1955.

MACURA J., KOTKOVÁ M., 1953: Vývoj azotobaktera v proudícím prostředí. Čsl. biol., 2: 41.

MÁLEK I., 1943: Pěstování mikrobů v proudícím prostředí, Čas. lék. čes., 82: 576.

MÁLEK I., 1950: Množení *E. coli* v proudícím prostředí: Biol. listy 31: 93.

MÁLEK I., 1952: Kultivace bakterií ve vícestupňovém proudícím prostředí. Čsl. biol., I: 18.

MÁLEK I., 1952: Kultivace azotobaktera v proudícím prostředí. Čsl. biol. I: 91.

MÁLEK I., 1953: Sporulation of Bacilli, VI. congresso internazionale di microbiologia, Roma, 1: 345.

MÁLEK I., 1955: O množení a pěstování mikroorganismů zvláště bakterií. NČSAV, Praha.

26

MÁLEK I., 1956: Protočnyj metod razmnoženija mikrobov. Mikrobiologija (Rus.), 25: 659.

MÁLEK I., BURGER M., HEJMOVÁ L., ŘIČICA J., FENCL Z., BERAN K., 1957: Asimilace cukrů a kyselin při kontinuálním zdrožďování neředěných sulf. výluhů. Čsl. mikrob., 2: 203.

MÁLEK I., VOSYKOVÁ L. a spol., 1953: Resistence bakterií pěstovaných v proudícím prostředí. Čsl. biol. 2: 68.

MÁLEK I., CHALOUPKA J., VOSYKOVÁ L., 1953: Sporulace bacilů, Čsl. biol. 2: 323.

MÁLEK I., VOSYKOVÁ L., 1954: Průtoková kultivace drožďárenských kvasinek. Čsl. biol. 3: 261.

MALMGREN B., HÉDÉN C. G., 1947: Studies of the Nucleotide Metabolism of Bacteria I—X, Acta Path. et Micr. Scand., XXIV, 412.

MALMGREN B., HÉDÉN C. G., 1952: Studies on the Cultivation of Micro-organisms on a Semi-industrial Scale. I. General Aspects of the Problem. Acta path. microb. Scand., 30: 223.

MAXON W. D., 1955: Continuous Fermentation. A Discussion of its Principles and Application. J. of Appl. Microbiol., 3: 110.

MONOD J., 1950: La technique de culture continue. Théorie et applications. Ann. Inst. Past., 79: 390.

MOYER H. V., 1929: A Continuous Method of Culturing Bacteria for Chemical Study. J. of Bact. 18: 59.

NORTHROP J. H., 1954: Apparatus for Maintaining Bacterial Cultures in the Steady State. J. Gen. Phys., 38: 105.

NOVICK A., 1955: Growth of Bacteria, Ann. Rev. of Microbiol., 9: 97.

NOVICK A., SZILARD L., 1950: Experiment with the Chemostat on Spontaneous Mutation of Bacteria. Proc. Nat. Acad. Sci., Wash., 36: 708.

NOVICK A., SZILARD L., 1951: Genetic Mechanisms in Bacterial Viruses. Experiments on Spontaneous and Chemically Induced Mutations of Bacteria Growing in Chemostat. Cold Spring Harbour Symposia Quant. Biol., 16: 337.

NOVOGRUDSKIJ D. M., 1950: K voprosu o vnutrividovych i mežvidovych vzaimootnošenijach počvennych mikroorganizmov. Agrobiologija 48: 5.

PASYNSKIJ A. G., 1957: Teorija otkrytych sistem i jejo značenije dlja biochemii. Usp. Sovr. Biol., 43: 263.

Patent Norddeutsche Hefeind. St. 48881, Kl. 66 16/01, 1934.

PERRET C. J., 1957: An Apparatus for the Continuous Culture of Bacteria at Constant Population Density. J. gen. Micr., 16: 250.

PIRFILEV, Leningrad: Personal Information.

PIRT S. J., 1957: The Oxygen Requirement of Growing Cultures of a Bacterium Species Determined by Means of the Continuous Culture Technique. J. gen. Microb., 16: 59.

POWELL E. O., 1956: Growth Rate and Generation Time of Bacteria, with Special Reference to Continuous Culture. J. Gen. Microb., 15: 492.

ROGERS L. A., WHITTIER E. O., 1930: The Growth of Bacteria in Continuous Flow of Broth. J. of Bact., 20: 127.

ROSENBERG M., 1956: Dynamics of the Breaking-down of Lysogenic Cells Irradiated by Ultra-violet Light. Fol. biol., Praha 2: 206.

ŠARKOV V. I., 1950: Gidroliznoje proizvodstvo. III. Goslesbumizdat.

SEVAGE M. C., Florey H. W., 1950: Induced Bacterial Antagonism. Brit. J. of Exp. Path., 31: 17.

ŠEVČÍK VL., 1952: Tvorba velkých forem u Bac. subtilis v proudícím prostředí. Čsl. biol., 1: 93.

SHERMANN I. M., ALBUS W. R., 1924: The Function of Lag in Bacterial Cultures. J. of Bact., 9: 303.

27

ŠVACHULOVÁ J., KUŠKA J., 1956: Studium metabolismu mykobakterií v proudícím prostředí. Rozhl. v tb., 16 : 488—491.

TAMYIA H., 1957: Mass Culture of Algae. Ann. Rev. of Pl. Phys., 8: 309.

UNGER E. D., STARK W. H., SCALF R. E., KOLACHOV P. J., 1942: Continuous Aerobic Process for Distiller's Yeast. Ind. Eng. Chem., 34: 1402.

UTENKOV M. D., 1941: Mikrogenerirovanie. Gos. izdat. ,,Sov. nauka'', Moscow.

VÁLYI-NAGY, T., CSOBÁN G., ZABOS P., 1954: Effect of Penicillin on the Nucleid Acid Metabolism of *Staphylococcus aureus*. Acta Microbiol. Acad. Sci. Hung, 2: 79—89.

VÁVRA J., 1958: Zařízení pro průtokovou kultivaci prvoků. Čsl. biol., In press.

VERBINA N. M., 1955: Priučenije drožžej k antiseptikam različnymi metodami. Trudy Inst. mikrob., IV: 54.

VICTORERO F. A., 1948: Apparatus for Continuous Fermentation. US Patent, 2, 450, 218.

VOLODIN A. P., 1952: K voprosu o prirode bakterialnych kolonij, Agrobiologija 2: 138.

WINSLOW C. E. A., WALKER H. H., 1939: The Earlier Phases of the Bacterial Cultures Cycle. Bact. Rev., 3: 147.

ZELLE M. R., 1955. Genetics of Microorganisms. Ann. Rev. Microbiol., 9: 1—20.

28

18

Reprinted from *Microbial Physiology and Continuous Culture* (Proc. 3rd Intern. Symp.), Her Majesty's Stationery Office, London, 1967, pp. 196–208

PROPERTIES OF MICROORGANISMS GROWN IN EXCESS OF THE SUBSTRATE AT DIFFERENT DILUTION RATES IN CONTINUOUS MULTISTREAM CULTURE SYSTEMS

ŘIČICA, J., NEČINOVÁ, S., STEJSKALOVÁ, E. AND FENCL, Z.

Institute of Microbiology, Czechoslovak Academy of Science, Prague

According to classical theory, the growth rate in the chemostat is controlled by the concentration of a certain limiting factor. In most continuous processes the microbial population is considerably limited and changes in the physiological state of the culture can be followed to a limited extent corresponding to the final stages of the batch growth curve. Not only the formation of products, but also many physiological processes of importance to subsequent phases of development of the culture occur under conditions where growth and multiplication are not limited by any substrate. In continuous cultivation such conditions can exist only when the dilution rate is so high that the growth rate approaches the maximum value. The use of such high flow rates makes it very difficult to maintain a steady-state in a single stage chemostat and it is impossible to exceed the critical dilution rate. In our opinion this disadvantage can easily be avoided by using a two-stage system with the addition of fresh medium to the second stage. As the rate of inflow of fresh substrate can be changed from very low values corresponding almost to zero up to very high values, the whole growth curve can be reproduced including the phase preceding the exponential phase. When compared with the turbidostat, the main advantage of this system is that both the dilution rate and the composition of the medium added to the second stage can be chosen at will. Thus changes can be made in the concentration of all components of the culture and the properties of the cells in the second stage can be compared with those in the first stage. Since the second stage is continuously seeded with cells from the first stage, all steady states are stable and there is no washing out of the culture; at most, it is diluted.

As early as 1958 at the first Symposium on Continuous Cultivation of Microörganisms held in Prague, Novick (1958) suggested the two-stage system with inflow of fresh substrate to the second stage as an interesting one from a theoretical point of view. Further theoretical development of the system was made by Herbert (1960) at the Symposium in London. However, insufficient experimental data were available for further theoretical considerations and for deciding on the advantages and disadvantages of this system, both for laboratory and industrial application.

MATERIALS AND METHODS

Organism. In all the experiments *Escherichia coli* B was used. Stock cultures were kept on agar slants at 4°C and subcultured every fortnight.

Media. For continuous culture the following media were employed:—

(1) Complex medium—(g/l) Bacto peptone (Difco), 100; Beef extract (Difco), 5·0; NaCl 50·0

(2) Mineral medium M56 (Monod *et al.* 1951)—(g/l) lactose 2·0; KH_2PO_4, 13·6; $(NH_4)_2SO_4$, 2·0; $MgSO_4.7H_2O$, 0·2; $CaCl_2$, 0·01; $FeSO_4.7H_2O$, 0·0005; pH 7·0.

Apparatus. Continuous cultures were run in laboratory fermentors with 1 litre working volumes; by connecting two fermenters in series a two-stage system was formed. The cultures were aerated and stirred. The rate of oxygen transfer was 80–100 millimoles $O_2/l./hr.$, temperature 37°C. Peristaltic pumps were used to add media to the culture vessels and to transfer culture from the first to the second stage.

Dry weight of bacteria. The concentration of bacteria by mass was determined by the following methods:

(i) By weighing the washed and dried cells from a suspension;

(ii) By measurement of optical density.

Dry mass concentration is given in mg./ml.

Protein. Protein in dried bacteria was estimated by the method of Lowry et al. (1951).

RNA and DNA. The determination of both nucleic acids was made by the method of Ogur and Rosen (1950), modified by Škoda et al. (1964).

Enzyme activity. The ability of the cells to form the enzyme β-galactosidase was followed directly in the sample taken from the cultivation vessel and with cells from the same sample washed and suspended in the mineral medium without the nitrogen source. 0·2% maltose solution was used for the carbon source. Enzyme formation was induced by shaking with 5 × 10^{-3}M thio-methyl-β-D-galactopyranoside (TMG) at 37°C. Samples were taken at regular short intervals before and after the addition of inducer. Synthesis of the enzyme was stopped immediately by diluting the sample into ice-cold phosphate buffer containing chloramphenicol (CAP). The final concentration of CAP was 100μg./ml. The samples were then shaken with toluene for 30 minutes at 30°C.

The formation of β-galactosidase in continuous cultures grown in mineral medium was induced directly by lactose during the cultivation, lactose being simultaneously inducer, carbon source and limiting substrate. Further synthesis of enzyme was stopped by means of CAP immediately after taking the sample. The activity of β-galactosidase was determined after washing and treating with toluene.

Determination of the specific activity of β-galactosidase was performed by the method of Lederberg (1950). As substrate for the enzyme o-nitro-phenyl-β-D-galactopyranoside (ONPG) was used and the yellow colour of the solution measured on a colorimeter at 420mμ.

THEORETICAL CONSIDERATIONS

According to the mathematical theory applied by Novick (1958) and Herbert (1960) for the characterisation of the steady state in the second stage with addition of fresh substrate, the concentration of microorganisms is given by the equation:

$$X_2 = \frac{F_1 X_1 / V_2}{D_2 - \mu_2},$$

where $F_1 X_1$ is the amount of organisms entering from the preceding stage, V_2 is the volume of the culture in the second stage, μ_2 is the specific growth rate and D_2 is the dilution rate. From this equation the specific growth rate μ_2 can be calculated as follows:

$$\mu_2 = \frac{D_2 X_2 - F_1 X_1 / V_2}{X_2}.$$

If the value of μ_2 is obtained under conditions of unrestricted growth, it represents the real value of the maximum specific growth rate μ_{max} which can be reached by the organism in a given medium.

Similarly the optimum dilution rate can be found at which the specific rate of synthesis of a certain product, i.e. the culture activity, reaches its maximum. From the equation derived for the concentration of the product P_2, supposing that a certain amount of product P_1 enters from the preceding stage:

$$P_2 = \frac{F_1 P_1 / V_2 - k_2 X_2}{D_2}.$$

The specific rate of product formation can be calculated:

$$k_2 = \frac{D_2 P_2 - F_1 P_1 / V_2}{X_2}.$$

From the above equations the dependence of the rate coefficients μ_2 and k_2 on the dilution rate D_2 can be seen. However, for a given dilution rate D_2 there is a variety of possible ratios of the individual flow rates to the two stages; thus, although $(F_1 + F_2)$ is constant, the ratio of $F_1 : F_2$ can be different and therefore all the components of the culture (cells, substrate and products) will also vary. So it is necessary to consider not only the quantitative but also the qualitative interrelation of the components in that the physiological state of the culture is influenced by the quality of the cells entering the second stage. For example, when using different rates in the first stage, the concentration of cells X_1 and product P_1 need not vary, but the physiological properties of the cells can be quite distinct. This fact is not generally taken into account in mathematical equations for mass balance. It is why we have followed not only the changes in the physiological state of the culture with different dilution rates in the second stage, but also its dependence on changes in the properties of the cells entering from the preceding stage.

The arrangement of the apparatus, as shown diagrammatically in Fig. 1, allowed not only maintenance of the quantitative ratios but also changes to be made in the properties of the cells in the first stage. The rubber tubing connecting the first and second stages passes through the peristaltic metering pump which maintains a constant flow of culture F_1 into the second stage. Using a low dilution rate in the first stage, the amount of medium F_0 entering the first stage was equal to F_1; at higher dilution rates F_0 was greater than F_1 and the excess culture flowed out from the first stage through the constant level overflow device. In both cases the quantity of organisms entering the second stage was $F_1 X_1 / V_2$ but, although it was quantitatively the same, the cells had different properties. Changes in the dilution rate in the second stage were made by altering the rate of medium addition F_2 to the second stage.

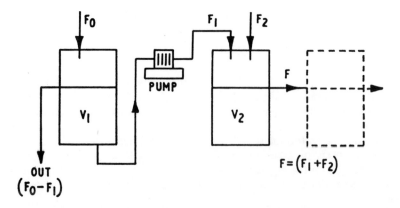

Fig. 1. *Diagram of multistage multistream culture system.* V_1, V_2 = culture volumes in first and second stages; F_0, F_2 = flow rates of fresh medium fed to first and second stages respectively (ml/hr.); F_1 = flow rate of culture from first stage fed to second stage by peristaltic pump (ml/hr.); F = overall flow rate ($F_1 + F_2$); D_2 = dilution rate (F/V_2).

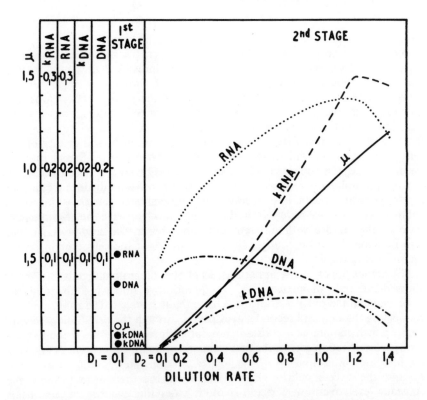

Fig. 2. *Changes in the physiological properties of* Escherichia coli *with various dilution rates in the second stage* (D_2) *of the two-stage multistream system.* Medium in both stages nutrient broth; $D_1 = 0.1$ hr.$^{-1}$, $D_2 = 0.1 - 1.4$ hr.$^{-1}$; $F_1/V_2 = 0.1$; μ = specific growth rate; k_{RNA} = specific rate of RNA formation; RNA = mg.RNA/mg. dry weight; k_{DNA} = specific rate of DNA formation; DNA = mg.DNA/mg. dry weight.

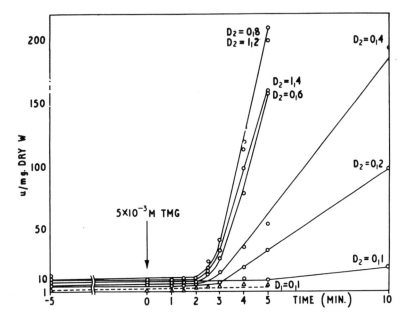

Fig. 3. *Specific activity of* β-*galactosidase of* E. coli B *grown at various dilution rates in the second stage of the two-stage multistream system.* Inducer (TMG, final concentration 5×10^{-3}M) was added directly into the sample taken from the culture vessel. U/mg. dry wt. = specific activity of enzyme.

RESULTS

In the first series of experiments, nutrient broth was the medium supplied to both stages. In the first stage the dilution rate D_1 was $0.1 \mathrm{hr^{-1}}$ and thus the volume ratio of the culture ($F_1 = 100 \mathrm{ml/hr}$) flowing into the second stage ($V_2 = 1000$ ml) was set at 0.1. In the second stage the dilution rate D_2 was changed, by degrees, within the range $0.1—1.4 \mathrm{hr^{-1}}$. Results are given in Fig. 2. In the first stage the values of the specific growth rate μ and the rate of nucleic acid formation k_{RNA}, k_{DNA}, as well as the content of nucleic acids (mg. RNA, DNA/mg. dry wt.) were very low. In the second stage, all values increased with increasing dilution rate D_2 and at a dilution rate of $1.2 \mathrm{hr^{-1}}$ RNA,. k_{DNA} and k_{RNA} reached their maxima; but the highest DNA content occurred within the much lower range $0.2—0.6 \mathrm{hr^{-1}}$. The specific growth rate μ would reach its maximum only at much higher dilution rates.

The ability of the cells to form β-galactosidase was also dependent on changes of D_2 (Fig. 3). Whereas the cells from the first stage had almost no ability to form enzyme, in cells taken from the second stage, this ability increased with increasing dilution rate. Maximum β-galactosidase activity expressed as values that were reached 5 minutes after the addition of inducer (final concentration 5×10^{-3} M TMG) was found within the range $D_2 = 0.8$-$1.2 \mathrm{hr^{-1}}$ (Fig. 4). The time taken for the first appearance of activity shortened as well, until a constant value of 2.5 minutes was reached.

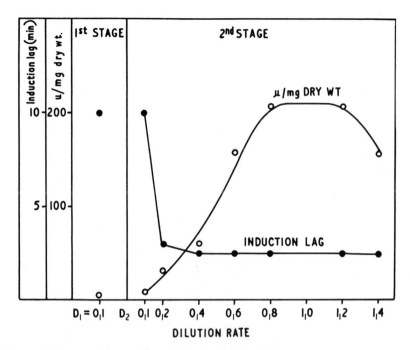

Fig. 4. *The influence of dilution rate upon specific activity and induction lag of the enzyme β-galactosidase* (From Fig. 3). U/mg. dry wt = the values for specific activity 5 minutes after the addition of inducer.

In order to follow the extent to which the properties of the cells from the first stage influenced the properties of the cells in the second stage, in the following experiments we increased the dilution rate of the first stage to $D_1 = 0.8\text{hr}^{-1}$. We used this dilution rate because we had found in previous experiments that the maximum enzymic activity was reached at $D_2 = 0.8\text{hr}^{-1}$ (see Fig. 4). At this increased dilution rate the concentration of cells in the first stage remained approximately the same as that at $D_1 = 0.1\text{hr}^{-1}$. The rate of flow of culture into the second stage was adjusted to keep the quantitative ratio of 0.1, and consequently F_1X_1/V_2, the same. With the increased dilution rate in the first stage, all rate coefficients (μ, k_{RNA} and k_{DNA}) and the RNA content were considerably higher than at $D_1 = 0.1\text{hr}^{-1}$ although the DNA content remained practically the same (Fig. 5). Also, in the second stage, higher values were reached and a high concentration of RNA was reached much sooner than in the previous experiment. The difference was also obvious with μ_2 and k_{RNA}, the values of which were very high, but even at the dilution rate $D_2 = 1.8\text{hr}^{-1}$ their maxima were not reached. Nevertheless, it is probable that in this case, RNA, k_{RNA} and μ_2 have their maxima at a certain higher and identical D_2. However, the influence of this increased dilution rate on the physiological state of the culture in the first stage and on the properties of the cells in the second stage is evident from their ability to form β-galactosidase (Fig. 6). Even in the first stage, the rate of enzyme synthesis after induction was rather high and equal to the rate attained in the second stage at $D_2 = 0.8\text{hr}^{-1}$. With increasing

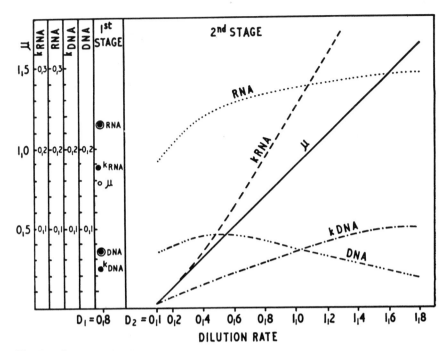

Fig. 5. *Changes in the physiological properties of* E. coli B *with various dilution rates in the second stage* (D_2) *of the two-stage multistream system.* Medium in both stages, nutrient broth; $D_1 = 0.8$hr.$^{-1}$, $D_2 = 0.1 - 1.8$ hr.$^{-1}$. $F_1/V_2 = 0.1$.

dilution rate in the second stage the specific activity of β-galactosidase increased and the induction lag shortened again to 2·5 minutes (Fig. 7). The specific activity of the enzyme after 5 minutes induction was several times higher than it was with a low dilution rate in the first stage. It is worth noting that the activity and rate of enzyme synthesis were ten times greater than in batch culture.

The rate of induced enzyme synthesis in washed cells suspended in mineral medium (M-56 without nitrogen source) was almost the same at all dilution rates (Fig. 8). The values of specific enzyme activity were influenced by the variable time-lag of induction. The curve of these values passes through a maximum that was reached at $D_2 = 0.8$hr^{-1} i.e. the same dilution rate as that used in the first stage (Fig. 9).

If the higher dilution rate ($D_1 = 0.8$hr.$^{-1}$) was applied to the first stage while the second stage was fed with medium containing lactose, the shape of some of the curves altered (Fig. 10). The RNA and DNA content reached the maximum at lower dilution rates, whereas their rates of synthesis k_{RNA} and k_{DNA} reached the maximum at higher dilution rates. The curves of k_{RNA} and μ were similar and both reached their maxima at the same dilution rate.

As can be seen from the curves, conditions can be chosen under which, for example, either the content of RNA or DNA, or the rate of nucleic acid synthesis (k_{RNA} or k_{DNA}) would be maximum. Using this culture

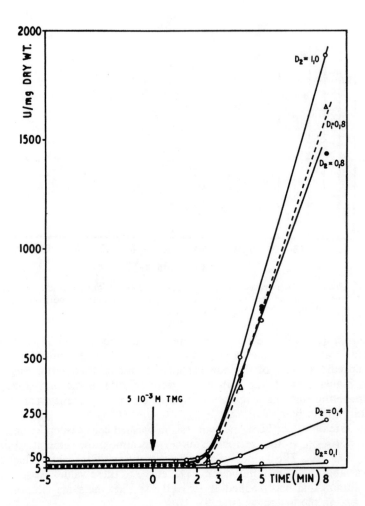

Fig. 6. *Specific activity of β-galactosidase of* E. coli *B in the second stage.* $D_1 = 0.8$ hr^{-1}, $D_2 = 0.1 - 1.0$ hr^{-1}; Medium, nutrient broth in both stages. Inducer (TMG, final concentration 5×10^{-3}M) was added directly into the sample taken from the culture vessel.

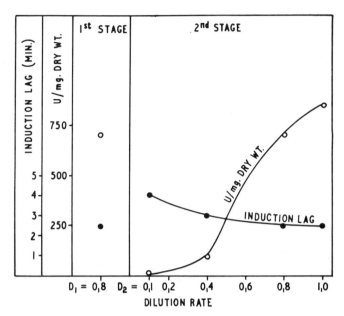

Fig. 7. *The influence of dilution rate upon the specific activity and induction lag of the enzyme* β-*galactosidase.* (From Figs. 5 & 6). U/mg. dry wt. = the values of the specific activity 5 minutes after the addition of inducer.

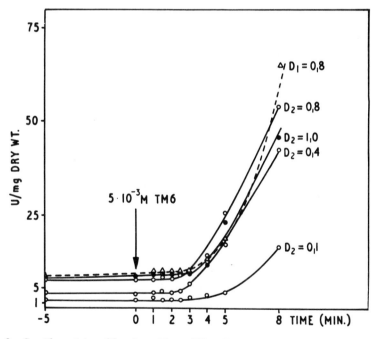

Fig. 8. *Specific activity of* β-*galactosidase of* E. coli *B grown at various dilution rates in the second stage of the two-stage multistream system.* (Culture conditions as Figs. 5 & 6). Before the addition of inducer (TMG, final concentration 5×10^{-3}M) the cells were washed and suspended in mineral medium without nitrogen and containing 0·2% maltose.

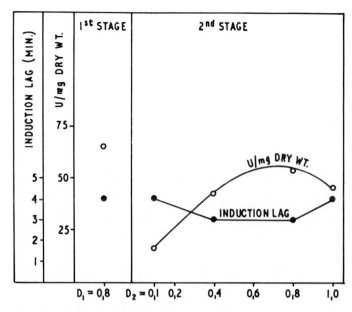

Fig. 9. *The influence of dilution rate upon the specific activity and induction lag of the enzyme* β-*galactosidase.* (From Fig. 8). U/mg. dry wt. = the values of the specific activity 5 minutes after the addition of inducer.

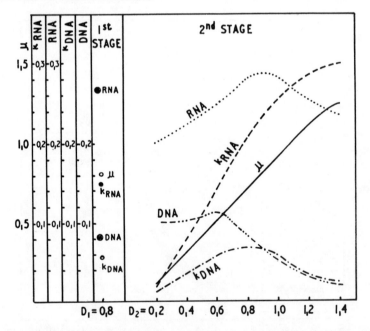

Fig. 10. *Changes in the physiological properties of* E. coli B *with various dilution rates in the second stage of the two-stage multistream system.* $D_1 = 0.8$ hr^{-1}; medium, nutrient broth. $D_2 = 0.1 - 1.4$ hr^{-1}; medium, mineral salts and 2% lactose. $F_1/V_2 = 0.1$; μ = specific growth rate; k_{RNA} = specific rate of RNA synthesis; RNA = mg.RNA/mg. dry wt.; kDNA = specific rate of DNA synthesis; DNA = mg. DNA/mg. dry wt.

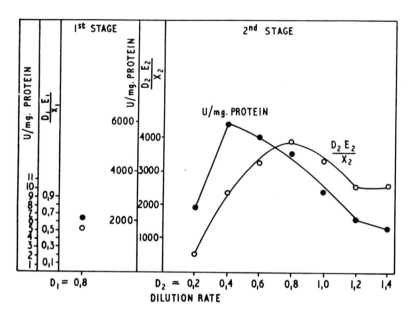

Fig. 11. *The influence of dilution rate upon the specific activity and rate of synthesis of the enzyme β-galactosidase.* $D_1 = 0·8 \text{ hr}^{-1}$; medium, nutrient broth; $D_2 = 0·1 - 1·4 \text{ hr}^{-1}$; medium, mineral salts and 2% lactose. The enzyme was induced directly by the lactose during growth. Specific activity of enzyme $= E/X$, (E = units of enzyme, X = mg. protein). Specific rate of enzyme synthesis $= D_1E_1/X_1$ and D_2E_2/X_2 in 1st and 2nd stages respectively since the amount of enzyme entering from the first stage is negligible.

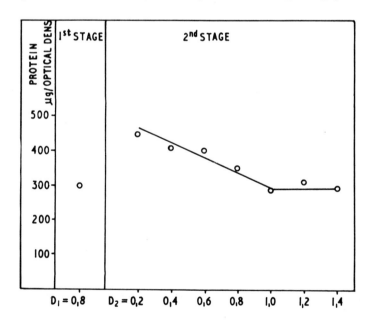

Fig. 12. *The influence of the dilution rate in the second stage upon the protein content per optical density unit of culture.* Culture conditions as for Figs. 10 and 11.

system a certain "partition" of the culture properties could be reached at suitable dilution rates in the second stage; these properties, and eventually their changes, caused by various factors, can be continuously compared with the values kept constant in the first stage. It is supposed that from $D_2 = 1\cdot4\mathrm{hr}^{-1}$ upwards, the rate coefficients k_{RNA} and k_{DNA} and μ as well as the RNA and DNA contents should be constant and that both the synthesis of RNA and the growth rate of organisms should be maximum. We are of the opinion that only under such conditions would the unrestricted growth of the culture be possible.

In the mineral medium the synthesis of β-galactosidase was induced by the lactose in the growth medium and the specific activity of the enzyme/mg. protein reached its maximum at $D_2 = 0\cdot4\mathrm{hr}^{-1}$. Further increase in this dilution rate decreased the ratio between the volume (F_1) of nutrient broth-grown culture entering the second stage and the increasing volume (F_2) of mineral medium also supplied to the second stage. This produced a parallel decrease in the specific activity of the enzyme (Fig. 11). In other words the proportion of organic nitrogen decreased and that of the inorganic increased. The maximum rate of enzyme synthesis was reached at $D_2 = 0\cdot8\mathrm{hr}^{-1}$.

With increasing dilution rate the protein content (mg. protein/optical density unit of the culture) decreased to a constant value as shown in Fig. 12. Even at high flow rates, this value was never lower than that in the first stage.

DISCUSSION

The results presented show that in the second stage the dilution rate, or more properly the increasing concentration of the substrate, had a considerable influence upon the physiological state of the microörganisms. Likewise, the physiological state of the culture in the first stage affected the properties of the cells growing in the second stage. By combining these two effects and by suitable composition of the medium the "partition" of the culture properties can be achieved. In our opinion this fact is very important from the experimental point of view, as conditions can be chosen that are optimum for a certain property. The possibility of continuously comparing the physiological state of the culture in the first and second stages and of following the influence of various factors is also of advantage. In evaluating the physiological state of the microörganisms a comparison should be made of several properties of the culture as the values referring to a single property might give an insufficient characterisation of the state.

If in the second stage a certain desirable physiological state is attained then it would be necessary to keep the first stage in a very similar physiological state. Only in this way would it be possible to attain the conditions under which the culture could be maintained in unrestricted or balanced growth for a long time. We suppose that the dilution rate D_2 at which both growth and synthesis of RNA reach their maximum should provide conditions for real balanced or unrestricted growth.

Unlike batch culture populations, populations in the steady-state, and especially those growing at higher dilution rates, are metabolically homogeneous. The cells are uniform in their life history since they originated under constant environmental conditions and prepared their progeny for

Q

the same conditions, and the rates of metabolic processes are increased. This is why in continuous cultures much higher enzyme activities, higher rates of product synthesis and more positive genetic response etc. are achieved. We are of the opinion that a multi-stage system with inflow of suitable fresh medium into the second stage can be very advantageous for studying the kinetics of biosynthesis of a given cell component (enzyme, metabolite, by-product or end-product) as well as for studying genetic problems such as the transfer of information.

REFERENCES

HERBERT, D. (1960). A theoretical analysis of continuous culture systems. In *Continuous Culture of Microorganisms*. Monograph No. 12. London: The Society of Chemical Industry.

HERBERT, D., (1961). The chemical composition of microorganisms as a function of their environment. In *Microbial Reaction to Environment*. *Symp. Soc. gen. Microbiol.*, **11**, 391.

LEDERBERG, J. (1950). The β-D-galactosidase of *Escherichia coli* strain K 12. *J. Bact.* **60**, 381.

LOWRY, M. O., ROSENBROUGH, N. J., FARR, A. L., & RANDALL, R. J. (1951). Protein measurement with the Folin phenol reagent. *J. biol. Chem.*, **193**, 265.

MONOD, J., COHEN-BAZIRE, G., COHN, M. (1951). Sur la biosynthèse de la β-galactosidase (lactase) chez *Escherichia coli*. La specificité de l'induction. *Biochim. biophys. Acta*, **7**, 585.

MÜLLER-HILL, B., RICKENBERG, H. V., WALLENFELS, K. (1964). Specificity of induction of the enzymes of the *lac*-operon in *Escherichia coli*. *J. molec. Biol.*, **10**, 303.

NOVICK, A. (Rev. BERAN, K.) (1958). Discussion part of the Symposium on the Continuous cultivation of microorganisms held in Praha in June 1958. *Folia microbiol, Praha*, **4**, 390.

OGUR, M., & ROSEN, G. (1150). The nucleic acids of plant tissues. I. The extraction and estimation of desoxipentose nucleic acid and pentose nucleic acid. *Archs Biochem.*, **25**, 262.

ŠKODA, J., ČIHÁK, A., ŠORM, F. (1964). Inhibition of the pyrimidine pathway by 5-azauracil, N-formylbiuret and its combination with 6-azauridine in Ehrlich ascites-bearing mice. *Coll. Czech. chem. Communs.*, **29**, 2389.

Editor's Comments on Papers 19 and 20

19 Powell et al.: *The Analogy Between Batch and Continuous Culture*

20 Elsden et al.: *Balanced and Restricted Growth*

In Paper 19 the editors of the 3rd (Porton) symposium on continuous culture comment upon the interrelationships of batch and continuous cultures and the physiological state. The discussion centers on asynchronous cultures and excludes any considerations of synchrony.

On the plateau of continuous growth studies, there is presently an area where confusion and controversy exist because of the different meanings and concepts that are conveyed by the use of such terms as "balanced" and "unbalanced" or "restricted" and "unrestricted" growth. An attempt to clear away this tangle arose at the 3rd (Porton) symposium of continuous culture and the remarks by the chairman and the editors of that meeting (Paper 20) are pertinent observations to a problem that concerns all studies of microbial growth.

The comments of the chairman and editors of the Porton symposium of continuous culture should be considered together with those of Herbert (Paper 14, p. 211), of Padilla and James (Paper 24, pp. 356 – 358), of Bu'lock (Paper 9, p. 141), and of the Copenhagen School (Paper 7, pp. 120 – 121).

It would appear that balanced growth is strictly a condition related to multiplication and common to open systems—whether synchronous or asynchronous—but alien to closed ones.

19

The Analogy Between Batch and Continuous Culture

E. O. POWELL, C. G. T. EVANS, R. E. STRANGE, and D. W. TEMPEST

EDITORIAL COMMENT

There is a certain community of thought among several of the authors contributing to this Symposium: the metabolic activity of a culture is being recognized as a measurable expression of its physiological state. Thus Ierusalimsky's discussion of the growth rate as depending on the RNA content of the cells and on the substrate concentration is a special case of Powell's dissection into 'potential metabolic activity' and a simple kinetic term. The two authors independently reach the same conclusion as to the applicability of the Michaelis-Menten equation. It is all the more important that work and thought of this kind should not be confused by the acceptance of naïve ideas belonging to an earlier stage of development.

The phases of a batch culture, described in so many textbooks, were readily recognized as presenting to us a spectrum of physiological states arranged in a temporal order. But then it came to be thought that some if not all of these states could be imitated and maintained permanently in the steady states of a simple chemostat at different dilution rates; and this is wrong, in general. There are two forms of continuous culture which do offer an exact parallel with batch culture. First, the turbidostat. It is possible, with attention to the supply of nutrients (including oxygen if necessary) to maintain the exponential phase of a batch culture for about 30 generations, during which the concentration of organisms increases 10^9- to 10^{10}-fold. In at least the later stages of this period we can be confident that growth is balanced and unrestricted. A population in the same physiological state, balanced and unrestricted, can be maintained indefinitely in a turbidostat if the population density is low; the growth rate is μ_{max}, the same as obtains in the batch culture during the exponential phase. Second, the ideal tubular fermenter. Herbert's (1961, in *Continuous Culture of Microörganisms*. Monograph No. 12. London: Society of Chemical Industry) discussion shows that the ideal tubular fermenter with feed-back can present the same spectrum of physiological states as does a batch culture; these states are arranged in a spatial and not a temporal order. The system as a whole is in a steady state, but at most points the organisms are not in balanced growth. The ideal tubular fermenter is difficult to realise in practice, but it can be approximated by a chain of simple stirred fermenters. (Powell & Lowe, 1964, in *Continuous Cultivation of Microörganisms*. Ed. Málek, Beran & Hospodka. Prague: N.Č.S.A.V.).

Apart from these two special examples, no exact analogy exists, and any supposed approximate parallelism must be discussed and not assumed. To continue with the simple chemostat: In any batch culture we can select two points, say E and L, the first during the transition from lag to exponential growth, the second during the transition from exponential growth to the stationary phase, and such that the corresponding growth rates (say μ_E and μ_L) are equal. Using the same medium and organism, we can also

Editor's Note: On page 210 of this reprint reference is made to (1) "Professor Elsden's contribution" and (2) "the experiments of Řičica and of Fencl"; the first appears as Paper 20 in this volume, the second as Paper 18.

work a chemostat at a dilution rate D such that
$$D = \mu_E = \mu_L < \mu_{max}$$
To which of the states E and L is the state of the chemostat to be supposed analogous? Clearly not E, because there nutrients are in excess; growth is unrestricted and accelerating, and the organisms are usually larger than they are during steady exponential growth at μ_{max}. State L is more like the chemostat condition, in that growth (now slowing down) is being restricted by lack of nutrients, but the organisms are becoming smaller, and this alone is evidence of a changing physiological state. The earlier E and the later L are taken in the batch culture, the lower is the growth rate; but the experiments of Herbert & Tempest (this Symposium) show that at very low growth rates the physiological state in the chemostat is quite unlike any state met with in the course of a batch culture. The difference between the two systems is best thought of in terms of heredity. We know that organisms exhibit hysteresis in response to a change in their environment; the proper-ties of an individual are determined in part by its ancestry, not only by its immediate surroundings. In the steady state of a chemostat, the ancestors of every organism for many past generations have all been subjected to the same environment. But we can safely assume that the more remote the ancestor, the smaller the part it plays in determining the particular properties of its ultimate descendents: the physiological state in the chemo-stat (under steady conditions) is decided by the dilution rate alone; the organisms may be said to have no history. In batch culture the environment is changing, and over much of the course of the culture it changes so rapidly that the reaction of the organisms at any one time is affected by the history of recent ancestors, which have been subjected to a different environment.

Still other difficulties arise in multi-stage systems—systems which have been used with a view to imitating batch culture conditions. These difficulties stem from the considerations sketched in the Editorial note succeeding Professor Elsden's contribution (p. 259). If in one stage of a multi-stage system we have an input of organisms from a previous stage at a rate D_1, and concentration x_0, and an addition of fresh medium at a rate D_2; and if in the effluent the concentration is $\tilde{x}$ then there is a net growth rate μ_{net} in that stage given by
$$\mu_{net} = D_2 - D_1 (x_0/\tilde{x} - 1)$$
This growth rate μ_{net} is not a simple property; it is a weighted arithmetic mean of the growth rates appropriate to the spectrum of physiological states represented in the fermenter. Similarly a net metabolic activity q_{net} with respect to a given substrate can be ascribed to the population; again it is a mean value. Corresponding to any physiological state we expect definite values of μ and q to exist; given experimentally determined values, we can infer little about the physiological state unless we know that the population is homogeneous.

All this is not to say that continuous cultures cannot be so designed and adjusted as to produce in a population a practical approximation to a required state. Some of the conditions used in the experiments of Řičica and of Fencl (this Symposium) evidently produced organisms closely similar to those found in the unrestricted growth of a batch culture; and Powell (this Symposium) suggests how the phenomenon of hysteresis might be put to rational use in the improvement of productivity.

20

Reprinted from *Microbial Physiology and Continuous Culture* (Proc. 3rd Intern. Symp.), Her Majesty's Stationery Office, London, 1967, pp. 255–257, 259–261

Balanced and Restricted Growth

S. R. ELSDEN, E. O. POWELL, C. G. T. EVANS, R. E. STRANGE, and D. W. TEMPEST

CLOSING SUMMARY

S. R. ELSDEN

Agricultural Research Council, Earlham Laboratory, Norwich

I offered to attempt to summarize this symposium as a means of avoiding writing a concluding paper. This offer was accepted by Mr. Tempest and I find myself with what is to me an impossible task. So widely have our discussions ranged that it is beyond me to achieve a "total synthesis". I propose, therefore, to limit my comments to a few topics which either have not been discussed in detail and which, in my view, merit attention or which interest me particularly.

I want first to draw your attention to apparatus. Two really new types of apparatus have been described to us. First the "oxystat"—what a horrible word this is!—and here we heard of two sorts, that of Dr. Herbert, and that developed by Dr. Pirt and his colleagues. The experiments on the effect of oxygen tension on the cytochrome system of *Aerobacter aerogenes* which Dr. Herbert discussed were most impressive. The other advance in engineering was the two-stage process described by our colleagues from Czechoslovakia, the use of which produced some exciting results. Here we had a means of achieving totally unrestricted growth over long periods of time. It surprised me that no one has discussed in any detail the turbidostat. Does this reflect the technical difficulties of maintaining optically clean surfaces in the monitoring cell? Or does this mean that it is not a useful instrument? Personally I think that is has considerable uses, for, in contrast to the chemostat, the cells produced have not been limited by any nutrient and they are thus of very considerable interest.

This brings me to the question of a 'balanced growth'. As always it is the greys which produce the most heated arguments and pride of place amongst the greys of science is terminology. On those few occasions when I was not immediately involved in the discussions of the meaning to be attached to the term 'balanced growth', I sat back fascinated at the heat engendered. This was the matter of the symposium which really exercised us—here was controversy and argument unconfined. At the risk of going over the ground once again, I would like to take advantage of my present position and comment further on the question. I have thought about the question a great deal and I am now convinced that the term is of little use and should be abandoned, at least as far as continuous culture in the chemostat is concerned. To use the term balanced growth implies that there exists an ideal cell, an international standard cell, to which cell others may be referred and compared.

My point is this. In the chemostat a whole range of conditions can be used under which an organism grows in a steady state. As far as one can judge for each condition, the cells produced are unique and under each condition the composition of the cells is fixed by the circumstances. Thus, the ribosome content reflects the dilution rate; the polysaccharide content reflects the C:N ratio and dilution rate, at least in *Escherichia coli*; and the acid phosphatase reflects the phosphate content of the medium. For each steady state there is a particular sort of cell produced, with unique proportions of cell constituents. I maintain, therefore, that each steady state in a chemostat, once achieved, produces a particular balanced growth, i.e. the proportions of cell constituents in the culture are constant so long as the particular steady state is maintained. Because of the multiplicity of terminology which can thus be imposed and hence of steady states which can be achieved, and therefore of balances which can be struck, I think the term balanced growth is meaningless in the context of continuous culture in the chemostat.

It seems to me to be much more profitable to think in terms of the limitations which are imposed upon the cell. Professor Postgate first made this point, and I think that it is a wise one. There are, first of all, what I propose to call external limitations. These are limitations which the operator is able to impose upon the cell. Second, there are internal limitations, i.e. limitations which the cell's genetic make-up imposes, e.g. the limiting specific growth rate, the limiting temperatures which the cell will grow at, the saturation constants for the various essential nutrients and so on.

External limitations are of two sorts. First, chemical, when the concentration of a particular nutrient affects the cell either in its enzymic make-up or structurally, as in the case of the effect of magnesium ions on the structure of ribosomes. Second, physical limitations, and here one has temperature, osmotic pressure, hydrogen ion concentration, and, unique to the chemostat, the doubling time which is entirely under the control of the operator.

So far I have considered growth in the chemostat. In the turbidostat, or in the apparatus described by Dr. Fencl, growth is not affected by the chemical limitations and proceeds as fast as the cell is able under the physical conditions imposed by the environment. It seems to me, therefore, that one has three sorts of cells: first, externally limited, chemically limited cells; second, externally limited, physically limited cells; third, in the turdidostat, internally limited cells. Now these are concepts rather than precise terms. If this line of argument is accepted then we need someone expert in terminology and well versed in the classics to consider these, and who better than Academician Málek, who has already demonstrated his skill in this direction.

Consideration of the 'balanced growth' question and the composition of cells has lead me to reflect upon the meaning of some of the analytical data which have been presented to us. Of the various macromolecular species the protein content of cells seems to vary least if one is to judge from the experiments described. But what does this mean? It means no more and no less than what the analytical methods employed are able to tell us. When I was a young man, protein used to be the acid precipitable total nitrogen times a magic number—6·25. We now use colorimetric methods and the magic number varies according to the method. The most popular method is a combination of the biuret colour test and the Folin test for phenols. That is to say it depends partly upon the number of peptide bonds and partly upon the tyrosine content of the material. Further, a standard has to be employed and this is usually bovine serum albumen which may or may not be relevant to the proteins of the microbe being analysed. The number one obtains by such methods is not, in my opinion, particularly meaningful in so far as total protein is concerned. Just as important, we do not know how many protein species a cell contains and because of this can never be sure that a constant total protein content means a constant mixture of proteins. There is, of course, ample evidence to show that the amount of a particular enzyme may vary over a wide range depending on the medium used; by contrast we know very little about the relative amounts of the broad mass of proteins under different conditions of culture. I realise that such measurements are difficult. But, with the advent of gel electrophoresis, either starch gel or acrylamide—it is possible to get a qualitative picture of the relative amounts of the proteins present in the soluble fraction of cells.

As I see it a series of numbers purporting to show that the protein content of a culture is constant is not meaningful. It would begin to be meaningful were these numbers accompanied by a demonstration that the proportion of the various proteins remained approximately the same. Such experiments have not, to my knowledge, been done. The trouble really is that it is not as easy a question as that about the protein content.

Professor Dawes and Dr. Hamlin have measured the amounts of certain enzymes in cells grown under interesting conditions, namely on two substrates, each of which is catabolized by its own particular pathway. This is a most interesting approach, and one of considerable practical importance. Both Dr. Pirt and Dr. Mateles pointed out that under industrial conditions organisms are usually grown on very complex media, i.e. instead of just two substrates a multiplicity of substrates. Further, nobody knows the effect of such conditions on the enzymic make up of the cell. Clearly to make such an analysis at this stage would be exceedingly difficult and it is for this reason that two-substrate studies such as that described by Professor Dawes and Dr. Hamlin are important for they serve as very useful models.

Dr. Pirt gave us a classification of end products of bacterial and fungal growth, which I found both apt and useful. Amongst this list were a group, produced in the main by fungi and streptomycetes, which are of considerable commercial importance, but the physiological significance to the producing organism is not clear. They appear to have two features in common. First, that they are produced after growth has ceased, produced in fact by cells which if not non-viable are certainly far from healthy. This production after growth stops was admirably brought about in Dr. Maxon's elegant analysis of the system producing neomycin. Second, they appear to be unrelated to any essential metabolite produced by the organism. In this latter respect these compounds contrast sharply with the excess production of vitamins or amino acids by certain organisms, for these are known metabolites, and with citric acid produced in large amounts by *Aspergillus niger* and other fungi. It is possible that compounds such as penicillin or neomycin are derived from as yet unknown essential metabolites, the production of which becomes uncontrolled when the organism stops growing—but we do no know why. I find it difficult to see how continuous culture can be applied in a rational way to the production of compounds whose physiological role is unknown. We can, of course, approach the matter by trial and error, but I doubt whether we would understand what we were doing. It would, therefore, be exceedingly interesting to know what these odd compounds, or their precursors, are doing in the cells which produce them. Once we have this information we shall be in a better position to rationalize their production.

EDITORIAL COMMENT:
Balanced and Restricted Growth

Most of the discussion stimulated by Professor Elsden's review was concerned with the meaning of a number of words and phrases related to or qualifying 'growth': 'balanced', 'restricted', 'external control' and their opposites. The difficulties felt by some of the speakers (including Professor Elsden himself) evidently arose from the ascription of different meanings to the same word. Now such difficulties are really trivial, and the editors propose to resolve them by *ex cathedra* pronouncement.

The expression 'unbalanced growth' was used by Cohen & Barner (1954, *Proc. natn. Acad. Sci. U.S.A.*, **40**, 885) of organisms in which important cell components were synthesized at very different rates, the resulting disparity in concentration ultimately leading to death. Campbell (1957, *Bacteriological Reviews*, **20**, 263) generalised this idea: 'growth is *balanced* over a time interval if, during that interval, every extensive property of the growing system increases by the same factor'. In particular, growth is balanced when all the molecular species composing a population of cells are increasing at the same rate; if in a closed system x_r is the concentration of the rth molecular species and μ_r is its growth rate,

$$\mu_r = \frac{1}{x_r} \frac{dx_r}{dt} ;$$

all the μ_r are alike equal to the growth rate μ of the population in balanced growth. It follows that then the chemical composition of the organisms is unchanging. But we would emphasize that the condition of balance in this sense (which we accept) implies nothing about the relative amounts of the different molecular species, nor does it imply a condition of normality or ideality in any sense.

We must recognise, as Campbell did, that the growth of one or a few cells cannot be balanced over an arbitrary time interval, because fluctuation in the component processes certainly occurs within the generation time of an organism (else it could not divide, and its shape would not vary). As indeed in chemical kinetics, we can speak of balanced growth only in a population so large that the erratic behaviour of individuals does not sensibly affect the measurable average properties.

For example, in a nitrogen-limited chemostat population, the steady-state growth is balanced; the organisms' environment has been unchanging for a long time relative to the generation time; there is no input of organisms from without, and the ancestors of every organism have been subjected to the same environment as itself. We cannot believe that, if some of the organisms were removed to a similar environment elsewhere, their properties or the properties of their progeny would change. Yet the organisms may contain 50% of their mass as polysaccharide. This is felt to be 'abnormal' by some who nevertheless are willing to grow organisms on agar plates, i.e. in circumstances where the growth is probably never balanced in our sense.

In the absence of definite precedent, the expressions 'free growth', 'restricted growth', 'external control' etc. are less easy to deal with. Since growth rates are finite, they may be said to be restricted, but this is a sterile tautology.

In another sense, growth is always free, because the experimenter cannot interfere directly with the internal workings of the cell, but can only leave it to a population to come to terms with the environment offered. In fact, growth rate is always determined both by internal and external factors. No matter how well we may tailor a medium to the needs of a given species, we expect, on the one hand, that internal processes will ultimately govern the growth rate attainable; on the other, we cannot be sure that there exists no growth factor which might be added to improve the medium we already have. To be useful, 'restricted' and 'unrestricted' must be confined to a narrower field. Now it is easy to devise, or to light upon accidentally, media in which the precise concentrations of the components are unimportant. Tryptic meat broth, for example, made in the conventional way will support sparse populations of many varieties of organisms at high growth rates; but the same growth rates are obtained if the medium is diluted manyfold with water. In the chemostat, the growth rate is less than that attainable by a sparse population in the medium supplied because the organisms consume all but a small fraction of at least one of the medium components. The ('external') control imposed by the experimenter is indirect; he fixes the dilution rate; the organisms either so change the environment as to match their resulting growth rate to the dilution rate, or they wash out. We think, then, that 'unrestricted growth' can be usefully defined only with respect to a medium in which the molecular species present can be stated, and then only at a given pH and temperature. We should say that growth in a medium of such-and-such composition is unrestricted if a sensible (but not necessarily wide) variation in the concentration of any of the components causes no change in growth rate; and that, if a change in growth rate is brought about by varying the concentration of some one component, and this is true of no other component, then growth is *simply restricted*. In practice, there can be some relaxation of the definition; we can be fairly sure that the useful components of tryptic meat broth are for many species present in excess, but not in such excess as to be toxic.

Apart from these semantic questions, there was one genuine difficulty.

Consider a two-stage continuous culture, the first stage a chemostat simply limited by glucose, say, the second having two inputs: one a slow feed from the first stage, the other a rapid supply of fresh medium. After a time, the system settles to a steady state. In the first stage, the quality and concentration of the organisms is then constant, and their growth is balanced. But while repeated samples from the second stage would also show no change in properties, growth in that stage is not balanced. Some few of the organisms present in the second stage will have ancestors which for several generations back have completed their life-span within the second stage. Their growth will be balanced, nearly enough; that is, their physiological state will be that appropriate to unchanging growth in the environment of the second stage. But most of the population (especially those whose inception occurred in the first stage) will be in transition between the physiological states appropriate to the low glucose concentration in the first stage and the much higher concentration in the second.

Thus the second fermenter in the system, with its population of organisms, can properly be said to be in a steady state, but the conditions within it cannot be permanently realised in a simple chemostat with no input of organisms. We need a distinctive term here, to avoid the implication that the organisms are in kinetic equilibrium with their environment. We propose to adapt a suggestion from Professor Postgate and say that the fermenter is in a *transitional steady state*.

Similarly, any geometrically fixed region of an ideal tubular fermenter contains a population whose properties do not change from sample to sample, but which is not in balanced growth. Every such region (and by an excusable extension, the whole fermenter) is in a transitional steady state.

Editor's Comments on Paper 21

21 Powell: *The Growth Rate of Microorganisms as a Function of Substrate Concentration*

In this paper, Powell, of the Porton group, who has been deeply associated with mathematical aspects of growth, reexamines and compares the Monod and related mathematical formulations and attempts to account for some types of physiological change in terms of them. Toward the end of the paper Powell begins an attempt to accommodate the problem of the physiological state. In later papers Powell has developed his thesis [see E. O. Powell, "Transient changes in the growth rate of microorganisms" in *Continuous Cultivation of Microorganisms* (Proc. 4th Intern. Symp.), I. Málek, K. Beran, Z. Fencl, V. Munk, J. Říčica, and H. Smrčková, eds., Academia, Prague, and E. O. Powell, *J. Appl. Chem. Biotechnol.*, **22**, 71 – 78 (1972)], but the approach is still limited by the difficulty of having to define in mathematical terms the enigmatic stumbling block of the physiological state.

21

Reprinted from *Microbial Physiology and Continuous Culture* (Proc. 3rd Intern. Symp.), Her Majesty's Stationery Office, London, 1967, pp. 34–56

THE GROWTH RATE OF MICROÖRGANISMS AS A FUNCTION OF SUBSTRATE CONCENTRATION

E. O. POWELL

Microbiological Research Establishment, Porton Down, Salisbury, Wiltshire

INTRODUCTION

The present day theory of continuous culture stems from and is still dominated by Monod's (1942, 1950) formulation of the dependence of growth rate on substrate concentration: the growth rate (μ) is given as a function of the concentration (s) of a single component of the medium by the relation

$$\mu(s) = \mu_m \frac{s}{K + s} \tag{1}$$

depending on parameters μ_m, the maximum growth rate, and K the saturation constant. (It is assumed that substrates other than that represented by s are held at constant concentration, or are present in excess). The great defect of all formulae such as (1) is that they represent the growth rate as depending only on the instantaneous value of s; they imply that there is no lag in the response of the growth rate to changes in s. In fact, the growth rate at a given time t depends not only on $s(t)$ but also on the history of the culture, in particular, on the way in which s has varied in the past. It is easy to find examples of both the approximate truth and the obvious falsehood of equations such as (1). For example, during the lag phase of a batch culture, the growth rate only slowly approaches the value it can attain during the exponential phase, even though during the transition s scarcely alters (and in fact falls slightly). On the other hand organisms seem to respond quickly to moderate changes in s, once growth is well established (Herbert, 1964).

Since, and largely as a result of, Monod's earlier work, the technique of continuous culture has so developed as both to permit and call for reexamination of the dependence of μ on s, but in a more limited sense. We can ask, not for a quite general formula for $\mu(s)$, but for a formula which is true under steady state conditions, when μ and s have remained unchanged for a long time relative to the mean generation time of the organisms concerned. Except in the last section, the formulae which I propose to discuss are to be understood as restricted to this sense, though they will often represent quite closely the transition from the exponential to the stationary phase of batch cultures. I assume, and I believe it to be true, that the growth rate in a given environment is intrinsically independent of population density; that the supposed effects of density on growth rate are due to changes in the environment brought about by the organisms themselves; for example, low pH or a deficit of oxygen. (Population density effects may nevertheless be important in practice; thus a great deal of power is required to provide full oxygenation for a fast-growing culture in a large vessel).

One other general defect of all these formulae must be mentioned: they represent the growth rate as asymptotic to a maximum value μ_m nearly attained at large values of s. For many nutrilites, there is indeed a wide range over which μ does not sensibly differ from μ_m but there are some obvious exceptions of practical consequence, and all sufficiently soluble nutrilites will suppress growth at very high concentrations. (For a brief discussion, see Powell, 1965). No quantitative description of the inhibition can yet be given; its causes are probably various (Dixon & Webb, 1964).

Hinshelwood (1946), in spite of his own contribution to the problem and his favourable discussion of Monod's work, long ago concluded that formulae like Monod's could be of little more than descriptive value. He was concerned with the transition stages of culture, but his conclusion remains true for formulae restricted to represent steady states, because of the great chemical complexity of the system through which μ depends on s. Meanwhile, Monod's hypotheses have been widely applied and have often proved practically satisfactory, but a number of examples of their failure have come to light; some at least of these can be explained biochemically and described quantitatively. In the following paragraphs I review three of the formulae which have been proposed for $\mu(s)$ and discuss a number of refinements intended to take account of particular phenomena whose nature is partly understood. Mathematical spatchcocking does not make for elegance, but if we take some simple formula as a basis, any rational modifications we can make will reduce in importance the quantitative defects of the original.

EQUATIONS OF GROWTH AND ASSIMILATION

Some general expressions

It will be necessary in the first place to distinguish (i) the growth rate of the organisms; (ii) the rate of consumption of a particular nutrilite, each as a function of the concentration of the nutrilite in question, and it will be supposed that other nutrilites are held at a constant concentration or are present in excess. Then we can write.

$$x\mu(s)$$

for the rate of production of new cell mass and

$$xq(s)$$

for the rate of consumption of the nutrilite (each per unit volume of culture), where x is the concentration of organisms by mass, s that of the nutrilite, $\mu(s)$ is the growth rate, and $q(s)$ is a 'metabolic coefficient' of the same dimensions (T^{-1}) as μ. In the steady state of a continuous culture, the derivatives dx/dt and ds/dt are both zero, so to avoid false statements (cf. Monod, 1950) while retaining a mnemonic form, I shall put where necessary

$$\left[\frac{dx}{dt}\right] = x\mu(s),$$

$$\left[\frac{ds}{dt}\right] = -xq(s)$$

to indicate the contribution from the activity of the organisms to the net derivatives.

The yield, Y, relative to a given substrate is the mass of organisms produced per unit mass of substrate consumed:

$$Y = -\frac{[dx/dt]}{[ds/dt]} = \frac{\mu(s)}{q(s)}. \tag{2}$$

In this form Y is given as a function of s, but in work with continuous cultures the growth rate is usually the independent variable (since at equilibrium it is equal to the dilution rate), and it is more convenient to express Y as a function of μ; this offers no difficulty when $\mu(s)$ is known.

Equation (2) shows that so far we need only two independent functions to describe the growth, not three. For historical reasons μ and Y have come into use, but the retention of q is more than a convenience. The known gross deviations from the simple Monod theory come about through variations in Y, that is, through peculiarities in the fate of the ingested substrate; there is thus some reason to regard q as a simpler function of s (at least outwardly) than are Y and μ.

It is to be remembered that although Y is defined with respect to a particular substrate whose concentration (s) is considered to be variable, it is not a function of s only; it depends also on the chemical nature of all other components of the medium even when they are present in excess. We can if we wish define and discuss (as Monod, 1950, did) a yield with respect to a second substrate under conditions when the growth rate is limited by the first. It is, however, the yield with respect to the limiting substrate which is of first importance in establishing the kinetics of continuous cultures.

In a chemostat at equilibrium, the mass concentration of organisms is given by (Herbert, Elsworth & Telling, 1956)

$$\tilde{x} = Y(s_R - \tilde{s})$$

where by convention the tilde letters indicate equilibrium values, and s_R is the limiting substrate in the ingoing medium. Or, if $\tilde{x}$ is to be expressed as a function of μ,

$$\tilde{x} = Y\{s_R - \tilde{s}(\mu)\} \tag{3}$$

where $\tilde{s}(\mu)$ is the function inverse to $\mu(\tilde{s})$. It is frequently difficult to measure $\tilde{s}$ accurately because of the awkward chemical nature of some substrates (particularly polyhydroxy compounds) and the very low concentrations encountered. In such cases the experimental determination of $\tilde{x}$ as a function of μ provides a test of a proposed relation between μ, $\tilde{s}$ and Y by means of an equation of the type of (3).

The saturation constant K occurring explicitly as a parameter in Monod's equations is also the 'half-rate constant': $\mu = \mu_m/2$ when $s = K$ if the yield is constant. In any case if the yield is variable, and for all other forms of growth equation, the half-rate constant does not naturally occur as an explicit parameter. I propose therefore to use H as a symbol for the half-rate constant (which is directly measurable) and to keep K for the parameter of a kinetic term of the Michaelis-Menten type (c.f. equation (12) for example).

Monod's Equations

Monod's expression for the growth rate has already been given (Equation (1)). For simplicity, he assumed that the yield Y is independent of s (or μ) though he discussed the possibility of its varying and in fact almost anticipated Herbert's (1958) introduction of 'endogenous metabolism' to account for variation of a certain kind: 'toute' dépense nutritive affectée d'un "coefficient d'entretien" appréciable ne saurait être indépendente du taux de croissance' (Monod, 1950).

With the assumption of constant Y we have, writing q_m for the maximum rate of consumption of limiting nutrilite,

$$q = q_m \frac{s}{K + s} = \frac{\mu_m}{Y} \cdot \frac{s}{K + s} \cdot \qquad (4)$$

Monod's equations are attractive in part because of their familiarity (through the Michaelis-Menten equation) and in part because of their relative mathematical simplicity; they lend themselves well to modification. They have been found defective mainly through the gross failure in certain circumstances of the assumption that the yield is constant. But there exists also a widely felt impression, frequently spoken of but not written about, that the variable μ of equation (1) approaches its asymptote too slowly to be a good representation of experimental data even in uncomplicated instances (Fig. 1). It will be seen later that there is some substance in this impression.

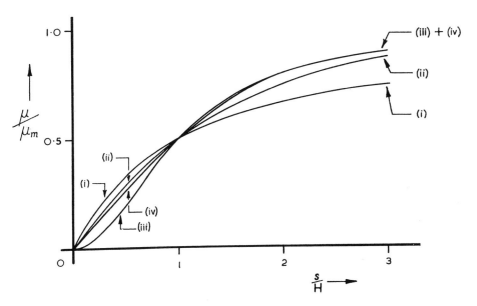

Fig. 1. Equations of growth: (i) Monod; (ii) Teissier; (iii) Moser, $r = 2$; (iv) Powell, $L = 10K$. Scaled to the same half-rate concentration $s/H = 1$. (iii) and (iv) intersect also at $s/H = 3$.

Equation (1) can be rearranged to

$$s = \frac{K\mu}{\mu_m - \mu}$$

and so under Monod's hypothesis the explicit expression for $\tilde{x}$ in a continuous culture (cf. (3)) is

$$\tilde{x} = Y \left\{ s_R - \frac{KD}{\mu_m - D} \right\} \tag{5}$$

since at equilibrium $D = \mu$.

Moser's Equations

Monod's equation (1) represents a rectangular hyperbola; varying the parameters produces changes of scale but the curve has always the same shape. Moser (1958) replaced the first power of s in (1) by an arbitrary power, obtaining a greater degree of flexibility:

$$\mu = \mu_m \frac{s^r}{H^r + s^r} . \tag{6}$$

Using this equation, he was able to graduate experimental data apparently with more accuracy than was provided by Monod's equation, but he gives no numerical comparisons of goodness of fit. An objection to (6) is that the gradient of μ at the origin is either zero (when r is greater than 1) or infinite (when r is less than 1); the latter condition in particular is not easy to account for chemically. In addition, the change from the first to a variable power of r in the equation is not a trivial but a major increase in complexity (Jeffreys, 1937); in the absence of a justifying hypothesis it seems better to investigate other methods of modifying equation (1).

In most of his graduations Moser's estimates of r were greater than 1. For $r > 1$ the curve of equation (6) approaches its asymptote more rapidly than does a rectangular hyperbola (Fig. 1) and is convex towards the s-axis near the origin. This latter feature is of interest in comparison with the algebraic effect of Herbert's (1958b) correction for endogenous metabolism and of a correction for viability discussed below.

Teissier's Equations

Teissier (1942) developed his equations from the consideration that the effect on the growth rate of an increment in s must be less, the nearer is μ to its maximum value μ_m. Making the hypothesis of simple proportionality, he wrote

$$\frac{d\mu}{ds} = \frac{1}{T} (\mu_m - \mu)$$

which gives on integration

$$\mu = \mu_m \left(1 - e^{-s/T} \right). \tag{7}$$

The half-rate constant H is $T \log 2$. The curve of equation (7) approaches its asymptote more rapidly than does a rectangular hyperbola (Fig. 1); apart from changes of scale, it always has the same shape.

Under Teissier's hypothesis, with the added assumption of constant yield, the explicit expression for $\tilde{x}$ in a simply-limited chemostat is

$$\tilde{x} = Y \left\{ s_R - T \log \left(\frac{\mu_m}{\mu_m - D} \right) \right\} . \tag{8}$$

Diffusion and Permeability

The assimilation of a substrate involves a sequence of chemical reactions of which some members may be common, and in some instances are known to be common, to a variety of species. In such instances we might expect the kinetics of the process to be similar for all the species concerned. Longmuir (1954) studied the uptake of oxygen in the presence of glucose by a range of microörganisms, and found that the apparent saturation constant (H) for oxygen differed from one kind to another; there was a marked positive association between H and the physical size of the organisms. He ascribed the differences to the concentration gradient of the substrate near the sites of its absorption. Longmuir's work prompted me to examine theoretically the possible effects of diffusion and cell membranes in restricting the access of substrates to the enzymes within the cell (I have already stated without details (Powell, 1958) the resulting modified Monod formula: equation (13) of this paper).

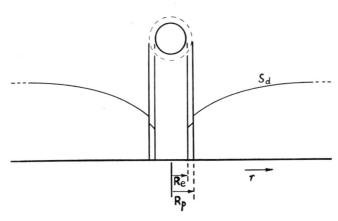

Fig. 2. Concentration of substrate (s_d) near a spherical organism as a function of radial distance r. R_e: radius of reactive surface; R_p radius of membrane.

We can obtain an appreciation of these effects from the simple spherically symmetrical system of Fig. 2. Here we have an amount of enzyme capable of converting at most G_m grams per second of substrate, disposed on a spherical surface of radius R_e. Outside this surface is a membrane of radius R_p and permeability P with respect to the substrate; if the concentrations of substrate on the two sides of the membrane differ by $\triangle s_d$, the rate of passage through the membrane is $-P \triangle s_d$ grams per square centimetre per second. We write s_e for the concentration of substrate near the enzyme, s for its concentration at a great distance, and we assume the system to be in a steady state. (On this scale, a close approach to diffusion equilibrium is attained within about 10 milliseconds for most substrates). Then Laplace's equation holds in the space between R_e and R_p and outside R_p: with spherical symmetry

$$\nabla^2 s_d = \frac{d^2 s_d}{dR^2} + \frac{2}{R}\frac{ds_d}{dR} = 0 \ .$$

Assuming the kinetics of the enzyme reaction to be of the Michaelis-Menten form, we arrive after some tedious algebra at

$$G_m \frac{s_e}{K + s_e} = (s - s_e) \left\{ \frac{4\pi \mathscr{D} P R_e R_p^2}{P R_p^2 + \mathscr{D} R_e} \right\}, \tag{9}$$

where $\mathscr{D}$ is the diffusion coefficient of the substrate. Because of the non-linearity of the enzyme kinetics, we cannot use the classical method of superposition of solutions in order to obtain the corresponding result for a more realistic geometry, but it is clear that access of substrate will be impeded more if the enzyme is distributed throughout a spherical volume instead of on the surface. We can readily obtain a rough impression of the extent to which diffusion and permeability may affect the flow of substrate. A fast-growing microörganism will double its weight in 20 minutes and at the same time consume say 3 times its weight of glucose. If we take the organism to be a sphere of radius 0.5μ, its mass will be about 5×10^{-13} gm and the corresponding G_m about

$$\frac{3 \times 5 \times 10^{-13}}{20 \times 60} \simeq 10^{-15} \text{ g/sec.}$$

If in the formula (9) we take P to be very large (or assume that there is no membrane), make $R_e = 0.5\mu$, and take $\mathscr{D} = 4 \times 10^{-6}$ cm²/sec we find approximately

$$s - s_e = (3 \times 10^{-6}) \frac{s_e}{K + s_e}. \tag{10}$$

For substrates such as glucose the apparent saturation constant is of the order of 10μg/cc. At this value of s, $s_e \simeq K$ and (10) gives

$$s_e = 8.5 \times 10^{-6},$$

about 15% lower than s. Thus diffusion alone can appreciably impede the access of substrates at low concentration, and inflate the apparent saturation constant. Its effect is greater, the smaller is s and the difference can only be heightened by the presence of an inert membrane enclosing the absorbing sites. (An active membrane, however, is itself a site of reaction).

We can now apply (9) to a whole culture without too much violence. We can say that if $q(s_e)$ is the rate of consumption of the substrate as a function of its *local* concentration s_e then

$$q(s_e) = (s - s_e) F \tag{11}$$

where the factor F depends on the geometry and physical characteristics of the organisms and their environment. In view of the various shapes, sizes and growth rates of individual organisms, as well as the usually unknown distribution of sites of reaction, each parameter implied in F can represent only some kind of average for the system.

Now if we adopt Monod's formulation for $q(s_e)$ with a constant yield, we have

$$q = q_m \frac{s_e}{K + s_e}.$$

On substitution into (11), this gives

$$s = \frac{q}{F} + \frac{qK}{q_m - q} = \frac{\mu}{YF} + \frac{\mu K}{\mu_m - \mu} \quad . \tag{12}$$

It is convenient to introduce a new constant L equal to q_m/F:

$$s = L\frac{q}{q_m} + \frac{qK}{q_m - q} = L\frac{\mu}{\mu_m} + K\frac{\mu}{\mu_m - \mu} \quad . \tag{13}$$

Then L has the dimensions of concentration; it is that value of s which would allow the organisms to grow at their maximum rate if K were zero and the enzyme reaction of zero order. The explicit expression for μ is

$$\mu = \frac{\mu_m (K + L + s)}{2L}\left[1 - \sqrt{\left\{1 - \frac{4Ls}{(K + L + s)^2}\right\}}\right]. \tag{14}$$

A rational approximation to this equation is of some interest. We have, since s, K and L are positive

$$(K + L + s)^2 > (L + s)^2 = (L - s)^2 + 4sL$$

$(L - s)^2$ is necessarily positive, *a fortiori* the last term under the root in (14) is less than unity, and expansion by the binomial theorem is always permissible. Retaining only two terms of the expansion, we find

$$\mu = \mu_m \frac{s}{K + L + s} \quad .$$

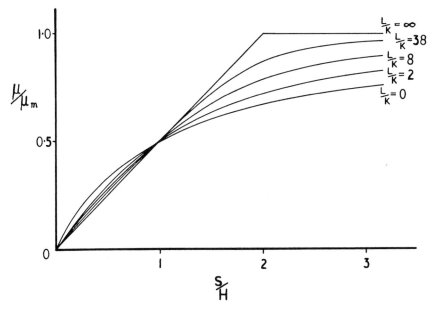

Fig. 3. The range of shapes of Powell's equation of growth.

If L is only a small multiple of K we return approximately to Monod's form, but with an apparent saturation constant $H, = K + L$, dependent in part on the diffusion and permeation of the substrate.

E

Equations (13) and (14) represent a hyperbola with one asymptote parallel to the s-axis and one at 45-90° to it. An example is shown in Fig. 1, and the possible range of shapes in Fig. 3.

By substituting (13) into (3) we find for the chemostat characteristic

$$\tilde{x} = Y \left\{ s_R - L \frac{D}{\mu_m} - K \frac{D}{\mu_m - D} \right\} . \tag{15}$$

Endogenous Metabolism and Viability

Bacterial cells from a growing culture, suspended in a medium devoid of an energy source will for a time consume their own substance; their mass growth rate is negative. Herbert (1958b) used this fact to account for the fall in Y observed at low values of dilution rate in carbon-limited cultures. He supposed that the endogenous metabolism proceeded at a constant rate at all values of μ, and he wrote

$$\left[\frac{dx}{dt} \right] = (\mu - k) \, x,$$

k being the rate constant of the endogenous process. It seems better, however, to keep μ to mean the observable growth rate $d \log x/dt$ and in the sequel I shall replace Herbert's k by μ_e.

The simplest way of representing quantitatively the effect of endogenous metabolism is to assume that the conversion of unit mass of substrate is potentially able to provide a constant mass Y_g of cell material. Then the growth rate would be $q(s) Y_g$ but for the loss μ_e due to a standing requirement for energy and replacement. In terms of our general equations

$$- [ds/dt] = xq$$

$$[dx/dt] = x\mu = x(Y_g q - \mu_e),$$

and the observed yield will be

$$(Y_g q - \mu_e)/q = Y_g - \mu_e'/q;$$

or, since

$$q = (\mu + \mu_e)/Y_g,$$

$$Y = Y_g \frac{\mu}{\mu + \mu_e} . \tag{16}$$

Under the assumption of constant yield, $\mu(s) \propto q(s)$; whatever algebraic form we assume for q, the effect of μ_e is simply to shift the corresponding curve downwards (Fig. 4; Marr, Nilson & Clark, 1963); the growth rate becomes zero at a small positive value of s. Below this point, the curve represents μ as negative, but negative values cannot be realised permanently; at sufficiently low s, the equilibrium value of μ is zero.

The rate of endogenous metabolism can be measured directly (Nilson, Marr & Clark, 1962; Marr, Nilson & Clark 1963). Its effect on the equilibrium values of x in continuous culture is striking; if (16) is substituted into (3) we find

$$\tilde{x} = Y_g \frac{(s_R - \tilde{s})D}{D + \mu_e}$$

—the concentration of organisms falls to zero towards the origin of D

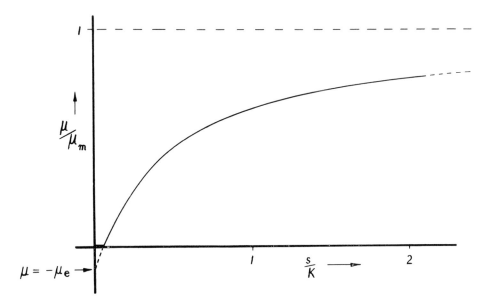

Fig. 4. Monod's equation with Herbert's correction for endogenous metabolism.

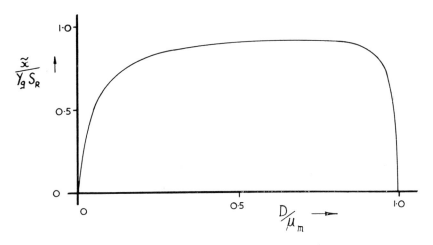

Fig. 5. Continuous culture characteristic according to the Monod—Herbert equation.

If we adopt Monod's form for q we have

$$\mu + \mu_e = Y_g q = Y_g q_m \frac{s}{K + s} \quad,$$

and the maximum observed value of μ (for which we retain the symbol μ_m) is

$$Y_g q_m - \mu_e \quad.$$

Hence

$$\mu = (\mu_m + \mu_e) \frac{s}{K + s} - u_e ; \tag{17}$$

equation (16) becomes

$$Y = Y_g \left\{ 1 - \frac{(K + s) \mu_e}{(\mu_m + \mu_e) s} \right\} ,$$

while in continuous culture (from (3))

$$\tilde{x} = Y_g \left\{ \frac{s_R D}{D + \mu_e} - \frac{KD}{\mu_m - D} \right\} . \tag{18}$$

Equation (18) does not agree with Pirt's (1965) expression purporting to represent the same characteristic (Pirt uses a 'maintenance coefficient', m, such that $m Y_g$ is identical with our μ_e). The reason for the discrepancy is that Pirt takes the equation (5) derived from Monod's hypotheses and substitutes $Y_g \mu/(\mu + \mu_e)$ for Y, forgetting that s is no longer proportional to $\mu/(\mu_m - \mu)$ and that the last term also has to be modified. With our notation, from (17),

$$s = K \frac{\mu + \mu_e}{\mu_m - \mu} \quad.$$

Under Teissier's hypothesis in the form

$$q = q_m \left(1 - e^{-s/T}\right)$$

the analogues of (17) and (18) are

$$\mu = (\mu_m + \mu_e)\left(1 - e^{-s/T}\right) - u_e; \tag{19}$$

$$x = Y_g \frac{D}{D + \mu_e} \left\{ s_R T \log \left(\frac{\mu_m + \mu_e}{\mu_m - D} \right) \right\} . \tag{20}$$

Equations (13) and (15), modified to take account of endogenous metabolism, became

$$s = L \frac{\mu + \mu_e}{\mu_m + \mu_e} + K \frac{\mu + \mu_e}{\mu_m - \mu} ; \tag{21}$$

$$\tilde{x} = Y_g \left\{ \frac{s_R D}{D + \mu_e} - L \frac{D}{\mu_m + \mu_e} - K \frac{D}{\mu_m - D} \right\} . \tag{22}$$

Microörganisms may exhibit an endogenous metabolism equally when their energy source is not a carbon compound. An example (*Nitrosomonas* sp.) due to Downing, Painter & Knowles (1964) is analysed below.

Herbert's assumption that μ_e is a constant independent of growth rate was made for simplicity. Because of the small ratio of μ_e to μ_m it is quantitatively of little importance whether we assume μ_e to be constant or to be a function of μ; but it would be of great biochemical interest to know whether or not the same mechanism of self-consumption operates in the presence as well as in the absence of a substrate.

Over most of the range of μ, nearly all the organisms produced in steady growth are viable; the index of viability, α (Powell, 1956) is usually greater than 0·99. But at very low growth rates the viability falls. Thus Herbert & Tempest (this symposium) working with a glycerol-limited continuous culture of *Aerobacter aerogenes*, found α to be only 0·67 at a dilution rate of 0·0035 hr^{-1} (corresponding to a mean effective generation time of about 8 days). Now continued multiplicative growth can only occur if $\alpha > \frac{1}{2}$; for smaller values the expected number of the immediate viable progeny of an organism is less than unity, and the culture must die out. We can expect the fall in viability to be reflected in the behaviour of continuous cultures at very low growth rates.

There is no simple relation between α, Y and the mass and number growth rates because in general non-viable organisms do not have the same mean size as viable organisms (Errington, Powell & Thompson, 1965), but we can suppose that the culture contains a fraction $V(\mu)$ depending on μ, of living organisms measured by weight. Then, neglecting μ_e for the time being and assuming a constant yield Y_g we have

$$\left[\frac{dx}{dt}\right] = x\mu = xVY_g q$$

since only the viable portion of the population contributes to the growth rate. Now $V(\mu)$ is a function which is nearly unity for all but very small μ, and vanishes with μ; again it matters little what form we choose for it. A simple and convenient choice is to put

$$V = \frac{\mu}{\mu + \mu_v}$$

where μ_v is a constant.
Then

$$\mu = \frac{\mu}{\mu + \mu_v} Y_g q$$

and the yield is

$$Y_g \frac{\mu}{\mu + \mu_v} .$$

This is exactly the same as equation (16) but for the substitution of μ_e by μ_v. That is to say, we can represent the effect either of low viability or of endogeneous metabolism by the same formulae, nearly enough. In cultures where both factors operate, the influence of endogenous metabolism is felt at growth rates much above those at which the viability is appreciably reduced; no further refinement of (16) is yet justified or practically necessary.

It will be noticed that the displaced curve of Fig. 4 resembles a curve of Moser's type with $r > 1$ (Equation (6) and Fig. 1).

There are several statements in the literature to the effect that there is a lower limit to the dilution rate D at which a chemostat can be worked; that at sufficiently low rates the organisms pass into a 'resting state' and are washed out. If the dilution rate of a culture is changed from a moderate to a very small value, the concentration of organisms begins to fall, and this circumstance probably accounts for the statements in question; the culture has to be run for a very long time to attain equilibrium at the new, lower, value of $\tilde{x}$. My colleagues Herbert and Tempest have failed to find a lower limit to D; if there is one, it is much less than any of the values that have been suggested.

Change in Composition

If the growth rate of a culture is limited by a substrate which (or part of which) is incorporated into the substance of the organisms, and if the composition of the organisms in respect of this substrate varies with growth rate, the yield must vary in the opposite sense. This is a general rule of which a good example is given by Tempest, Hunter & Sykes (1965) for magnesium-limited cultures. The proportion of ribonucleic acid (RNA) in cells of *Aerobacter aerogenes* varies with growth rate in an approximately linear manner (Herbert, 1958b), and there is a fixed ratio between the amounts of the RNA synthesised and the magnesium assimilated. We can then say that the amount of magnesium required per unit mass of cells is also a linear function of μ, say proportional to

$$\beta\mu/\mu_m + (1-\beta);$$

the yield must be

$$\frac{Y_m/\mu_m}{\beta\mu + (1-\beta)\mu_m}$$

where β is a constant and Y_m the yield when $\mu = \mu_m$. If for simplicity we adopt Monod's form for $q(s)$ we find

$$\mu = \frac{Y_m\mu_m}{\beta\mu + (1-\beta)\mu_m} \cdot q_m \cdot \frac{s}{K+s} \tag{23}$$

and

$$\mu_m = Y_m q_m.$$

Since (23) is quadratic in μ, it is conveniently written in the inverse form:

$$s = \frac{K\mu}{1+\beta}\left(\frac{1}{\mu_m-\mu} - \frac{\beta^2}{\mu_m+\beta\mu}\right),$$

whence in continuous culture

$$\tilde{x} = \frac{Y_m\mu_m}{\beta D + (1-\beta)\mu_m}\left\{s_R - \frac{KD}{1+\beta}\left(\frac{1}{\mu_m-D} - \frac{\beta^2}{\mu_m+\beta D}\right)\right\}.$$

I have verified that this last equation is in qualitative agreement with the results of Tempest, Hunter & Sykes (1965), but the data are not extensive enough to permit of a more rigorous test.

The Secretion of Polysaccharides

Holme & Palmstjerna (1956) found that *Escherichia coli* secreted a large amount of a glycogen-like substance when growing under conditions of nitrogen starvation in the presence of an excess of carbon source. Holme

(1958) later found this to be true also in nitrogen-limited continuous cultures, and noticed that the yield with respect to nitrogen was greater, the lower the dilution rate. It now appears that the phenomenon is a fairly general one among bacteria and yeasts under limitation by nitrogen, sulphur and phosphorus, but not magnesium (Herbert, 1958b; Herbert & Tempest, unpublished). The organisms behave as if the polysaccharide synthesis proceeds at a constant rate almost independent of their own growth rate. At very low growth rates however, the yield does not continue to rise indefinitely because relatively more and more of the incoming carbon substrate is required by the still finite demand of endogenous metabolism; this fact greatly complicates the kinetics.

We can suppose the gross mass concentration, x, of organisms to be made up of a portion y of living matter and a remainder $x - y$ of polysaccharide. If then we retain s for the concentration of limiting substrate and write z for that of carbon source, we can write down and solve a system of equations for the equilibrium between s, z. x and y in the chemostat. Again. I have verified that the resulting system can be in qualitative agreement with the available data, but the number of parameters is large and some *ad hoc* assumption has to be made about the bifurcation in the metabolic pathway of the carbon substrate. The problem is best approached by a piecemeal analysis of the component processes; this is being done by my colleagues Herbert & Tempest.

CURVE FITTING

One object of this paper is to compare the goodness of fit of the proposed equations of assimilation and growth to existing data in those instances where the yield is grossly affected only by endogenous metabolism. Six equations were tested; they are restated here, each with a reference symbol, for ease of reading:

Monod's hypothesis (Equation 'M'; q may replace μ):

$$\mu = \mu_m \frac{s}{K + s}$$

Teissier's hypothesis (Equation 'T'; q may replace μ):

$$\mu = \mu_m \left(1 - e^{-s/T}\right)$$

Powell's hypothesis (Equation 'P'; q may replace μ):

$$\mu = \frac{\mu_m(K + L + s)}{2L} \left[1 - \sqrt{\left\{1 - \frac{4Ls}{(K + L + s)^2}\right\}}\right]$$

Monod's hypothesis, with endogenous metabolism (Equation 'MH'):

$$\mu = (\mu_m + \mu_e) \frac{s}{K + s} - \mu_e$$

Teissier's hypothesis, with endogenous metabolism (Equation 'TH'):

$$\mu = (\mu_m + \mu_e) \left(1 - e^{-s/T}\right) - \mu_e$$

Powell's hypothesis, with endogenous metabolism (Equation 'PH'):

$$\mu = \frac{(\mu_m + \mu_e)(K + L + s)}{2L} \left[1 - \sqrt{\left\{1 - \frac{4Ls}{(K + L + s)^2}\right\}}\right] - \mu_e$$

The observations in the literature have been obtained by a variety of methods; some series are very ragged; those derived from batch cultures give at best an average value of μ over a small range of s. Probably only the original authors could assess the weights to be assigned to their observations. I have therefore thought it best to treat all the observations on the same footing, and to apply the method of least squares directly to them in the conventional way. The arithmetical procedure involved is sufficiently illustrated by equation MH. If δ_r is the residual of an observation μ_r at $s = s_r$,

$$\sum \delta_r^2 = \sum \mu_r^2 + (\mu_m + \mu_e)^2 \sum \left(\frac{s_r}{L + s_r}\right)^2 + \mu_e^2 \sum 1$$

$$- 2(\mu_m + \mu_e) \sum \frac{\mu_r s_r}{K + s_r} + 2\mu_e \sum \mu_r$$

$$- 2\mu_e(\mu_m + \mu_e) \sum \frac{s_r}{K + s_r}$$

and μ_m, μ_e and K have to be chosen so that $\Sigma\delta_r^2$ is a minimum. By differentiating with respect to the three in turn, we find

$$\mu_m = \sum \left\{\frac{(\mu_r + \mu_e)s_r}{K + s_r}\right\} \bigg/ \sum \left(\frac{s_r}{K + s_r}\right)^2 - \mu_r;$$

$$\mu_e = \left\{\mu_m \sum \frac{s_r}{(K + s_r)^2} - \sum \frac{\mu}{K + s_r}\right\} \bigg/ K \sum \frac{1}{(K + s_r)^2} \; ;$$

and for K,

$$0 = \sum \frac{\mu_r s_r}{(K + s_r)^2} + \mu_e \sum \frac{K s_r}{(K + s_r)^3} - \mu_m \sum \frac{s_r^2}{(K + s_r)^3} \; .$$

The system has to be solved by iteration, beginning with approximate values of the parameters obtained graphically. Within each cycle of the iteration, the third member of the set has itself to be solved by iteration. The bulk of arithmetic is so large as to demand the use of an electronic computer. The iterations were considered to be complete when successive estimates of the parameters simultaneously differed by less than 1 part in 10^4. The goodness of fit was estimated from the mean square residual per degree of freedom (number of observations minus number of parameters).

The relative goodness of fit of the three types of equation to 13 sets of data is indicated in Table 1. Equations MH, TH, PH were applied where the limiting substrate was also the principal energy source. Two anomalies were found (Schaefer, 1948; Monod, 1949), both associated with *M. tuberculosis* and glucose. All six fittings gave inadmissible (i.e. negative) estimates of μ_e. Considering the accuracy of the data and the smallness of the ratio μ_e/μ_m it is not surprising that this should happen by chance; however, the equations M, T, P (corresponding to $\mu_e = 0$) were fitted instead. The results of Downing, Painter & Knowles (1964) also call for special mention. Their data consist of measurements of oxygen consumption at different concentrations of ammonia. When therefore equations MH, TH, PH are applied, we should expect the estimates of 'μ_e' to be negative,

Table 1. *Goodness of fit of equations of assimilation and growth. The equations are ranked according to the magnitude of the mean square residual per degree of freedom, rank 1 corresponding to the best fit.*

			Equation					
Author	*Organism*	*Substrate*	*M*	*T*	*P*	*MH*	*TH*	*PH*
Cohen & Monod, 1957	*E. coli*	*	3	2	1			
Dagley & Hinshelwood, 1938	*A. aerogenes*	PO_4^{3-}	3	2	1			
Downing, Painter & Knowles, 1964	*Nitrosomonas* sp.	NH_3				2	3	1
Monod, 1942	*E. coli*	glucose				3	2	1
,,	,,	lactose				3	1	2
,,	,,	mannitol				3	2	1
Monod, 1949	*M.tuberculosis*	glucose	3	2	1			
Novick, 1958	*E. coli*	arginine	3	2	1			
Novick & Szilard, 1950	*E. coli* (i)	tryptophane	3	1	2			
	E. coli (ii)	tryptophane	3	2	1			
Schaefer, 1948	*M.tuberculosis*	glucose	1	3	2			
Wyss, 1941	*E. coli*	(i)†	1	3	2			
,,	*E. coli*	(ii)†	3	2	1			

*o-nitrophenyl-β-D-galactoside
†p-aminobenzoic acid in competition with (i) 100µg/cc (ii) 250µg/cc of sulphanilic acid.

because endogenous metabolism adds to the oxygen consumption; and this was found to be the case. (The estimates are actually not of μ_e itself, but of a quantity proportional to it). From correspondence with Dr. Painter I understand that the estimates of 'μ_e' are not trustworthy, because his cultures may have contained a little ammonia besides that intentionally added; the effect of a constant excess, however would be merely to displace the curve and would not prejudice the comparison of the fitted equations.

Monod's equations give the worst fit in 10 of 13 comparisons, indicating that the rectangular hyperbola does indeed approach its asymptote too slowly to represent the observations well. In 9 comparisons equations P or PH are superior to the rest, and this is some evidence that the correction for diffusion and permeation is worth making. Equations T and TH might still be preferred for their simplicity; they are in fact less convenient algebraically.

PHYSIOLOGICAL STATES AND TRANSIENT CHANGES

Even though we may understand few details of the process they represent, accurate formulae for $q(s)$ and $\mu(s)$ are always of practical value in work with continuous cultures. In fact, from a theoretical point of view the formulae we already have are less jejune than they seem.

All our expressions for $q(s)$ are of the form $q_m.S(s)$ where $S(s)$ is a simple function of s. We know, however, that the physiological state of micro-organisms changes with growth rate. Let us recognise this by writing

$$q(s) = q_a(s). S(s) \qquad (24)$$

where $q_a(\infty) = q_m$ and $S(\infty) = 1$; in this dissection of q we imply that q depends directly on s through $S(s)$ and indirectly through $q_a(s)$, which is a measure of the organisms' activity and related to their physiological state. It is to be understood that $q_a(s)$ is defined for steady growth at constant

s; we may call it the *potential metabolic activity* (with respect to a given medium and limiting substrate). If we could hold the physiological state constant while changing s we should expect q to vary directly as $S(s)$, attaining the value q_a for large s. We cannot in fact do this, but we have some indication from experiment as to how q_a changes. Tempest & Herbert (1965) measured the oxidation rate of a variety of substrates by washed *Torula utilis* cells grown at different rates in a chemostat. Where the test substrate was the same as the limiting substrate in the chemostat, the measurements may be accepted as proportional to q_a at least approximately.

We can now illustrate the application of (24) by assuming, say, that q_a is a linear function of μ; this is roughly true of some, but not all, of Tempest & Herbert's examples. Let

$$q_a = q_m \left\{ \beta \frac{\mu}{\mu_m} + (1-\beta) \right\} \tag{25}$$

so that $q_a = q_m$ for $\mu = \mu_m$.

Then if we follow Monod and put

$$S(s) = \frac{s}{K+s}$$

we have, introducing a term μ_e for endogenous metabolism,

$$q = q_m \left\{ \beta \frac{\mu}{\mu_m} + (1-\beta) \right\} \frac{s}{K+s} \; ;$$

$$\mu = Y_g q - \mu_e; \tag{26}$$

$$\mu_m = Y_g q_m - \mu_e \, . \tag{27}$$

On eliminating q and q_m from these equations we find

$$\mu = (\mu_m + \mu_e) \frac{s}{s + K \mu_m / \{(1-\beta)\mu_m - \beta\mu_e\}} - \mu_e \, .$$

This is of exactly the same form as (17), but the apparent saturation constant is increased; we cannot determine K and β separately from measurements of the curve alone. Similarly if we replace the q_m of (12) by q_a as given by (25), we find using (26) and (27) that (21) is reproduced in form, with $K\mu_m/\{(1-\beta)\mu_m - \beta\mu_e\}$ instead of K.

We can go still further than this. Herbert (1958b) originally proposed that the effect of endogenous metabolism should be represented by a simple constant term in the growth equations. Tempest & Herbert's (1965) results on the consumption of oxygen by washed cells seem to show, however, that endogenous metabolism varies with growth rate (in their examples, in a linear manner). If in our equations (17) and (21) we replace μ_e by $\beta\mu_e - (1-\beta)\mu$, we find as before that the form of the equations is reproduced, but K is replaced by βK.

In sum, if our observations are represented adequately by the equations derived from either Monod's or Powell's hypothesis both the potential metabolic activity and the endogenous metabolism may vary linearly with growth rate, and it will not be possible to detect the variation from the measured curves alone.

Equation (24) suggests a method of approaching the problem of transient changes in growth rate. So far we have insisted that our growth equations refer always to the steady state, so that, in particular, q_a is a definite function of s. When s varies with time, the metabolic coefficient q is a function of time (t) determined, however, by all the values taken by s at times prior to t. (Naturally we expect q to be influenced by the values of s in the recent past more than by those of the remote past). That is, q is a *functional* (Volterra, 1931) of $s(t)$, and we can write

$$q(t) = Q \left[\begin{array}{c} \zeta = \infty \\ s\,(t - \zeta) \\ \zeta = 0 \end{array} \right] . S(s) . \tag{28}$$

Suppose that at time t_0, Q has the value Q_0, and s is changed from s_0 to s_1. We should then expect q to change from $Q_0 S(s_0)$ to $Q_0 S(s_1)$ immediately, because the cell's complement of enzymes will not have altered appreciably; thereafter there will be a slower change as Q adjusts itself to the new conditions. If the value s_1 is maintained, Q will gradually approach the equilibrium value $q_a (s_1)$. Q may be called the *metabolic activity functional*.

An example based on the simplest possible assumptions will show that (27) is in qualitative agreement with the few data we have. Let us take

$$S(s) = \frac{s}{K + s}$$

and assume that Y is constant; let us assume also that Q is such that its rate of change at time t is proportional to the difference between itself and the equilibrium value $q_a(s)$ corresponding to the s obtaining at t, i.e.

$$\frac{dQ}{dt} = \lambda \left\{ q_a(s) - Q \right\} \tag{28}$$

or

$$Q = \lambda e^{-\lambda t} \int_{-\infty}^{t} e^{\lambda \zeta} q_a\{s(\zeta)\} d\zeta. \tag{29}$$

Then, in a chemostat,

$$\frac{ds}{dt} = D(s_R - s) - xQs/(K + s). \tag{30}$$

To this equation we must add, in general

$$x = \exp \int_{-\infty}^{t} (YQS - D)\, d\zeta$$

obtained from

$$\frac{1}{x} \frac{dx}{dt} = \mu - D = YQS - D, \tag{31}$$

but we can avoid this complication if we keep s_R constant and consider only changes in D. Suppose then that the chemostat is running in a steady state at dilution rate D_0. We have $x_0 = Y(s_R - s_0)$ and $s_0 = KD/(\mu_m - D_0)$. The dilution rate is now suddenly changed to a new, say a higher, value D_1. We shall have at first, as s begins to increase, $Q = q_a(s_0)$ and

$$\frac{ds}{dt} = D_1 (s_R - s) - Y (s_R - s) q_a(s_0)s/(K + s)$$

and if Q adjusts itself very slowly in comparison with D (i.e. if $\lambda \ll D$) the culture will approach a condition where, nearly,

$$\frac{ds}{dt} = 0 \text{ and } s = \frac{KD_1}{Yq_a(s_0) - D_1} . \tag{32}$$

But the true equilibrium for the new condition is

$$\tilde{s} = s_1 = \frac{KD_1}{Yq_a(s_1) - D_1}$$

and because q_a is an increasing function of s, s_1 is less than the value of s given by (32). That is to say, s will at first overshoot its true equilibrium value, approaching s_1 gradually as Q approaches $q_a(s_1)$ (Fig. 6). We can argue similarly about x; if D_1 is not near μ_m the permanent change in x will be slight, but there will be a marked dip during the transition. These conclusions accord with the findings of Mateles, Ryu & Yasuda (1965) and with experiments now being conducted by Tempest. Herbert's (1964) observations of the rapid response of a continuous culture to small periodic additions of the limiting substrate are now reconciled with the observations of Mateles *et al* (1965): if we make a sudden change in s, there will indeed be an immediate response from the organisms, but only to an extent accorded by the factor $S(s)$ in (27).

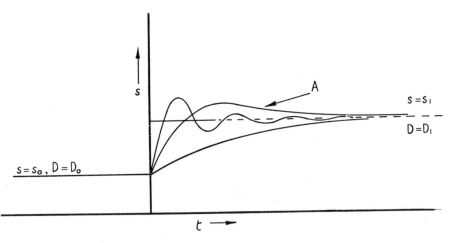

Fig. 6. Transient changes in s when the dilution rate in a chemostat is increased from D_0 to D_1. Curve A has been observed experimentally.

If we differentiate equation (30) with respect to time, and eliminate Q and Q' with (28) and (30), keeping s_R constant so that $x = Y (s_R - s)$ we arrive at an ordinary differential equation of the second order and second degree which it is quite feasible to solve by numerical methods:

$$\frac{d^2s}{dt^2} + D\frac{ds}{dt} + \lambda q_a(s)\frac{(s_R - s)\,Y}{K + s}$$

$$-\left\{(s_R - s)D - \frac{ds}{dt}\right\}\left\{\lambda + \frac{ds}{dt}\left(\frac{1}{s_R - s} + \frac{1}{s} + \frac{1}{K + s}\right)\right\} = 0.$$

I have verified that this equation can be brought into rough agreement with the figures of Mateles *et al* (1965). Its solution can take three forms, however, depending on the relative values of the parameters and the way q_a changes. By applying the Poincaré criteria (see e.g. Stoker, 1950) it is found that (i) s may change monotonically from s_0 to s_1; (ii) there may be a single overshoot; (iii) s may approach s_1 by way of a damped oscillation (Fig. 6).

Equation (29) gives the form of Q under the assumptions of the foregoing example. We may rewrite it as

$$Q = \int_0^\infty \lambda e^{-\lambda\eta} q_a\,\{s(t - \eta)\}\,d\eta.$$

In this expression we notice that $\lambda e^{-\lambda\eta}$ is a frequency function; it is always positive and

$$\int_0^\infty \lambda e^{-\lambda\eta}\,d\eta = 1.$$

Thus Q is a weighted mean of the values of q_a which have obtained in the past, and $\lambda e^{-\lambda\eta}$ is the weighting function. This suggests that we may be able to represent Q more generally as a simple linear functional

$$\int_0^\infty f(\eta)q_a\,\{s(t - \eta)\}\,d\eta$$

in which $f(\eta)$ is a weighting function to be determined experimentally—it may well be true that Q is not so strongly dependent on values of s in the present and immediate past (as when $f(\eta) = \lambda e^{-\lambda\eta}$) as in a rather more remote past; i.e. $f(\eta)$ may have a maximum for $\eta > 0$. But the real nature of Q is still a matter for future investigation.

The Outlook

In the equation

$$\mu = Q.Y.S \tag{33}$$

three problems are implied: how does each of Q, Y, S depend on s?

(i) We have a partial and practically satisfactory answer to the simplest: the form of $S(s)$.

(ii) We know how Y varies with s in a few particular cases, but we do not yet know that Y is not itself a functional of $s(t)$ under non-equilibrium conditions.

(iii) The nature of Q is almost entirely unknown. The potential metabolic activity (i.e. the limit $q_a(s)$ of Q) we know to vary with s but very few acceptable measurements have been made.

The formation of microbial products, as distinct from biomass, offers analogous problems which so far have hardly been systematically formulated (but see Fencl, 1966), and which are likely to be even more complex—as the contributions to this symposium alone will show. By analogy with (33) we can write

$$\frac{1}{x} \left[\frac{dp}{dt} \right] = R. \, \alpha. \, S$$

where p is product concentration and α is product yield. The functional R depends on the physiological state of the organisms, but it is not identical with Q because we know that some products can be formed without accompanying growth.

A knowledge of the form of the activity functionals Q and R is in my opinion likely to prove of great practical value, and in particular to lead to a more rational and efficient use of multi-stage systems. For example, certain antibiotics and toxins are formed within a short range of physiological states which are attained only transiently in the growth of a batch culture. We can imagine a two-stage system in which the first stage is used only to bring the organisms into the appropriate state; the effluent passes into the second stage, which is also fed with additional substrate, or another appropriate substrate. In the second stage, provided the activity functionals change slowly enough, we then have the advantage of high physiological activity coupled with high substrate concentration.

REFERENCES

COHEN, G. N. & MONOD, J. (1957). Bacterial permeases. *Bact. Rev.*, **21**, 169.

DAGLEY, S. & HINSHELWOOD, C. N. (1938). Physicochemical aspects of bacterial growth. I. Dependence of growth of *Bacterium lactis aerogenes* on concentration of medium. *J. chem. Soc.*, 1930.

DIXON, M. & WEBB, E. C. (1964). *Enzymes*. 2nd Edition. London: Longmans, Green & Co. Ltd.

DOWNING, A. L., PAINTER, H. A. & KNOWLES, G. (1964). Nitrification in the activated sludge process. *J. Proc. Inst. Sew. Purif.*, Part 2, p.3.

ERRINGTON, F. P., POWELL, E. O. & THOMPSON, N. (1965). Growth characteristics of some Gram-negative bacteria. *J. gen. Microbiol.*, **39**, 109.

FENCL, Z. (1966). Theoretical Analysis of Continuous Culture Systems. In *Theoretical and Methodological Basis of Continuous Culture of Microörganisms*. Ed. Málek, I. & Fencl, Z. Tr. Liebster, J. Prague: Nakladatelství Československé Akademie Věd.

HERBERT, D. (1958a). Continuous culture of microörganisms: some theoretical aspects. In *Continuous Cultivation of Microörganisms* (Proceedings of the 1st International Symposium). Ed. Málek, I. Prague: Nakladatelství Českslovenské Akademie Věd.

HERBERT, D. (1958b). Some principles of continuous culture. In *Recent Progress in Microbiology*. Ed. Tunevall, G. Stockholm: Almqvist & Wiksell.

HERBERT, D. (1964). Multi-stage continuous culture. In *Continuous Cultivation of Microörganisms* (Proceedings of the 2nd International Symposium) Prague: Nakladatelství Československé Akademie Věd.

HERBERT, D., ELSWORTH, R. E. & TELLING, R. C. (1956). The continuous culture of bacteria; a theoretical and experimental study. *J. gen. Microbiol.*, **14**, 601.

HINSHELWOOD, C. N. (1946). *The Chemical Kinetics of the Bacterial Cell.* Oxford: The University Press.

HOLME, T. (1958). Glycogen formation in continuous culture of *Escherichia coli* B. In *Continuous Cultivation of Microörganisms* (Proceedings of the 1st International Symposium) Prague: Nakladatelství Československé Akademie Věd.

HOLME, T. & PALMSTJERNA, H. (1956). Changes in glycogen and nitrogen-containing compounds in *Escherichia coli* B during growth in deficient media. *Acta chem. Scand.*, **10**, 578.

JEFFREYS, H. (1937). *Scientific Inference.* Cambridge: The University Press.

LONGMUIR, I. S. (1954). Respiration rate of bacteria as a function of oxygen concentration. *Biochem. J.*, **57**, 81.

MARR, A. G., NILSON, E. H. & CLARK, D. J. (1963). The maintenance requirement of *Escherichia coli. Ann. N.Y. Acad. Sci.*, **102**, 536.

MATELES, R. I., RYU, D. Y. & YASUDA, T. (1965).. Measurement of unsteady state growth rates of microörganisms. *Nature, Lond.*, **208**, 263.

MONOD, J. (1942). *Recherches sur la Croissance des Cultures Bactériennes.* Paris: Hermann et Cie.

MONOD, J. (1949). The growth of bacterial cultures. *A. Rev. Microbiol.*, **3**, 371.

Monod, J. (1950). La technique de culture continue: théorie et applications. *Annls Inst. Pàsteur, Paris*, **79**, 390.

MOSER, H. (1958). *The Dynamics of Bacterial Populations Maintained in the chemostat.* Washington, D.C.: The Carnegie Institution.

NILSON, E. H., MARR, A. G. & CLARK, J. D. (1962). The maintenance requirement of *Escherichia coli. Bact. Proc.*, **62**, 37.

NOVICK, A. (1958). Experimentation with the chemostat. In *Recent Progress in Microbiology.* Ed. Tunevall, G. Stockholm: Almqvist & Wiksell.

NOVICK, A. & SZILARD, L. (1950). Experiments with the chemostat on spontaneous mutations of bacteria. *Proc. natn. Acad. Sci., U.S.A.*, **36**, 708.

PIRT, S. J. (1965). The maintenance energy of bacteria in growing cultures. *Proc. R. Soc.*, (B), **163**, 224.

POWELL, E. O. (1958a). An outline of the pattern of bacterial generation times. *J. gen. Microbiol.*, **18**, 382.

POWELL, E. O. (1958b). In Symposium VI of *Recent Progress in Microbiology.* Ed. Tunevall, G. Stockholm: Almqvist & Wiksell.

POWELL, E. O. (1965). Theory of the chemostat. *Lab. Pract.*, **14**, 1145.

SCHAEFER, W. (1948). Recherches sur la croissance du *Mycobacterium tuberculosis* en culture continue. *Annls Inst. Pasteur, Paris*, **74**, 458.

STOKER, J. J. (1950). Nonlinear Vibrations. New York: Interscience Publishers, Inc.

TEISSIER, G. (1942). Croissance des populations bactériennes et quantité d'aliment disponible. *Rev. sci., Paris*, **80**, 209.

TEMPEST, D. W. & HERBERT, D. (1965). Effect of dilution rate and growth-limiting substrate on the metabolic activity of *Torula utilis* cultures. *J. gen. Microbiol.*, **41**, 143.

TEMPEST, D. W., HUNTER, J. R. & SYKES, J. (1965). Magnesium-limited growth of *Aerobacter aerogenes* in a chemostat. *J. gen. Microbiol.*, **39**, 355.

VOLTERRA, V. (1931). *Theory of Functionals.* Ed. Fantappiè, L.; tr. Long, M. London: Blackie & Son, Ltd.

WYSS, O. (1941). The nature of sulphanilamide inhibition. *Proc. Soc. exp. Biol. Med.*, **48**, 122.

EDITORIAL COMMENT

Endogenous metabolism and maintenance

'Endogenous metabolism' means, primarily, the rate at which organisms consume their own mass in the absence of an energy source. The misgivings felt about the use of the expression in any other circumstances clearly arise from our lack of a general theory of the processes and pathways involved. Herbert (1958b) originally proposed to regard endogenous metabolism as operating at a constant rate over the whole range of growth rates in a given medium, but later found (Tempest & Herbert, 1965) that in fact it varied. Whether it is variable or constant (in the primary sense), we do not know that the same processes occur in the presence of the energy source under steady state conditions—no permanently realisable steady state is characterised by a negative growth rate. On the other hand we do know that the efficiency of utilization of the substrate varies with growth rate. In the present state of knowledge, then, it might seem better to avoid any unwarranted implication and to work instead with a 'maintenance coefficient' (Pirt, 1965; 'coefficient d'entretien', Monod, 1950) which for uniformity we may write as q. Then

$$Y_g(q - q_e) = \mu; \quad q = \frac{\mu + q_e Y_g}{Y_g},$$

$$Y = \frac{\mu}{q} = Y_g \frac{\mu}{\mu + Y_g q_e} = Y_g \frac{q - q_e}{q}$$

and $Y_g q_e$ is arithmetically the same as the μ_r used by Powell, whatever its interpretation. These equations embody the simplest possible assumptions to account for the fall in yield at low growth rates. They do not necessarily imply that q_e is independent of s (or μ). Their use ought, at least for the time being, to be restricted to conditions of balanced growth.

Without any direct knowledge of mechanism, we can say that q_e is the rate at which a nutrilite is being consumed in processes which do not serve to increase biomass, processes calling for an expenditure of energy or for material replacement or both; at the same time μ_r is the amount by which the growth rate is reduced below a potential maximum because of this diversion.

Editor's Comments on Papers 22 Through 26

22 **Scherbaum and Zeuthen:** *Induction of Synchronous Cell Division in Mass Cultures of* Tetrahymena piriformis

23 **Helmstetter and Cummings:** *Bacterial Synchronization by Selection of Cells at Division*

24 **Padilla and James:** *Continuous Synchronous Cultures of Protozoa*

25 **Dawson:** *Continuous Phased Culture – Experimental Technique*

26 **Goodwin:** *Synchronization of* Escherichia coli *in a Chemostat by Periodic Phosphate Feeding*

In the middle 1950s Scherbaum and Zeuthen's synchronization of free living populations of the protozoan *Tetrahymena pyriformis* was a landmark development in the study of growth that came from the biological fringe of the discipline. Employing similar alternations of temperature, Hotchkiss [*Proc. Natl. Acad. Sci. (U.S.)*, **40**, 49–55 (1954)] synchronized bacteria (*Bacillus subtilis*) soon afterward, and then followed a spate of other methods.

The methods now available may be classified by relating them to closed and open systems, as in asynchronous cultures, and show the changing and steady-state characteristics of those methods.

Further developments from the papers reproduced here include Zeuthen [*Exptl. Cell Res.*, **68**, 49–60 (1971)], Helmstetter and Cummings [*Biochim. Biophy. Acta*, **82**, 608 –610 (1964)], Goodwin [*European J. Biochem.*, **10**, 515–522 (1969)], and Dawson [*J. Appl. Chem. Biotechnol.*, **22**, 79–103 (1972)].

In this new dimension of growth new problems arise, as, for example, the discrimination of artifact production, the development of new methods for harvesting, immobilizing, examining, and analyzing the cells. Elsden's remarks (Paper 20, p. 310) concerning the pertinence and meaning of analytical data are of special significance for studies of microbial growth at the cellular level. In this area, the simple empirical parameters of protein, RNA, lipid, and other classes are no longer adequate or useful except in a superficial manner; spectral details of their compositions, qualitative and quantitative, are required now. It is in this area of experimental endeavor that future developments in growth studies presently lie dormant.

22

Reprinted from *Exptl. Cell Res.*, **6**(1), 221–227 (1954)

INDUCTION OF SYNCHRONOUS CELL DIVISION IN MASS CULTURES OF *TETRAHYMENA PIRIFORMIS*

O. SCHERBAUM[1] and E. ZEUTHEN

Laboratory of Zoophysiology, University of Copenhagen, Denmark [2]

Received October 22, 1953

Wᴛʜ the ultimate scope of making the mitotic cycle accessible to the study with standard chemical and physiological techniques, we have during the past year endeavored to produce synchronism of cell division in mass cultures of *Tetrahymena piriformis*, Lwoff's strain.

Tetrahymena was selected because it can be grown in pure culture in proteose-peptone media and—sole among animal cells—also in fully defined media as reported by Kidder and Dewey (4). One possible way of synchronizing the growth activities of this organism would be to place a barrier at a certain point in the cell cycle, say by using inhibitors known or suspected to inhibit specific phases. Colchicin and Aminopterin were tried, with meso-inositol (5) and folic acid, or folic acid derivatives, as suggested releasing agents. We failed because the inhibitors proved non-toxic, as earlier reported (3, 4).

The effect of temperature has been tried, however, starting out from a different theory. Ephrussi (2) many years ago reported that the Q_{10} is different for the different phases of cell division in the sea urchin egg. It was, therefore, considered that exposure to lowered temperature, for a time shorter than the mitotic cycle at the low temperature, with subsequent return to optimum temperature, might tend to bring closer separate stages having different Q_{10}'s. After the lapse of another cycle at optimum temperature repetition of the treatment should tend to make the group larger, etc., till synchronism in the whole culture had been produced. Indeed, shifts of temperature $28° \rightleftharpoons 7°C$ had some effect, but disturbing factors seemed to interfere. These experiments, however, led the way to successful experiments in which the temperature was periodically raised to sublethal levels. This paper reports such experiments. We are not yet going to offer full interpretation of our findings; we think, however, that the results are best understood, not on the basis of different Q_{10}'s as discussed above, but on

---

[1] Supported by a grant of "Statens Almindelige Videnskabsfond".
[2] Address: Juliane Maries Vej 32, Copenhagen.

the basis of the assumption that sublethal temperatures act by blocking —with a considerable recovery time at optimum temperature—a specific step in the cell cycle.

EXPERIMENTS

The cells were grown as pure cultures in 2 per cent proteose peptone (Difco) + 1 per mille liver fraction L (Wilson Laboratories); salts were added according to Kidder and Dewey (4). Under conditions of ample supply of oxygen, in single cell cultures grown in capillaries on 0.5–1.5 µl, medium population densities may reach 1.2 million cells/ml on this medium (personal communication from Mr. H. Thormar). However, there is evidence that true exponential multiplication does not continue beyond population densities of about 100,000 cells/ml or perhaps less. In the experiments to be reported we have therefore worked with very dilute cultures (5,000–50,000 cells/ml), offhand assumed to be in the log-phase of growth. We used two types of culture flasks: either salt cellars (1 ml culture medium, covering 4.5 cm², average depth 2.2 mm, bottom slightly hollow) or $^1/_2$ l, flat-bottomed flasks (10 ml culture medium covering 57 cm², depth 1.75 mm). In the salt cellars the glass walls are thicker (5–6 mm) than in the big flasks (about 1 mm). No shaking was used. The flasks were dipped into a water bath which was regulated to $\pm$ 0.1° C. The temperature of the bath could be shifted up and down by the use of a signal watch which at predetermined times switched the current from one thermoregulator to another. Adjustment of the temperature in the bath always occurred with some delay (8 min. for an increase from 28° to 34° C, and 12 min. for a drop from 34° to 28° C), and within the cultures themselves the delay was in addition influenced by the amount of culture medium and by the thickness of the glass. The figures corresponding to those given for the bath were 20, resp. 17 minutes for the salt cellars, and 15 resp. 14 minutes for the $^1/_2$ l flasks. Usually, cultures were inoculated from stationary phase cultures (kept at 23–24° C) and grown at 28° C; after the end of a lag the division index was followed; it represents the ratio between number of cells in division and number of cells not dividing. As division cells were counted all stages from the onset of cytoplasmic constriction to separation of the daughter cells. The division index was determined on 100 µl samples fixed in 5 per cent formaldehyde; every point represents the counting of about 300 cells. We found the division index to vary from about 0.05 to about 0.10. Single cells growing in 0.5–1.5 µl of the medium here used were observed to divide every 135 minutes at 28° C ($\pm$ 20 min.). Division times of 6.8–12.5 min. have been reported (1). The division index calculated from the two set of data is 0.05–0.095, which corresponds nicely with our observations (cf. Fig. 1).

Upon transfer from 28°–29.5° C (which shall hereafter, without special qualification, be called "optimum temperature", cf. however (6)) to a sublethal temperature (32°–34° C)) the division index tends to decrease and soon it gets very low. However, at the high temperature there is a partial or a complete block for growth (6). So we suggest that sublethal temperature prevents growth (e.g. syntheses), it also prevents new cells from entering a division, but it does not prevent a division, once initiated, from running to completion.

Fig. 1.
Division index in a culture of *Tetrahymena piriformis* treated for $6^1/_2$ hours with intermittent heat shocks. The uppermost curve shows the switch of the temperature as indicated by signals on the watch, respectively transfer into bath (1 h) and removal from bath (8 h). Proteose - peptone culture, 10 ml in flask. Total initial population about 400,000.

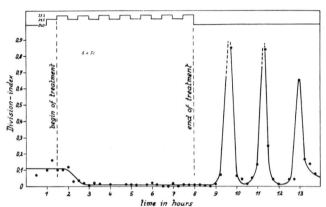

However, we expose cells to sublethal temperatures only for a short time and then, for another short period, we transfer them to the optimum temperature. The cells then resume growth but they do not, within this short time, begin to divide. Before they recover the mechanisms which in normal cells operate to switch cells from synthesis·(growth in mass) to division they are exposed to another short temperature shock and thereafter again transferred to optimum temperature and so on. With this treatment of intermittent heat shocks (we might say: "intermittent fever") the cells grow bigger and bigger (cf. Fig. 2 a, b). After several hours they are returned to constant optimum, or usually just to room temperature (the dishes were placed under a binocular microscope for continuous observation). About $1^1/_2$ ($\pm$ 10 minutes) hours at this temperature a burst in division activity was observed (Fig. 2 c). As many as 85 per cent of the cells may be in division simultaneously. From now on, through a few mitotic cycles, the divisions appear to be synchronized (cf. Fig. 1). We have observed three successive peaks, with respectively 85, 83 and 64 per cent of the cells dividing simultaneously. In between there are minima in which only about 2 per cent of the cells are in division.[1] The peaks appear to be closer in time than indicated by the duration of the normal mitotic cycle (at 24° C: 1.7 hours, as compared with a normal of 2.5–3 hours).

---

[1] Here we may add that in the experiments with lowered temperature we found a low first peak in division activity, but after this division rate returned, more or less, to the average for an untreated culture, e.g. no rhythmicity was induced.

Experimental Cell Research 6

Obviously, in the type of experiment reported there is a considerable number of variables which have not yet been examined in sufficient detail. In attempting to establish conditions of temperature treatment which will produce the highest possible degree of synchronism in the cultures we have made the following observations:

Time of treatment: 6–10 hours. Shorter treatment tends to produce lower peaks of division activity and longer treatment (16 hours) results in expressed distortion of the cells.

Temperature during treatment: so far we have obtained the best results if we let the temperature shift between 29° and 33° C (32.3°–33.7° C in the different cases) with a period of 1 hour, e.g. $^1/_2$ hour high, and $^1/_2$ hour low. Two-hour periods were rather ineffective, the division index was not considerably reduced during treatment, and not very high after. Thirty-minutes periods were more effective than 2-hours, but still inferior to 1-hour periods. We have also tried to change the ratio: time at high temperature over time at low temperature, however, with no beneficial effects. Increasing the ratio tended to make the cells more distorted, decreasing the ratio resulted in cell multiplication during treatment and in less piling up of divisions.

During the heat treatment the average cell size increases, e.g. about 3-fold in 7–8 hours. This is illustrated in the photomicrographs, Fig. 2 a, b, c. The treated cells (b) become big and, compared with the controls (a), relatively thick (pear-shaped). Also, they may become somewhat distorted. In addition to what has been said above we add that for a minimum of distortion to develop it may be essential that 1) the population density is so low (order 5,000–50,000/ml) that true exponential multiplication would have been possible had we not exposed the cells to temperature treatment, that 2) oxygen is not limiting in any part of the culture, and 3) that the temperature changes are not too abrupt. Even if there are some distorted cells by the end of heat treatment, all cells assume a perfectly normal look after one or two cycles at optimum or room temperature. It should be commented that no—or at least very few—cells are killed either during or after the heat treatment. Actually, during treatment cell numbers increase very slowly, and about as much as could be expected from the very low, however real, division index found during treatment.

The increase in cell size was followed by volume measurements of single cells in a compression chamber a.m. Scholander, Claff and Sveinsson (7). While the average cell volume increases as shown in Fig. 3, curve $\times$, the relative size differences, initially observed, are retained. The average protein content was studied with an unconventional (and not yet well enough

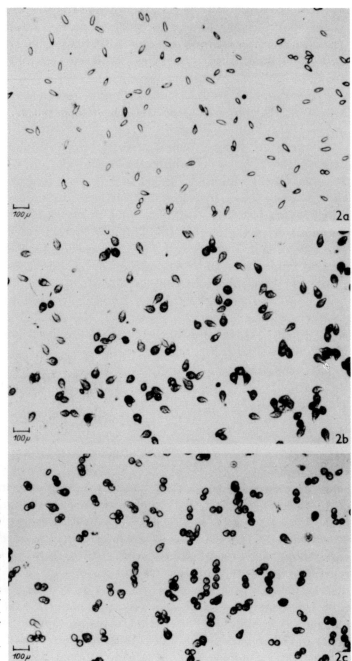

Fig. 2. Exp. 29. 4. 1953. Proteose - peptone culture, 10 ml in flask. Total initial population 115,000 cells. Individuals alive when pictures taken.
a: untreated culture (temp. 24° C) several hours after inoculation.
b: cells treated 9 hours at 28° ⇌ 34° C ($^1/_2$ hour high, $^1/_2$ hour low). Photographed $^1/_2$ hour after return to 24° C.
c: same after 100 minutes at 24° C.

checked) method in which the amount of fixative picked up by cells fixed (for 3–5 days, maximum weight after 3 days) in 3 per cent phosphotungstic acid was measured as the reduced weight of the fixed cells. A Cartesian diver balance (8) floating in the fixative was used. The growth during heat

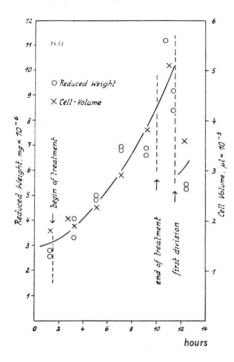

Fig. 3.
Reduced weight (RW) per cell, after 7–8 days of fixation in 3 per cent phosphotungstic acid. RW measurements carried out in the fixative (density: 1.0226, water at same temperature 1.0000). RW of unfixed cells not measured; however, cells fixed for 4 hours weigh only 40 per cent of same cells fixed for three days or more. × cell volumes measured as area × height in a compression chamber, height 10 μ.

treatment is not only cytoplasmic; also the macronuclear volume increases very considerably. After recovery from heat treatment, e.g. at optimum or room temperature, the cells tend to become smaller with each new maximum of division. That is, now the situation is reversed relative to the one during heat treatment: cell divisions are outbalancing the growth activities. For *Tetrahymena* it has previously been reported (9) that big cells can be harvested from stationary phase cultures. Upon transfer to fresh medium these cells resume growth, but they divide faster than they double in mass, with the result that in the early generations cells get smaller and smaller. Carefully traced growth curves for single cell cultures (9) were interpreted to indicate that the growth in mass (mass measured by the amount of respiration!) does not keep up with the divisions because the initial big cell grows exponentially with only part of its body; the remainder does not grow at all and becomes diluted away upon the increasing number of cells in the progeny. It is a situation like this which we believe pertains to the giant cells produced by the heat treatment.

SUMMARY

This preliminary report shows that, using intermittent heat treatment of mass cultures of *Tetrahymena*, 85 per cent of the cells can be induced to undergo division simultaneously. Three successive maxima of divisions have been observed. The cell populations used represent total cell numbers of anything between 5,000 and $^1/_2$ million corresponding to dry weights of the order 0.03 to 5 mg. The work is continued with attempts to synchronize much larger quantities of material, first on proteose-peptone media, then on synthetic media.

REFERENCES

1. BROWNING, I., VARNEDOE, N. B., SVINFORD, L. R., *J. Cellular Comp. Physiol.*, **39**, 371 (1952).
2. EPHRUSSI, B., *Protoplasma*, **I**, 105 (1926).
3. HALBERSTAEDTER, L., and BACK, A., *Nature*, **152**, 275 (1943).
4. KIDDER, G. W., and DEWEY, V. C., Biochemistry and Physiology of Protozoa. Ed. by A. Lwoff, p. 323, 1951.
5. MURRAY, M. R., DE LAM, H. H., and CHARGAFF, E., *Exptl. Cell Research*, **2**, 165 (1951).
6. PHELPS, A., *J. exptl. Zool.*, **102**, 277 (1946).
7. SCHOLANDER, P. F., CLAFF, C. L., and SVEINSSON, S. L., *Biol. Bull.*, **102**, 157 (1952).
8. ZEUTHEN, E., *Compt. rend. Lab. Carlsberg, Sér. chim.*, **26**, 243 (1948).
9. —— *J. Embryol. and exptl. Morphol.*, **1**, in press (1953).

23

Reprinted from *Proc. Natl. Acad. Sci. (U.S.)*, **50**(4), 767–774 (1963)

BACTERIAL SYNCHRONIZATION BY SELECTION OF CELLS AT DIVISION

By Charles E. Helmstetter* and Donald J. Cummings

NATIONAL INSTITUTE OF NEUROLOGICAL DISEASES AND BLINDNESS, BETHESDA, MARYLAND

Communicated by Raymond E. Zirkle, August 26, 1963

Since Maruyama and Yanagita[1] first described a filtration method for obtaining synchronously dividing bacteria, a number of reports[2–5] have appeared utilizing this approach. Basic to the filtration technique were the assumptions that the bacteria played a passive role during filtration and that synchronous growth was achieved by the principle of selection by size. However, in the course of investigations to improve the reliability and increase the yield of the technique as used by Helmstetter and Uretz,[4] a new principle for bacterial synchronization became apparent. This principle involves (1) the fact that a bacterium will bind to a variety of surfaces and (2) the ability of the cell to divide while bound to these surfaces. If a population of bacteria could be irreversibly bound to a surface while growth medium flowed past the surface, the only cells which would appear in the medium would be those new daughter cells which were not involved in the attachment of their parent cells to the surface. Since the cells bound to the surface would be growing, the unbound sister cells which elute from the surface would be representative of the youngest[6] cells in a log phase culture. Under the proper conditions, new daughter cells could be removed for extended periods of time from the population growing on the surface, and these cells would grow synchronously.

The purpose of this report is to describe a technique for the continuous removal of new daughter cells from a growing culture and to present evidence that this technique operates by the principle stated above.

Materials and Methods.—Bacteria and growth conditions: The organism used was *Escherichia coli* strain B/r (ATCC 12407). The minimal medium contained NH$_4$Cl, 2 gm; Na$_2$HPO$_4$, 6 gm; KH$_2$PO$_4$, 3 gm; NaCl, 3 gm; MgSO$_4$, 0.013 gm; Na$_2$SO$_4$, 0.011 gm; and glucose, 5 gm in 1 liter of distilled water. In preparation for each experiment, an inoculum of the bacteria from a nutrient agar slant was grown to saturation in nutrient broth at 37°C. A 0.1-ml sample from this culture was inoculated into 100 ml of the minimal medium and incubated for 24 hr at 37°C. A 0.2–1.0-ml sample of a 100-fold dilution of the 24-hr culture was then inoculated into 5–10 liters of minimal medium and incubated with aeration at 37°C. After 17–20 hr the culture was in exponential growth at a titer of 2×10^7 to 2×10^8 bacteria/ml. One liter of this culture was used to inoculate the synchronization apparatus, and the remainder was filtered through a 0.65-μ-pore, 152-mm-diameter Millipore filter and used for washing and eluting the cells in the synchronization apparatus. This filtered medium will hereafter be called conditioned medium.

Synchronization: A schematic diagram of the synchronization apparatus is shown in Figure 1. It was essentially a two-section stainless steel funnel coupled to a recirculating device. The binding surface for the cells lay on a fine screen which was clamped between the upper and lower sections of the funnel. The binding material used was Whatman cellulose anion exchangers in the paper form which were cut to 15-cm-diameter circles. The exponential phase cells were bound to the anion exchanger by passing the culture through the paper under low pressure (1–3 lb/in²) so that the flow rate was approximately 100 ml/sec.[7] Most of the bacteria passed through the exchanger and were discarded. The anion exchanger was then washed as is described in detail below. Elution was carried out in a continuous fashion by recirculating the eluent with a Sigmamotor OV-22 Kinetic-Clamp pump. Cells were prevented from re-entering the anion exchanger by passing the recirculating eluent through a 0.65-μ-pore, 152-mm-diameter Millipore filter. A few aminoethyl-cellulose papers were always placed below this Millipore filter to stabilize any effects on the growth medium caused by the introduction of ion-exchange material. The elution rate was maintained reasonably constant since "bound" cells would only remain bound under steady-state flow conditions. The elution was controlled either by adjusting the height of the fluid column above the anion exchanger or by setting the pumping rate in a closed system. Samples were taken directly from the eluent and not the recirculating pool. All of these operations were performed in a 37°C room.

For studies on the properties of this system (Figs. 2–5), the number of ion-exchange papers was varied between 1 and 8, and these were inoculated with 1 liter of an exponential phase culture at 6×10^7 bacteria/ml unless otherwise specified. The exchanger was then washed with 1 liter of conditioned medium. The wash step was performed by applying and releasing a differential pressure at approximately 350-ml intervals until all of the washing fluid had passed through the exchanger. The flow rate was 100 ml/sec. The elution medium for these experiments consisted of a mixture of one part conditioned medium and one part fresh minimal medium unless otherwise specified.

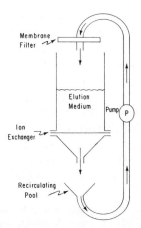

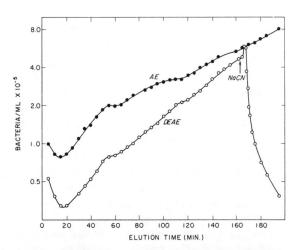

Fig. 1.—Schematic diagram of the synchronization apparatus. While the exact specifications were not critical, the two-section funnel was 15 cm in diameter at the position of the exchanger and had a capacity of 5 liters.

Fig. 2.—Elution of bacteria from 8 aminoethyl-cellulose papers (AE) and 4 diethylaminoethyl-cellulose papers (DEAE). After elution of the DEAE papers for 163 min, NaCN was added to the eluent to give a final concentration of 0.005 M (arrow). The elution rate was 600 ml/min.

For optimal yield of synchronized bacteria the apparatus was operated with one aminoethyl-cellulose paper as the binding surface. The total number of cells passed through the paper for the binding operation depended upon the concentration of eluted cells desired. This step was usually performed using 1 liter of 1.2×10^8 bacteria/ml, but has been performed with titers ranging from 2×10^7 to 2×10^8 bacteria/ml with essentially the same results. In general, the more thoroughly the paper was then washed, the better was the subsequent selection of new daughter cells. Optimal selection was obtained after performing three 1-liter washing steps as described above and inverting the paper after each liter passed through. If the washing steps were reduced or eliminated, the selection was poorer but the yield was much greater. Elution was always performed with the paper inverted relative to its position at inoculation. After the paper was inverted for elution, a 0.65-μ-pore, 152-mm-diameter Millipore filter was placed above the paper to prevent backflow of cells, and elution with conditioned medium was begun. Samples were collected for synchronous growth analysis after elution had proceeded for 40 min, and were incubated with vigorous shaking.

Counting procedure: In order to avoid any effects of plating procedures on the presumed quality of synchrony,[4] all bacterial counts and size distributions were determined using a Coulter Counter Model B with a particle size distribution plotter. The counter was operated with a 30-μ-diameter orifice, maximum amplification, and a 1/current setting of 0.707. Coulter Counter readings were maintained in the 2–20 $\times$ 10^3 range and were performed either directly in the eluent or after dilution in normal saline which had been filtered through a type HA Millipore filter. No corrections were made in the data for background or coincidence counts.

Results.—The synchronization technique will be analyzed first by presenting the characteristics of the selection principle, and second by describing the results obtained under optimal conditions for continuous collection of new daughter cells. Analysis of the technique was best accomplished by observing the elution of bacteria from the anion exchanger for extended periods of time. Figure 2 shows the elution pattern during a few hours of recirculation of growth medium through a stack of 8 aminoethyl-cellulose papers (curve AE) and 4 diethylaminoethyl-cellulose papers (curve DEAE). During the first 15 min of elution, the number of cells eluted decreased with time, presumably because of the washout of unbound and weakly bound cells. After this initial period, the bacteria in the eluent were small in size, but this selection became poorer with time. To account for the increase in bacterial concentration with elution time, it was concluded that the cells were growing on the anion exchanger. This conclusion was supported by the observation that if cell growth were inhibited, the appearance of cells in the eluent was suppressed. When NaCN was added to the recirculating eluent (curve DEAE), the number of cells removed from the anion exchanger rapidly decreased. The requirement of growth on the exchanger for elution of bacteria is also shown in Figure 3, where the concentration of salts in the eluent medium was altered. This alteration impeded growth of the cells on the exchanger, and resulted in a rapid decrease of cell number in the eluent. When elution was continued for a sufficient time for the cells to adapt to the altered growth medium, recovery to approximately the original elution titer occurred.

Although the multiple-paper experiments illustrated that growth during elution was necessary for selection, the withdrawal of new daughter cells was not optimal under these conditions. If this selection is related to the division of cells bound to the exchanger, then selection should be enhanced by reducing the number of papers. Figure 4 shows the results of the elution of bacteria from 8, 4, and 1 aminoethyl-cellulose papers. As the number of papers was reduced from 8 to 1, the brief plateaus observed at intervals of a generation time became more pronounced. The elution pattern from 1· aminoethyl-cellulose paper occurred stepwise with the generation time because the barriers to removal of the new daughter cells were minimized, and this resulted in a constant efficiency of removal. In the multiple-paper experiments some of the cells eluted from the upper papers became bound in

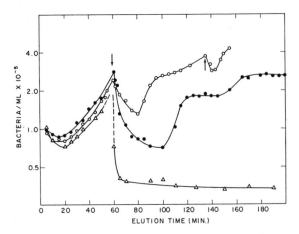

Fig. 3.—Effect of alteration of the composition of the eluent medium on elution of bacteria from 8 aminoethyl-cellulose papers. After 58 min of elution (first arrow), the minimal salts concentration was increased in the eluent in three separate experiments so that the final concentration was 1.6(○), 2.3(●), and 3.0(△) times the original. After 135 min of elution (second arrow), the eluent in the experiment described by the upper curve (○) was diluted 1:1 with distilled water. The elution rate was 600 ml/min.

the lower papers. As a result, the efficiency of removal increased with time and obscured the stepwise elution pattern characteristic of a single paper. Therefore, interpretation of these results is best accomplished by considering the single-paper experiment. After 35 min of elution the bacteria in the eluent from a single paper were primarily new daughter cells (see Fig. 7) and remained so for several hours of elution. The shape of the elution pattern between the steps depended upon the age[6] distribution of the bacteria initially bound to the exchanger. The flatness of the plateau regions, which was observed in experiments with various high efficiencies of removal, indicated that the cells bound to the paper were uniformly distributed in age. The number of cells which remained on the paper after each round of division depended on the efficiency of removal of the new daughter cells. Therefore, if less than 50 per cent of the new daughter cells was removed, then the concentration of cells in the eluent had to increase at intervals of a generation time.

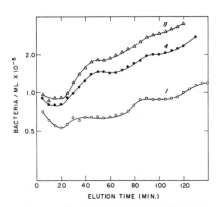

Fig. 4.—Elution of bacteria from 8 (△), 4 (●), and 1 (○) aminoethyl-cellulose papers. The elution rate was 600 ml/min.

As shown in Figure 5, the efficiency of removal of new daughter cells depended on the ionic strength of the eluent relative to that of the binding medium. Curve *A* in Figure 5 shows that it was possible to approach the condition of constant elution titer for an extended period of time by inoculating an aminoethyl-cellulose paper with a culture which had been diluted 1:10 with distilled water and eluting with normal minimal medium. This result was also obtained by eluting with medium containing approximately twice the salt concentration of the normal medium used for inoculation of the paper. However, in both cases the constant elution pattern was only approached, and the concentration of eluted cells began to rise after about 2 hr of elution. Curve *B* in Figure 5 shows that when the paper was eluted with medium which had been diluted 1:2 with distilled water relative to the inoculation medium, the elution pattern was similar to that obtained in the multiple-paper experiments.

While the elution behavior has been described at length, it remains to be shown that the eluted cells grew synchronously. The degree of synchronous growth of bacteria in samples taken at any time during elution depended upon the elution

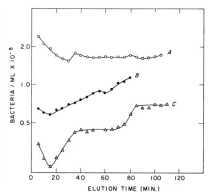

FIG. 5.—Elution of bacteria from 1 aminoethyl-cellulose paper as a function of the minimal salts concentration of the eluent relative to the inoculation medium. (*A*) The paper was inoculated with a log phase culture after it had been diluted 1:10 with distilled water (to 3×10^7 bacteria/ml), washed as described in *Materials and Methods* with 1 liter of a 1:10 dilution of the conditioned medium, and eluted with conditioned medium from the culture prior to dilution. The elution rate was 600 ml/min. (*B*) The paper was inoculated and washed in the normal fashion as described in *Materials and Methods*, but eluted with a 1:1 dilution of the conditioned medium in distilled water. The elution rate was 200 ml/min. (*C*) The paper was inoculated, washed, and eluted with the same conditioned medium. The elution rate was 600 ml/min.

medium used and the degree of binding of the cells to the paper. Figure 6 illustrates the synchronous division obtained with and without washing the aminoethyl-cellulose paper after inoculation. Both samples for synchronous growth analysis were taken after 40 min of elution, but the quality of synchrony was essentially the same in samples taken over 1 hr after this time. The synchronously dividing cells were collected without further manipulation from a growing culture as they fell off the

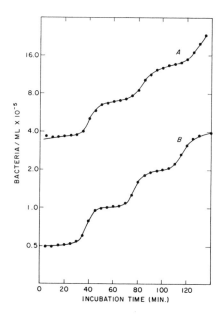

FIG. 6.—Synchronous growth under conditions for optimal degree of synchrony (*B*) and near maximum yield of bacteria (*A*). Both were performed in the same manner, as described in *Materials and Methods*, except that: (*B*) the paper was washed after inoculation and the bacteria were counted directly in growth medium, and (*A*) the paper was unwashed and the bacteria were counted after dilution in normal saline. The flow rates for elution were set at approximately 150 ml/min by maintaining a 1-liter head above the paper. In both cases, a sample of the eluent was collected for 1 min, and bacterial concentration was determined during incubation at 37°C with vigorous shaking.

351

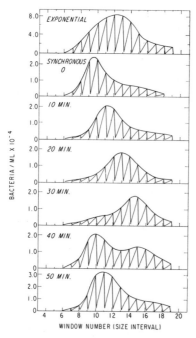

FIG. 7.—Size distribution of cells during synchronous growth (0–50 min) compared to the size distribution of an exponential phase culture (exponential). The exponential size distribution was determined on a culture at 1×10^8 bacteria/ml after dilution in conditioned medium. The size distribution of cells during synchronous growth was determined directly in the growth medium by withdrawing samples from the culture at the times indicated during growth. Time 0 for synchronous growth shows the size distribution of cells in the eluent after 40 min of elution. Size is recorded as direct window number (threshold interval) reading from the Coulter Counter particle size distribution plotter. The counting time for each size interval was set at 8 sec.

exchanger, and therefore no lag in growth occurred. The doubling time during synchronous growth was always shortest when the paper was inoculated at high titer and then thoroughly washed. Figure 7 shows the size distribution of cells taken at the times indicated during synchronous growth. As can be seen, the cells immediately after elution were small in size relative to the exponential phase population and increased in size during the interdivisional period. Changes in cell size during synchronous growth have also been reported by Lark and Lark.[3] At division (40 min) the cells again became small in size and this pattern repeated in succeeding generations.

Discussion.—A procedure for continuous withdrawal of synchronized bacteria from a growing culture has been described. The results show that the selection is related to an ionic binding between the bacteria and the anion exchanger,[8] and the subsequent removal of the youngest cells. The evidence supports the assumption that the majority of cells removed from the culture on the exchanger consists of those new daughter cells which were not involved in the binding. That the selection is based on this principle, and not simply on the removal of small cells due to their weaker or unusual binding properties, is indicated by the fact that growth on the exchanger is necessary for the elution of cells. Under steady-state elution conditions, cells which were not growing could not be removed regardless of their size. In addition, the stepwise elution curves obtained in the single-paper experiment indicate that the bound cells were uniformly distributed in age. Although the exchanger was inoculated with a log phase population, a uniform age distribution could arise on the paper by a counterbalancing of cell number with binding ability. While Powell[9] has shown that the younger cells are more frequent in a log phase culture, we have evidence that the larger cells bind more readily to the exchanger. If an unusually weak binding stage exists during growth of *E. coli* (as shown for HeLa cells in the synchronization technique of Terasima and Tolmach[10]), a pronounced dip would have been expected in the elution curve at intervals of a generation time. Finally, it was not possible to obtain a negative slope in the elution curve, as would be expected if random removal of small cells were involved, unless growth on the exchanger was stopped by artificial means. At best, a constant elution pattern could only be approached. It must be assumed, therefore, that a constant elution pattern corresponds to 100 per cent efficiency of withdrawal of the new daughter cells which are capable of being removed. If this is the case, then only one cell of a new sister pair falls off the paper, and this must be the sister cell which was not involved in the binding. There can be, of course, some cells on the exchanger that are so completely bound that neither sister cell elutes at division, and also cells so weakly bound that both sisters elute at, or any time after, division.

The results indicate that the assumed size fractionation technique as used by Helmstetter and Uretz[4] also operated in the fashion described here. The fact that selection by size was not involved in that technique is supported by a number of observations. First, as the number of papers is decreased in a pile constructed for filtration, the quality of selection improves. Second, the results remain essentially the same when different bacterial strains of widely varying sizes are filtered and eluted from the same size paper pile. Third, the concentration of cells eluted from the pile depends on the ionic strength of the eluent. Fourth, when various ion-exchange papers are used in the pile, there is considerable variation in the results although the porosity of the papers does not vary significantly; that is, when the filtration procedure is performed with a strong anion exchanger (diethylaminoethyl-cellulose), most of the cells remain on the pile. When a cation exchanger (carboxymethyl-cellulose or cellulose phosphate) is used, most of the cells pass through the pile. A quantitative analysis of the properties of ion-exchange materials in relation to micro-organisms has been reported by Rotman.[11]

The procedure described here has a number of interesting properties:

(1) The concentration of selected bacteria does not present a serious problem because the apparatus readily produces any concentration desired. Since selection is independent of elution rate and concentration is inversely proportional to elution rate, a concentrated sample can be obtained by reducing the elution rate. In the experiments described, the flow rates were deliberately adjusted so that bacterial counts could be performed directly in the eluent, but the apparatus yields 10^7-10^8 bacteria per minute in any volume desired. Concentration can also be enhanced by increasing the area of the anion exchanger, washing the paper with fresh minimal medium prior to inoculation, or by passing a larger total number of cells through the paper at inoculation. Finally, high titers can be achieved by delaying collection of selected cells until the concentration in the eluent has risen to the desired value under conditions of less than 100 per cent efficiency of removal.

(2) It is not necessary to collect a single large sample of the eluent for synchronous growth experiments. Since the quality of selection remains essentially constant for long elution times, it is only necessary to collect small samples at intervals and incubate these separately. In this manner, large quantities of cells of all ages are readily available.

(3) The synchronization apparatus is actually a growth vessel where the titer of the growing cells can be held nearly constant for some time by continuously removing half of the youngest cells.

(4) Mother-daughter relationships can be studied where this relation is identifiable if the mothers remain on the paper and the daughters elute.

(5) For some studies of the effects of various agents as a function of cell age, it is not necessary to grow a synchronous culture at all. If the agent is added directly

353

to the eluent, changes in the elution pattern of the new daughter cells will directly reflect the age at which the cells on the exchanger are affected by the agent.

We wish to thank Drs. David H. Rammler and Lawrence R. Fowler for many helpful discussions.

* Present address: University Institute of Microbiology, Copenhagen, Denmark.

[1] Maruyama, Y., and T. Yanagita, *J. Bacteriol.*, **71**, 542 (1956).

[2] Abbo, F. E., and A. B. Pardee, *Biochim. et Biophys. Acta*, **39**, 478 (1960).

[3] Lark, K. G., and C. Lark, *Biochim. et Biophys. Acta*, **43**, 520 (1960).

[4] Helmstetter, C. E., and R. B. Uretz, *Biophys. J.*, **3**, 35 (1963).

[5] Nagata, T., these Proceedings, **49**, 551 (1963).

[6] We shall use the concept of age in describing the stage of development of a cell during an interdivisional period.

[7] For the work reported here the funnel apparatus was fitted with a cap which was connected to an air pressure line for the binding and washing operations. However, these operations could be performed satisfactorily with vacuum.

[8] Although the best results were obtained with aminoethyl-cellulose, other binding materials were also used. Ordinary cellulose filter paper, glass or rubber tubes, and various ion exchangers all gave some selection but lower yields.

[9] Powell, E. O., *J. Gen. Microbiol.*, **15**, 492 (1956).

[10] Terasima, T., and L. J. Tolmach, *Exptl. Cell Research*, **30**, 344 (1963).

[11] Rotman, B., *Bacteriol. Rev.*, **24**, 251 (1960).

24

Reprinted from *Methods in Cell Physiology*, Vol. I, Academic Press, Inc., New York, 1964, pp. 141–157

Chapter 9

Continuous Synchronous Cultures of Protozoa[1]

G. M. PADILLA[2] AND T. W. JAMES

Gerontology Branch, National Heart Institute, National Institutes of Health, PHS, U. S. Department of Health, Education and Welfare, Bethesda, and the Baltimore City Hospitals, Baltimore, Maryland and Zoology Department, University of California, Los Angeles, California

I. Introduction

One natural outgrowth of the development of synchronous cultures of various cell types has been the establishment of a system which can be maintained for an indefinite number of generations without loss of cell division synchrony. We have called such systems "continuous synchronous cultures." This may be considered a misnomer since the term "continuous culture" is usually applied to a system in which a continuous inflow of

---

[1] This research has been supported in part by National Science Foundation Grant G-19297.

[2] Present address: Biology Division, Oak Ridge National Laboratory, Oak Ridge, Tennessee.

sterile medium and a concomitant washout of cells results in a steady state population in the growth vessel. The chemostat (Novick and Szilard, 1950), bactogen (Monod, 1950), and turbidostat (Myers and Clark, 1944) are typical examples of this type of continuous culture, although they differ widely from each other. In the present situation, a "continuous synchronous culture" is a system which is regularly permitted to expand from one population level to another in a stepwise fashion by a synchronous burst of cell division. The culture is then rapidly diluted to the population density level existing before the synchronous burst took place. Such systems can be shown to have certain properties common to a chemostat, irrespective of the method used to induce cell division synchrony. However, since dilution is intermittent rather than continuous, the system is continuous only in the long term sense of its operation.

Development of such culturing methods also serves to provide a partial answer to a criticism which is often leveled against synchronized systems. The criticism hinges on the concept of "balanced growth" which stipulates that growth of a cell can be considered "balanced" only if all the major constituents of a cell double during each cell cycle (Barner and Cohen, 1956). In many synchronized cultures this does not occur, particularly if they are dependent on batch culture techniques.

Through the various phases of the growth cycle in batch cultures, cells continuously adapt to an ever changing environment as the cell population increases. Such adaptations may be reflected in the biochemical profile of cellular constituents, so that cells in the early logarithmic phase of growth differ sharply from those in the deceleratory phase (Buetow and Levedahl, 1962). If synchrony-inducing forces are applied to such batch cultures, it is not surprising to find cells exceeding the prescribed limits of balanced growth, and in fact, fluctuations in the median values may be amplified and intensified through synchrony. Through the adoption of continuous culture techniques, the cells are offered a more stable environment which, in part, results in repetitive synchronous bursts whose characteristics are easily duplicated with a substantial degree of precision each time the cycle is repeated. In fact, if the continuous system is well controlled, the data from one cycle can be superimposed on that of another cycle with confidence.

A. Selection for Synchronized Cells

The use of stepwise dilution procedures in a continuous synchronized culture provides an added dividend that is not immediately evident but that is analogous to a chemostat's tendency to select for short generation time organisms. In a chemostat the washout rate determines the mean generation time of the cells. It has been shown experimentally (Novick and Szilard, 1950) as well as on mathematical and statistical grounds (Moser, 1958), that there is a selection for the shortest mean generation time in the culture. For example, short generation time mutants are selected (Novick and Szilard, 1951). In a continuous synchronized culture with stepwise dilution, the synchronized generation time is determined by the length of the cyclical program which brings the cells into simultaneous division. If halving the volume by dilution is carried out once each cycle (preferably as soon as the burst of division is completed), the number of cells at any fixed time in the cycle will be the same from cycle to cycle, but the proportion of cells that obey the cycle to those that do not will increase progressively. In other words, those cells which do not divide will be diluted or washed out of the culture more rapidly than those which do divide, and unless more nondividing cells are generated in the course of time, the culture will consist exclusively of dividing cells.

This is, of course, an oversimplified view of what is actually occurring. In practice, some small percentage of disobedient cells will be produced at a constant rate, but if their rate of production is much lower than the rate of production of obedient cells, i.e., synchronized cells, they will be reduced by repeated dilutions to a small proportion of the total population. This model offers an opportunity to determine the size of the nondividing class of cells and to change experimentally its dimensions through imposition of cyclical regimens of varying severity. The time of dilution is of some importance, but this will depend on the method by which synchrony is obtained. As a rule, some time in the nondividing period of the cell cycle will be favorable in bringing about this type of selection.

B. Balanced Growth

Another serious criticism often leveled against the product of synchronization is that the cells are not "normal" because of the temporary physiological states (transients) introduced by the synchronizing techniques or procedures being employed (Abbo and Pardee, 1960). For example, a temperature cycle may induce synchronous division by distorting the pattern of biochemical activities that are present in a "normal" cell, i.e., one that is grown at a constant temperature (Scherbaum and Zeuthen, 1954; Zeuthen, 1958). There is no complete answer to this criticism, since this is most likely what happens. Yet, the importance of the question can be appraised by a second one: Are "normal" cells essential to understanding the patterns of biochemical events that are important to cell division? Obviously, if cell division can occur in synchronized systems, a knowledge of just what the distortions are may provide an insight into the essential activities for the process of cell division. The discomfort which arises in some minds is undoubtedly associated with the difficulty in understanding the kinetics of physiological transients. Furthermore, the aim of many investigators who have attempted to use synchronized cultures has not been to understand the process of cell division, and the supposition may have been made that other aspects of the cell cycle are synchronized to the same degree to which the division activities are controlled. The introduction of continuous culture techniques into the problem will provide greater reproducibility in the synchronized systems and will help circumvent some of these difficulties (James, 1961a).

II. Light-Induced Synchrony in *Euglena gracilis*

Recently, the green flagellate, *Euglena gracilis*, has been added to those photosynthetic cell types that can be experimentally induced to divide synchronously by means of a repetitive light-dark cycle. We shall limit our discussion of the system to its continuous culture features, since the various other attributes of the system have been well described elsewhere (Cook, 1960, 1961, 1962; Cook and James, 1960). The discussion below summarizes some of these findings.

A. Development of the Light-Dark Cycle

The light-dark cycle which brought about repetitive synchrony in *Euglena* was derived from a series of considerations encompassing not only the growth of this cell as a function of light exposure but also its biochemical profile and kinetic responses following cessation of illumination (Cook, 1960). In particular, the basis for constructing a repetitive 24-hour cycle rested in great measure on the minimum energy (light) requirements for growth and cell division for *Euglena* at 20°C, cultured on the chemically defined Cramer-Myers (1952) medium (Table I). This medium, which is identical to that also used for *Astasia longa*, consists primarily of inorganic salts chelated by citrate, to which vitamins B_1 and B_{12} are added. In addition, as shall be discussed in more detail later, a source of reduced sulfur is required for maintenance of synchrony. With *Euglena*, this need was filled by cysteine and methionine added to final concentrations of 6.4×10^{-4} M and 10^{-5} M, respectively (Cook, 1960). Both vitamins and SH-bearing compounds were aseptically added to pre-autoclaved medium.

TABLE I

Growth Medium[a]

Compound	Mg/liter
$(NH_4)_2HPO_4$	1000
KH_2PO_4	1000
$MgSO_4 \cdot 7H_2O$	200
Na citrate $\cdot 2H_2O$	645
$CaCl_2 \cdot 2H_2O$	265
$Fe_2(SO_4)_3 \cdot XH_2O$	3.0
$MnCl_2 \cdot 4H_2O$	1.8
$CoCl_2 \cdot 6H_2O$	1.3
$ZnSO_4 \cdot 7H_2O$	0.4
$Na_2MoO_4 \cdot 2H_2O$	0.2
$CuSO_4 \cdot 5H_2O$	0.02
Vitamin B_1 (thiamine $\cdot$ HCl)	0.02
Vitamin B_{12}	0.01
Distilled water to one liter	
Na acetate	5000
pH (Acetate)	6.8

[a]Modification of basal medium of Cramer and Myers (1952).

Repeated determinations revealed that *Euglena* grows at 20°C under a saturation intensity of incandescent white light of 130 foot-candles with an average generation time of about 20 hours. If during exponential growth the light is suddenly turned off, a proportion of the population continued to divide for about 6.2 hours. These would be, of course, cells that had been continuously exposed to light in the preceding 13.8 hours of their life time, the other cells being unable to divide (Cook, 1960). They thus mobilized, in that period of time, sufficient energy to complete their generation. If now a population of cells were to be exposed to alternating periods of light and darkness, in which the same relationship is maintained, i.e., 13.8 hours of light followed by 6.2 hours of darkness, one would expect that initially only a small proportion of cells would complete division after the first light shut-off. As the cycle is repeated, and if there occurs little or no depletion of "stored" energy, a greater proportion of cells would complete cell division in the dark period. Eventually, the entire population would behave as a unit and synchrony would result. The results shown in Fig. 1 indicate that this is indeed the case. For a 24-hour cycle the light period was set at 16 hours and the dark period at 8 hours. At population densities below 5000 cells per milliliter 90–95% of the cells complete division at each burst; this is shown in the upper portion of the

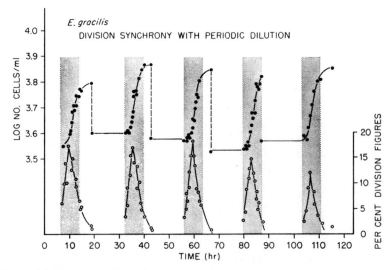

FIG. 1. Light-induced division synchrony in *Euglena gracilis*. The upper curve is a plot of the logarithm (to the base 10) of the population density. The 8-hour dark periods are represented by the shaded areas. The light periods are of 16 hours duration. The broken lines indicate dilution with fresh salt medium at the temperature of the culture (20°C). The lower curve shows the proportion of cells in recognizable fission. (From Cook, 1962; reproduced by permission of the publishers).

graph as the logarithm of the number. The division index, i.e., the fraction of cells in cytokinesis, indicates that the system is highly repetitive, the peak value being achieved regularly in the middle of the dark period.

B. Continuous Culture Techniques

The results shown in Fig. 1 also typify the continuous aspects of synchrony in *Euglena*. Continuity, which is interrupted only by accidental contamination, is achieved in this system with a culture vessel of simple design. The cells are grown in a Spinner Flask (Bellco Glass Co., Vineland, New Jersey), which is a cylindrical vessel equipped with two ports and a magnetically operated stirring bar. Stirring assures adequate gas exchange as well as uniform illumination. The culture is diluted daily in the early portion of the light period, taking care that the inflowing sterile medium is at the same temperature as the parent culture. Samples and excess cells are removed by a siphon suitably equipped with a contamination-preventing mantle at the delivery tip. As shown by the cell counts, performed on an electronic cell counter (Coulter Co., Hialeah, Florida), this method of sampling is adequate.

The primary deficiency in this system is the present inability to synchronize *Euglena* at high population densities. This difficulty, however, is offset by the finding that samples from separate cycles can be pooled and continuous patterns of change indicating balanced growth can be derived from such samples (Cook, 1962). This point again illustrates one of the advantageous features of a continuous synchronous culture.

III. Temperature-Induced Synchrony in *Astasia longa*

A. Growth vs. Temperature

One would expect that since growth is a complex of chemical and physical processes, a cell's response to temperature shifts would assume a variety of forms. In some instances a given function may be inhibited by reduced temperature (e.g., respiration) while an apparent compensatory change will be seen in the opposite direction (e.g., increase in size). Indeed, the parameter under examination may be said to alter the quality of the event being measured. Thus in attempts to derive an over-all formulation of stress vs. response, temperature effects on cell division and growth must be considered in general terms before examination is made at the molecular or even particulate level.

To derive a synchrony-inducing temperature cycle, it becomes necessary to limit one's choice of the range and duration of the temperature periods of the cycle, as well as its periodicity. To this end, the primary measure of a cell's response to temperature is its growth rate. At the same time one should also determine the biochemical profile of the cell at each steady state level of growth, so that with standard calculations based on the kinetics of cellular proliferation (Hutchens *et al.*, 1948) an estimate can be made of a cell's capability or effectiveness as a function of temperature. This may be in terms of substrate utilization as well as rates of synthesis of one or all the major components of protoplasm. To be sure, these determinations need not forecast a successful synchrony but will presage the temperature cycle best suited for its attainment.

In *A. longa* a variety of parameters were examined as function of ambient temperature (James and Padilla, 1959) and the results provided sufficient clues for the subsequent formulation of a synchrony-inducing temperature cycle (Padilla and James, 1960). In essence, it was found that *Astasia* is twice as effective at 25°C than at 15°C in its rate of dry weight production, even though it divides 3.3 times faster at the higher than at the lower temperature (James and Padilla, 1959). This was meant to imply that for a 24-hour temperature cycle, ignoring for the moment the effects of repetitive temperature shifts, the 15°C portion of the cycle must be twice as long as the 25°C portion in order for the cells to produce equivalent amounts of protoplasm in that span of time. In addition, as the 15°C-grown *Astasia* were considerably larger than their 25°C counterparts, the cold cells upon shifting to a higher temperature *could* operate their larger mass of respiratory machinery to greater advantage under temperature cycling than in a constant environment. As it turns out, one of the early temperature programs, set within these conjectural limits, gave a reasonably good synchrony provided two other criteria were met: (a) the temperature transition periods were rapid and (b) the medium was either organic (e.g., 2% proteose peptone) or supplemented with SH-bearing compounds (Padilla and James, 1960). We shall consider the question of the medium later.

The above criteria were met through the development of improved culture vessels, controlling units, and provision for optimum conditions of growth (Padilla, 1960; Blum and Padilla, 1962). The most adequate temperature program to date consists of a 17.5-hour period at 14.5°C and a 6.5-hour period at 28.5°C. With this cycle, highly repetitive synchrony is achieved and is characterized by a close phase relationship of cellular events to the temperature program (Padilla and Cook, 1963). With the development of continuous culture techniques, to be described below, the system approaches the qualifications of a model for a single cell.

B. Development of Continuous Culture Techniques

As with *E. gracilis*, maintenance of continuous cultures in synchrony rests simply with the adoption of techniques for serial dilution. The first problem one faces is the construction of a culture vessel that allows large yields of cells growing at optimum conditions.

1. Culture Apparatus

Figure 2 shows the type of culture vessel found to be most adequate in the continuous synchrony of *A. longa* (Blum and Padilla, 1962). An almost identical unit has been used by Wilson for studies on the oxygen consumption of synchronized cultures of *A. longa* (Wilson, 1962; Wilson and James, 1963). The vessel consists of a cylindrical Pyrex glass jar 10 inches in diameter and 12 inches high, to which is clamped a $\frac{1}{4}$-inch stainless steel plate as a cover. The jar is separated from the cover by rubber and gauze gaskets which prevent contamination and avoid undue strain on the glass surface. In lieu of these materials, gaskets of autoclavable inert plastic foam are now available (Gay-Mar Industries, Buffalo, New York).

Several ports are drilled through the plate cover for the following items:

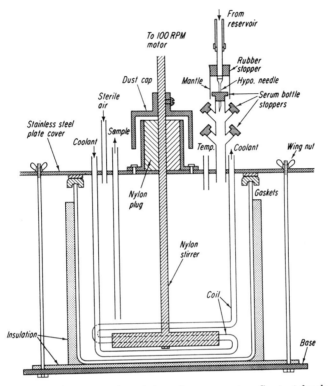

FIG. 2. Schematic representation of the culture apparatus. See text for details.

two $\frac{3}{8}$-inch OD stainless steel tubes of the cooling coil; inlet and outlet ($\frac{1}{4}$-inch OD, stainless steel tubes) for sterile air, flowing at a minimum rate of 3.5 liters per hour and 20 pounds per square inch (a lower rate is permitted if the air is bubbled through the medium); a stainless steel sampling tube ($\frac{1}{4}$-inch OD) connected to an automatic pipetter (Brewer Co., Baltimore, Maryland); an autoclavable thermistor probe (Yellow Springs Co.) for recording the temperature; a stainless steel inoculation tube ($\frac{1}{2}$-inch OD) provided with multiple stoppered inoculation ports; and lastly, a central opening ($\frac{1}{2}$-inch bore) for the stirrer collar. As shown in Fig. 2, the stirrer shaft, consisting of a $\frac{3}{8}$-inch diameter nylon rod (Zytel type, Du Pont and Co., Wilmington, Delaware), passes through a cylindrical nylon plug which is enclosed in a brass housing attached to the plate cover. The two nylon surfaces are self-lubricating and will withstand repeated steam sterilization. An aluminum cap covers the stirrer and its collar. A 3-inch steel or nylon blade completes the stirring mechanism.

Several methods are available for securing air-tight seals to all these ports and the plate cover: welding, cementing with heat-resistant epoxy resins, etc. The most advantageous method involves the use of stainless steel compression-type tube fittings which are available from industrial suppliers in a variety of sizes and forms. If other materials are contemplated in the construction of the culture apparatus, a recent study itemizing those materials found to be inert toward algae (Dyer and Richardson, 1962) may serve as a guide.

The culture vessel is inoculated with a hypodermic syringe. Dilution is also accomplished via a hypodermic needle attached to a tube passing through a rubber stopper (cf. Fig. 2). A glass mantle, consisting of a 25-mm glass tube section, covers the pierced serum bottle stopper and allows for repeated inoculations. An overhead reservoir, consisting of a 20-liter aspirator bottle (Corning Co., Corning, New York) contains sterile medium which is added by gravity flow. An in-line solenoid valve (not shown) permits unattended dilution of the culture. Heat-labile compounds, such as vitamins and SH-compounds, are sterilized by filtration and may be added directly to the culture vessel. If a more permanent setup is desired, the empty culture vessel and reservoir may be pre-sterilized and complete medium added and cold-sterilized by passage through high-capacity ceramic filter candles (Selas Co., Spring House, Pennsylvania) or membrane filters (Pall Corp., Glen Cover, Long Island, New York).

2. Controlling Units

A program clock (Paragon Electric Co., Two Rivers, Wisconsin) equipped with dual sets of trippers independently activates a sequence-delay timer and an 8 pole, double throw latching relay. The sequence timer (Herbach and Rademan, Philadelphia, Pennsylvania) consists of a slow moving clock motor whose shaft turns several adjustable cams. These cams sequentially depress circumferentially mounted microswitches which emit intermittent pulses to an automatic pipetter coupled to a fraction collector (Gilson Medical Electronics, Middleton, Wisconsin). The pipetter samples the culture four times and delivers each 15-ml sample to test tubes in the rotating fraction collector table. The first three samples are discarded since they serve to flush out the connecting lines and only the fourth sample, which is rapidly mixed with 0.2 ml of Bouin's fixative previously placed in each tube, is retained for counting. Cell counts are performed on properly diluted aliquots with an electronic cell counter (Coulter Co., Hialeah, Florida) previously calibrated for this cell type.

The latching relay governs the phases of the temperature cycle by alternatively directing the flow of water through the coils in the culture vessel from either a "hot" or a "cold" water bath. Paired solenoid valves (Automatic Switch Co., Florham Park, New Jersey) control the inflow and outflow of each bath. In addition the temperature level of each bath is also controlled through the switching of the relay: during the cold period a pair of thermoregulators sets the cold bath at the desired cold period temperature (in this case 14.5°C) and permits the hot bath to rise to a temperature higher than that of the warm period. During the warm period a second pair of thermoregulators is put into operation and the cold bath is now set to drop below 14.5°C while the hot bath is kept at 28.5°C (the temperature of the warm period). Thus, by allowing the hot bath to be warmer and the cold bath to be colder than the ultimate temperature levels of the cycle (when their water is not being circulated through the culture), an artificially higher temperature gradient is established at each temperature shift and the extra heat capacity of each bath is employed to reduce the transition periods considerably. We have found that by this device even two 5-liter cultures can be rapidly cooled and heated with conventional size water baths (Forma Scientific, Inc., Marietta, Ohio) of 20- to 30-gallon capacity. Additional refinements can be introduced in each bath so that the temperature constancy is kept well below ±0.1°C.

3. SYNCHRONY STABILITY AND REPETITIVENESS

A direct result of conducting these studies against a well-defined chemical background has been the finding that very subtle but necessary synchrony stabilizing agents are some reduced sulfur compounds (Padilla and James, 1960; James, 1960). Among several, thioglycollic acid (2-mercaptoacetic acid) is the most effective. This substance appears to perform at least two interrelated functions: (a) it permits *Astasia* to respond to the synchrony-inducing temperature cycle, and (b) it stabilizes the synchronized population so it does not drift out of phase with the temperature program. To be sure, stability in the synchronized system derives from a rigorously applied temperature cycle, as well as maintenance of optimum conditions for growth (such as proper aeration) (Blum and Padilla, 1962). But even with the temperature cycle outlined above, thioglycollic acid is required for the system to operate satisfactorily, as shown in Fig. 3.

The upper portion of Fig. 3 shows six consecutive cycles for cells repeatedly diluted with fresh medium but kept at a constant thioglycollic acid concentration (5×10^{-5} M). The lower portion shows a parallel culture "synchronized" in the absence of this compound and similarly diluted. It is obvious that without thioglycollic acid the bursts of division gradually drift into the cold period with a resultant departure from complete cell doublings at the end of each warm period. Moreover, the slope of the log-linear portion of each burst decreases, possibly indicating a progressive lengthening of the time required for cytokinesis. In essence, without a reduced sulfur source continuous synchrony is lost, and the

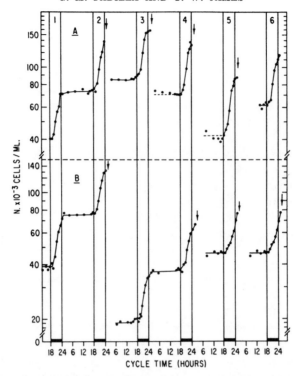

FIG. 3. The effect of thioglycollic acid on the synchrony pattern of *Astasia longa* grown on Cramer-Myers medium. The upper portion of the graph (A) shows the behavior of a culture to which thioglycollic acid was added to a concentration of 5×10^{-5} M at the time of dilution shown by each arrow. The lower portion of the graph (B) shows a parallel culture repeatedly diluted without addition of thioglycollic acid. The ordinate gives the population density and the abscissa shows the cycle time. The cold period (14.5°C) was 17.5 hours long and the warm period (28.5°C), indicated by the bar, was of 6.5 hours duration.

quality of repetitiveness decreases in the system. The metabolic function of thioglycollic acid itself is not yet known, but the rapid uptake at a time coincident with the burst of division suggests it is closely linked to the process of cell division (James, 1961b).

The question naturally arises as to whether repetitiveness of cell division is reflected in the biochemical activities of the synchronized cells. In other words, does synchrony occur in cellular processes other than cytokinesis? In partial answer to this question the protein and dry weight contents of *Astasia*, assayed colorimetrically and gravimetrically, were examined in consecutive synchronous cycles. The results of such determinations are shown in Fig. 4. The upper portion of the graph shows the pattern of

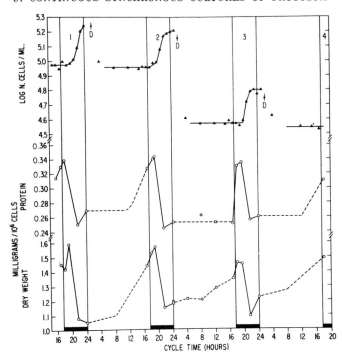

FIG. 4. Total protein levels and dry weights of synchronized *Astasia longa* during four consecutive cycles. The abscissa shows the cycle time, with the 6.5-hour warm period (28.5°C) indicated by the black bar. The upper tracing (▲——▲) represents the synchrony pattern expressed as the logarithm of the population density. The middle and lower curves relate to the total protein content and dry weights expressed in milligrams/10⁶ cells. Total proteins were determined by the Lowry method (Lowry *et al.*, 1951). Between the second and third warm periods several samples of cells were stored at 2°C overnight before being assayed. The corrected values of such determinations are indicated by the open squares. All other determinations were performed on fresh samples and represent mean values of triplicate aliquots.

synchrony and the middle and lower tracings indicate the changes in the protein and dry weight contents of the cells.

In spite of some variation in the timing of the onset of cell division, the pattern of protein synthesis appears unaffected. The cells regularly attain maximum protein level of approximately 0.34 mg/10⁶ cells in the first 2 hours of the warm period, and a minimum level of 0.25 mg/10⁶ cells in the last 2 hours of the warm period. This value is not one half of the maximum level since the interval of cytokinesis and minimum protein production are not coincident throughout the entire population of cells, some protein being synthesized during the burst of division (Blum and

Padilla, 1962). In any event, it is clear that protein synthesis is repetitively synchronized and closely parallels the cytological events.

Changes in the dry weight content, on the other hand, show some fluctuations, but in amplitude rather than periodicity. As with the protein content, the maximum dry weight of the cells is achieved in the first 2 hours of the warm period, but with each ensuing warm period this level decreases slightly. In the fourth warm period, only a portion of which is shown, the maximum level rises again, showing a return to the previous level.

To determine the patterns of synthetic activity occurring in the cold period, suggested by the broken lines, samples of cells were removed, stored in culture medium at 2°C overnight, and assayed for their total protein and a dry weight content. Cells stored at this temperature for 8 hours or longer lost as much as 25% of their protein content. This loss was in part reflected in the decreased dry weight of the cells which was found to be constant for the duration of the cold storage. The corrected values are shown in Fig. 4 only to suggest the over-all patterns of change. It is not clear at this time why cells should lose so much material when stored at low temperatures.

The above results clearly indicate that in the over-all aspects, *A. longa* is finely attuned to its temperature program. As reported elsewhere, such a phase relationship relates also to the time course of nucleic acid synthesis, which, as in the case of proteins, occurs in the latter portions of the cold period and most of the warm period, exclusive of the time of cytokinesis (Blum and Padilla, 1962). It is to be emphasized, however, that these values still relate to an *average* cell and not a single cell. As such, however, they are amenable to statistical considerations. In this sense then, fluctuations do not necessarily reflect deviations of a single cell from its norm, but rather represent the retention of variability usually associated with cell populations (James, 1961a). If the population is approaching the behavior of a single cell, is its synchrony dependent on the continual imposition of the temperature cycle? How will a synchronized population of cells behave when placed at constant temperature? The results shown in Fig. 5 indicate that for some time, at least, the temperature-induced alignment is retained by *A. longa*.

In this particular experiment, the temperature cycle was interrupted at the end of the second warm period and the cells were then kept at the higher temperature (28.5°C). The cell population, upon completing its division, paused slightly and then resumed cytokinesis in a phased manner. This phased pattern persisted for approximately two generations, after which the characteristic log-linear rate of growth was resumed, with a generation time of approximately 11 hours. Although not clear at the

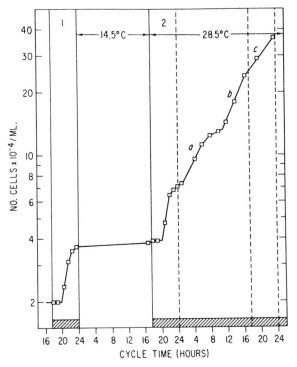

FIG. 5. Oscillatory pattern of growth of synchronized *Astasia* released from the temperature cycle. The curve shows the behavior of cells subsequent to their release from the temperature-inducing cycle to the temperature of the warm period (28.5°C). The dotted vertical lines indicate the times at which the ensuing temperature shifts would have occurred. In the first burst the cells doubled in number in 2.5 hours, in the second burst in 3.0 hours. Following release they showed three distinct log-linear rates of division with slopes equivalent to the following generation times: $a = 6.5$ hours, $b = 6.5$ hours, and $c = 11$ hours.

present time, a variety of feedback mechanisms must be involved in the retention of such an alignment and may be related to some form of control which is lost by the cells after two cell divisions at constant temperature.

IV. Concluding Remarks

The inherent quality of repetitiveness that is seen in the synchrony of *Euglena* and *Astasia*, it is hoped, does not rest with a uniqueness of these organisms, but rather represents the ability of the cell cycle to adapt to an externally applied cycle of stresses. If one considers a cell lineage as a discontinuous series of events grossly subdivided into an interphase and a mitotic phase, synchrony derived by these methods becomes but an alignment of cellular activities. In terms of developmental patterns, synchrony is the segregation of age classes, which in turn reflects temporally distributed cellular events.

It is apparent that at the present time synchrony in *Euglena* and *Astasia* cannot be operationally described in molecular terms. But it is also clear that since under these systems the cells are able to grow and divide, this presents an experimenter with a well-defined model of cellular growth and division for further experimentation (Padilla and Blum, 1961, 1962). That the present approach does not represent a cul de sac is to be determined by future studies, particularly in terms of whatever useful information can be derived from such synchronized systems. It is hoped, moreover, that the present approach will be extended to other cell types by those investigators least fearful of adventure.

REFERENCES

Abbo, F., and Pardee, A. B. (1960). *Biochim. Biophys. Acta* **39**, 478.

Barner, H. D., and Cohen, S. S. (1956). *J. Bacteriol.* **72**, 115.

Blum, J. J., and Padilla, G. M. (1962). *Exptl. Cell Res.* **28**, 512.

Buetow, D. D., and Levedahl, B. H. (1962). *J. Gen. Microbiol.* **28**, 578.

Cook, J. R. (1960). "Light-Induced Division Synchrony of *Euglena gracilis* and the Effects of Visible Light on Related Cells." Doctoral dissertation, Univ. of California, Los Angeles.

Cook, J. R. (1961). *Plant Cell Physiol.* **2**, 199.

Cook, J. R. (1962). *Biol. Bull.* **121**, 277.

Cook, J. R., and James, T. W. (1960). *Exptl. Cell Res.* **21**, 583.

Cramer, M., and Myers, J. (1952). *Arch. Mikrobiol.* **17**, 384.

Dyer, D. L., and Richardson, D. E. (1962). *Appl. Microbiol.* **10**, 129.

Hutchens, J. O., Podolsky, B., and Morales, M. F. (1948). *J. Cellular Comp. Physiol.* **32**, 117.

James, T. W. (1960). *Ann. N. Y. Acad. Sci.* **90**, 550.

James, T. W. (1961a). *Ann. Rev. Microbiol.* **15**, 27.

James, T. W. (1961b). *Pathol. Biol. Semaine Hop.* **9**, 510.

James, T. W., and Padilla, G. M. (1959). *In* "Proceedings of the First National Biophysics Conference" (H. Quastler, ed.), p. 694. Yale Univ. Press, New Haven, Connecticut.

Lowry, O. H., Rosebrough, N. J., Farr, A. L., and Randall, R. J. (1951). *J. Biol. Chem.* **193**, 265.

Monod, J. (1950). *Ann. Inst. Pasteur* **79**, 390.

Moser, H. (1958). *Carnegie Inst. Wash. Publ. No.* **614.**

Myers, J., and Clark, L. B. (1944). *J. Gen. Physiol.* **28**, 103.

Novick, A., and Szilard, L. (1950). *Proc. Natl. Acad. Sci. U. S.* **36**, 708.

Novick, A., and Szilard, L. (1951). *Cold Spring Harbor Symp. Quant. Biol.* **16**, 337.

Padilla, G. M. (1960). "Studies on the Synchronization of Cell Division in the Colorless Flagellate *Astasia longa*." Doctoral dissertation, Univ. of California, Los Angeles.

Padilla, G. M., and Blum, J. J. (1961). *First Ann. Meeting Am. Soc. Cell Biol., Chicago, Illinois, 1961*, p. 160.

Padilla, G. M., and Blum, J. J. (1962). *Second Ann. Meeting Am. Soc. Cell Biol., San Francisco, California, 1962*, p. 41.

Padilla, G. M., and Cook, J. R. (1963). *In* "Synchrony in Cell Division and Growth" (E. Zeuthen, ed.). Wiley (Interscience), New York. In press.

Padilla, G. M., and James, T. W. (1960). *Exptl. Cell Res.* **20**, 401.

Scherbaum, O., and Zeuthen, E. (1954). *Exptl. Cell Res.* **6**, 221.

Wilson, B. W. (1962). "Controlled Growth and the Regulation of Metabolism of the Flagellates *Astasia* and *Euglena*." Doctoral dissertation, Univ. of California, Los Angeles.

Wilson, B. W., and James, T. W. (1963). *Exptl. Cell Res.* In press.

Zeuthen, E. (1958). *Advan. Biol. Med. Phys.* **6**, 37.

25

Reprinted from *Continuous Cultivation of Microorganisms* (Proc. 4th Symp.), Academia: Publishing House of the Czechoslovak Academy of Sciences, Prague, 1969, pp. 71–85

CONTINUOUS PHASED CULTURE – EXPERIMENTAL TECHNIQUE*

P. S. S. DAWSON

National Research Council of Canada, Prairie Regional Laboratory, Saskatoon, Saskatchewan, Canada

Introduction

Almost twenty years ago the theory of continuous culture was introduced (Monod, 1950; Novick and Szilard, 1950) – an important event for microbiology – for it enabled growth and metabolism in an organism to be studied in the culture as a steady state, that is, in a condition of equilibrium.

From this revolutionary development, it was learned that growth of a culture is not a predestined sequence of change but is dependent upon the system in which the growth takes place: it can be changing in the closed system of the traditional batch culture, or constant in the steady state of the open system of continuous flow, or chemostat culture. With the latter technique it also became evident that growth could be appreciated in terms of the growth rate, or related to the doubling time, so that an expression of growth as a parameter of the cell became possible.

Despite these advances, however, the individual performance of the cell was still unrecognized; only an average picture was obtained of the cells in the culture, growing in random distribution throughout their cycle of reproduction, or cell-cycle.

This unresolved condition gives incomplete satisfaction, for surely it is necessary to know the precise behaviour of a cell rather than its overall performance

How can Microbiology be built upon firm foundations or on an absolute basis, if microbiologists continue to ignore the fundamental units involved in their cultivation systems? The individual status of the microbe should be respected, and understood, not treated as an empirical statistic to be hopefully used and exploited. The individual behaviour and functions of the cell must be known. After all, in medicine, the doctor has to treat his patient as an individual and not as a statistical entity.

Growth in terms of the cell

An individual microbe, growing in a culture, does not differ in its behaviour according to whether it finds itself in a batch, continuous or any other culture. It will duplicate regardless of the method; for it is known as an experimental fact, that within reasonably prescribed limits, the growth constants of Monod apply equally well in batch or continuous systems (Málek and Fencl, 1966).

A simple interpretation of Monod's classical studies of growth (Monod, 1942, 1949), indicates that an organism requires a basic amount of nutrient or ration to duplicate itself, and the speed at which this is performed depends upon the concentration of nutrient available. Under the same environmental conditions an organism will use the same amount of nutrient to reproduce itself in either system, but the manner of its subsequent development will then depend upon the system.

In the closed system of batch culture, as growth proceeds, the initial stockpile of nutrient rations is progressively and selectively removed, successive generations modify the medium, so that the growth rate and cell metabolism of the culture changes.

In the open system of the continuous culture, on the other hand, a constant condition or steady state operates, and the in-coming medium enables the cells to grow at a uniform rate indefinitely, and with an overall constant metabolism. In the continuous culture the steady stream of nutrient rations enables the randomly dividing cells to reproduce at their individual but similar doubling times.

In both batch and continuous methods the cells are at all stages in the reproductive cycle of the cell - so that, at any instant, an average randomized condition exists in the culture. Whilst this randomized condition changes temporally in the batch culture it does remain constant in a continuous culture.

Now, if instead of having a randomly dividing population in a chemostat, the cells are "in synchrony" and the total rations are supplied at one time, then the cells should grow and divide simultaneously at the end of the doubling period. If another ration of nutrient medium per cell arrives then, and is similarly repeated at subsequent intervals, a continued synchrony should be maintained in the culture. Furthermore, by altering the periodicity of the dosage of medium the growth rate could be changed. Such a system has been used by us and is termed Continuous Phased Growth.

72

Experimental

Continuous phased growth

The arrangement illustrated in Fig.1 shows the device that may be used to produce such a continuously phased culture: one in which the cell cycle may be investigated at any chosen growth rate.

In this method (Dawson, 1965) the continuous supply of medium to a chemostat is interrupted and collects in a dosing vessel placed ahead of the culture vessel. A volume of medium, equal to that of the culture, collects in this vessel during a doubling time and when complete automatically discharges into the culture. There it is intimately mixed,doubling the culture volume and halving the cell density; one half portion of this diluted culture is then removed and the cycle repeats.

It will be noted that in the method a repeated dilution of the growing culture with fresh medium is made at a ratio of 1:1, and takes place at regular intervals which correspond to the doubling time of the cells. The volume of culture in the apparatus is halved after each fresh nutrient addition to avoid an otherwise exponential increase required in the size of the system.

The half portion removed is used experimentally: this is transferred to a similar culture vessel which functions as a second stage and operates under the same conditions; the culture is sampled for analysis at regular intervals over the cycle period there.The changes observed in the culture reflect the changes taking place in the cell during the cycle. The same changes are taking place concurrently in the first stage which serves to continue phasing.

Results and discussion

Changes during the cell cycle

Fig.2. shows such a series of observations made on <u>Candida utilis</u> in phased growth on a 1% glucose mineral salts medium during a cycle time of 4 hours at 28 $^{\circ}$C (Dawson, 1965). It can be seen that growth doubles during the cell cycle: the cell density and dry weight of cells increase gradually over the cycle, but the cell number doubles towards the end of the cycle. Changes in morphology are outlined at the bottom of the slide.

There is a statistical distribution of doubling times about a mean value in continuous culture (Powell, 1958a,b) - so that 100% syn-

73

chrony cannot be expected in phased growth. In practice about 70–80%
of the cells are found to be in synchrony. A sequence of photomicrographs (Fig.3a–e) shows the changes taking place in a phased culture
during intervals of sixty minutes in a four-hour-cycle.

The oxygen uptake of the culture is shown (Fig.4) as a trace determined by paramagnetic analysis of the effluent gas from the culture vessel (Phillips,1963): and below, a series of similar traces
taken in succeeding cycles shows the reproducibility of the phasing
method as portrayed by the pattern of O_2 uptake.

In Fig.5 it can be seen that at different growth rates, doubling
of growth takes place in a manner similar to that in the four-hour-
cycle (Müller and Dawson, 1968). However, as the doubling time lengthens, the cell population increases, and exhaustion of the nitrogen
from the culture occurs earlier in the cycle. An inverse relationship
is apparent in Fig.6 between the amount of nitrogen initially available to the cell and the length of the cycle or doubling time. This
is in accord with the empirical relationship $[\mu = \mu_{max}(S/S + K_S)]$,
established by Monod (1949), which showed that the rate of growth depends upon the concentration of the substrate limiting to growth in
the culture.

Exhaustion of the controlling nutrient partway through the cycle
does not prevent the cells from completing their cycle, and as
phased cultures repeatedly maintain their existence (average runs of
6–8 months duration), it follows that at the point in the cell cycle
when limiting nutrient has been exhausted, the phased cells in the
culture have already satisfied their requirements for that nutrient
in the cycle. Thus, growth is limited in the culture by the amount
of nutrient added at the start of the cycle and, as long as this
balances the ration needed for each cell to double, the equilibrium
of phased growth is maintained.

With less nitrogen available to a cell, a reduced capacity for
protein and nucleic acid synthesis could be expected, and hence, following a smaller synthesizing mechanism within the cell, a longer
growth period before the cell can complete its reproduction: that is,
a longer doubling time. This could indeed serve as a basis for the
control of growth rate in the nitrogen limited cells.

Fig.7 illustrates the changes observed in protein and RNA con-
----
where:
μ = specific growth rate, μ_m = maximum specific growth rate, S = concentration
of substrate limitihg to growth , K_S = saturation constant.

74

tent of the cells during nitrogen limited growth at different cycle times. It will be noted that at short cycle times (less than 4 hours) when nitrogen is available to the cells throughout their cycle, both protein and RNA increase in a linear manner, but at longer cycle times (greater than 4 hours) when nitrogen exhausts during the cycle, the increases in protein and RNA contents are curtailed partway through the cycle. It is suggested that during this latter period in nitrogen-limited cells the cell is probably undergoing a turnover to complete its cycle (Müller and Dawson.1968).

Phased cultures are useful tools for examining metabolism and growth of a cell during the cell cycle at different growth rates: that is, for different conditions of growth (Dawson, 1966; Dawson and Craig, 1966).

Changes during the post - cycle

During the latter stages of the growth sequence in batch cultures, there occur changes of a secondary nature; metabolic developments that are not associated with positive aspects of growth. In continuous culture this is the area of multistage investigation - and it is not necessary to stress here the importance of physiological aspects of growth (Málek and Fencl, 1966), though perhaps one might suggest that emphasis on the cellular basis is lacking still.

A further development of phased culture, however, permits the post-cycle changes being examined, that is of developments in the cell following the cell cycle (Dawson, 1967a). This is described below. At the end of a cell cycle - the cells in the phased culture can remain in the second stage culture vessel without further addition. Any changes taking place subsequently in that culture must occur at the expense of unexhausted materials remaining; that is, of unused constituents in the medium, together with extracellular and intracellular products formed by the cells during the cell cycle.

Growth in phased culture is controlled by the limiting condition of a given nutrient, in relation to cycle time, as for example, by the nitrogen source - so that other constituents of the medium, including the carbon source, are present during the cell cycle in non-limiting concentrations. The amounts of these non-limiting constituents remaining at the end of the cycle period will depend upon the quantities added initially in the fresh medium, and consequently their relative proportions, or balance, will be important in directing the later, or post-cycle development. It follows, therefore, that some

75

measure of control for post-cycle development can be arranged by using suitable quantities of non-limiting components in the composition
of the fresh medium added initially, and by varying these proportions
as required.

In Fig.8 the post-cycle extension of the cell cycle on 1% glucose mineral salts medium is shown, and below this the effects of varying the concentrations of the carbon source to 3% and 5% whilst maintaining the other medium constituents unchanged. The cycle time was 4
hours and the temperature 28 ^{O}C.

In all three media, growth (measured as cell density, dry weight
and viable count) doubled over the cell cycle. (a) In 1% glucose medium - growth was limited by the carbon source: glucose was exhausted
after 3 hours in the 4-hour-cell cycle, nitrogen assimilation ceased
at that point. After 4 hours in the post-cycle period the viable count
started to fall. (b) In 3% glucose medium - growth was nitrogen limited:
nitrogen exhausted two-thirds of the way through the cell cycle. Glucose was not used up until 3 hours of the post-cycle had elapsed:past
this point growth remained constant. (c) In 5% glucose medium - a
slightly higher growth exhausted nitrogen halfway through the cell
cycle.Glucose now lasted into a long post-cycle period during which a
small gradual increase in growth occurred.

In Fig.9 the changes taking place in the major carbohydrate components of the cells are illustrated. During the cycle period, in all
three media, no accumulation of glycogen occurred in the cells while
nitrogen was being utilized; but when this ceased carbon assimilation
continued in the presence of glucose in the culture and glycogen was
formed (as in the 3% and 5% experiments). Upon exhaustion of glucose
from the culture, certain further changes occurred; for example,glucan
and acetic soluble glycogen remained unchanged, but mannan and the
alkali and perchloric acid soluble glycogen fractions decreased.

Cell wall components, mannan and glucan, were produced whenever
growth (measured as dry weight of cells) was increasing in the culture,
but as shown in Fig.10, the ratio of mannan/glucan formed depended
upon the conditions prevailing in the culture.

These results, in greater detail, have been discussed elsewhere
(Dawson, 1967a) in consideration of primary and secondary aspects of
metabolism, but here they serve to illustrate how important the overall composition of the medium is for controlling and directing growth
during the cell cycle and of development in the post-cycle.

The three experiments were all performed at one growth rate
(C_T = 4 hours), they could be repeated systematically at others, and

76

furthermore at different temperatures, or using a different set of
nutrient or environmental factors. Thus, the phasing technique is a
useful surveying method for the temporal metabolic and other growth
patterns of the cell and post-cycles.

Other aspects of continuous phased growth

The phasing technique lends itself to other experimental applications: interruptions (for example of O_2 supply) or additions (of nutrient supplements, precursors or inhibitors) may be made to the culture in the experimental stage (II) at any time - so that in vivo effects may be followed in the culture but on the basis of the cell.

We have used C_1, $C_{3,4}$ and C_6 labelled glucose, for example, to
show in preliminary experiments that the contributions of the Emden-
Meyerhof and pentose phosphate pathways play varying roles during the
cell cycle and at different doubling times (Dawson and Westlake, 1965).

On the applied level, the phased technique offers new prospects
for microbial fermentations - not only does the system permit of analysis during one cycle, prior to harvesting in the following ones, of
a single or different components; but it also permits of harvesting
enzymes or other metabolites produced transiently during the cell
cycle in yields that are unattainable, or undetectable, by other methods using random populations (Dawson and Hampton, 1965).

As a method for continuous production of biomass, the technique
permits a 50% increase on chemostat methods when judged on the capacity
of the culture vessel used; as the replacement time in the phasing
technique is the doubling time of the cells, not one and a half times
the doubling time ($D_r/\log_e^2$), as in the chemostat (Dawson, 1965).

The technique has been used for cultivating several types of organism (Dawson, 1967b; Kurz et al., 1968), other than the food yeast,
Candida utilis, used solely to illustrate the potentiality of phased
cultures in this paper: the results obtained with these different organisms which include various bacterial species such as Bacillus subtilis, Pseudomonas and Azotobacter species, follow the same general
patterns observed for the yeast.

These experiments have shown that microbial growth can be examined in a continuous culture on the basis of the cell: that the phased
condition or synchrony can be reproduced indefinitely; that the cell
metabolism is changing throughout the cell cycle; and that the pattern
of change varies and is characteristic of the doubling time. These
results illustrate the initial scope and possibilities of the technique.

77

Since 1950, the advantages to be derived from examining micro-bial growth and metabolism in the controlled and reproducible systems of continuous culture have been increasingly apparent; but, at the same time, the inability to translate batch culture operations dir-ectly into successful chemostat practice remains.This inability to bridge the gap could arise, on the one hand, from assuming a constant performance from a cell during the cell cycle; and on the other hand, from ignoring repression effects in a chemostat. Repression is likely to be present in a chemostat culture due to the continual incoming presence of fresh medium whilst these effects are removed during the temporal sequences of the batch growth.

Growth, as controlled at different rates in a chemostat, is be-lieved by some to be unnatural and alien to that obtained on various media during exponential phase growth in batch culture: likewise, distinctions are also made between the batch and continuous cultures of synchronous and synchronized growths. Problems such as these are fundamental to an understanding of growth phenomena, and the technique of continuous phased growth could be used to examine these more close-ly.

The method of continuous phased growth is far from being per-fected,but using the technique to unravel the behaviour and performance of the cell helps to understand and control the cell. This in turn leads to improvements in technique and application.

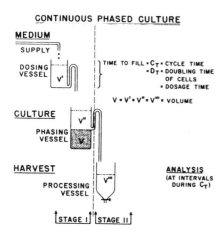

Fig.1.
Continuous phased culture: prin-ciples of operation.

78

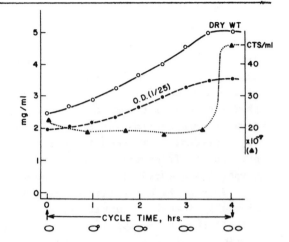

Fig.2.
Changes taking place in phased culture of <u>Candida utilis</u> growing in carbon limited medium with a doubling time of 4 hours.

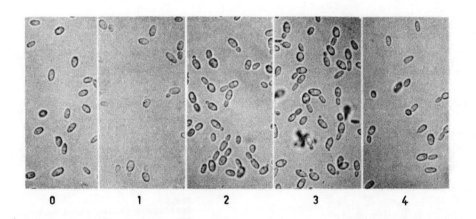

Fig.3.
Phased growth of <u>Candida utilis</u>: photomicrographic record of 4-hour-cycle. The sequence shows cells growing aerobically on glucose limited medium, (0) just after dosing and then at intervals of one hour (1, 2, 3 and 4) during the cycle.

79

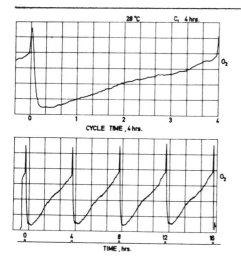

Fig.4.
Phased culture of Candida utilis:
oxygen uptake of cells during a
single 4-hour-cycle (upper trace)
and four consecutive cycles of
4 hours (lower trace).

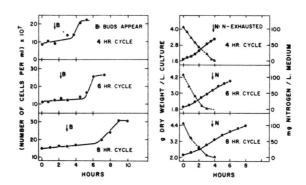

Fig.5.
Phased division of Candida utilis growing under nitrogen limitation at different growth rates: viable counts of cells; increase of dry weight of cells (-●-),
and utilization of growth limiting substrate (-▲-). The arrow means N-exhausted,
B-buds appear respectively.

80

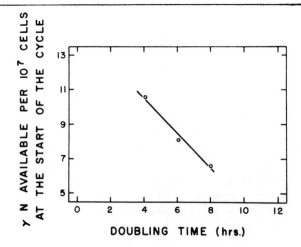

Fig.6.
Control of growth rate,
the amount of nitrogen
available to the cells
at different doubling
times.

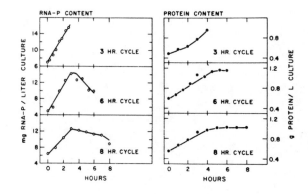

Fig.7.
Phased division of Can-
dida utilis growing un-
der nitrogen limita-
tions at different
growth rates: changes
in RNA-P and protein
content of cells during
the cycle.

81

382

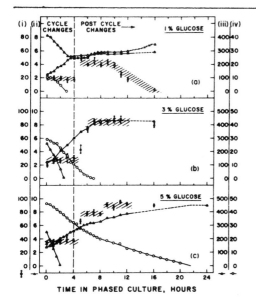

Fig.8.
Changes in the culture during phased
growth of <u>Candida utilis</u> on (a) 1 % ,
(b) 3 % , and (c) 5 % glucose medium,
cycle time 4 hours. ● cells x 10^7/ml:
-▲- dry wt. cells mg/ml: -△- N μ g/ml:
-O- glucose mg/ml.

Fig.9.
Changes in carbohydrate composition of <u>Candida utilis</u> cells
during phased growth on (i) 1%,
(ii) 3%, and (iii) 5% glucose
medium, cycle time 4 hours. -●-
total carbohydrate, -△- glucan,
-O- mannan, -□- glycogen.

↟glucose exhausted, ↡nitrogen
exhausted.

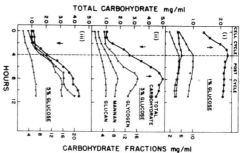

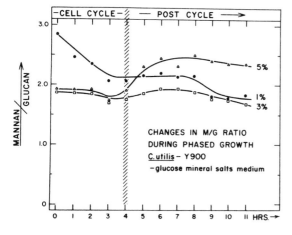

Fig.10.
Changes in mannan/glucan ratio
of <u>Candida utilis</u> cells during
phased growth on 1% . 3% , and
5% glucose medium.

82

383

References

Dawson,P.S.S.:

Continuous phased growth,with a modified chemostat. Can.J.Microbiol.11:893,1965.

Dawson,P.S.S.:

Dynamic aspects of growth and metabolism in Candida utilis. Nature 210:375,1966.

Dawson,P.S.S.:

Changes in carbohydrates composition of Candida utilis during and following the cell cycle. Abstract No.280, 4th FEBS, Oslo, 1967.

Dawson,P.S.S.:

Continuous phased growth B. subtilis. (Unpublished information), 1967.

Dawson,P.S.S., Hampton,A.N.:

Changes in acid soluble nucleotide constituents of Candida utilis during phased growth. (Unpublished information), 1965.

Dawson,P.S.S., Westlake,D.W.S.:

Patterns of labelled CO_2 evolution in C. utilis. (Unpublished information),1965.

Dawson,P.S.S, Craig,B.M.:

Lipids of Candida utilis: changes with growth. Can. J. Microbiol. 12:775, 1968.

Kurz,W.G.W., Dawson,P.S.S., Blakley,E.R.:

A comparative study in vivo of enzyme activities in batch, continuous and phased cultures of a Pseudomonad grown on phenylacetic acid. (Paper in preparation). 1968.

Málek,I. and Fencl,Z.(eds.) :

Theoretical and Methodological Basis of Continuous Culture of Microorganisms. Publ.House Czechoslov.Acad.Sciences,Prague, Academic Press, New York and London 1966.

Monod,J.:

Recherches sur la Croissance des Cultures Bactériennes.Masson et Cie,Paris 1942.

Monod,J.:

The growth of bacterial cultures. Ann. Rev.Microbiol. 3:371, 1949.

Monod,J.:

La technique de culture continue. Ann. Inst. Pasteur 79:390, 1950.

Müller,J., Dawson,P.S.S.:

The operational flexibility of the phased culture technique, as observed by changes in the cell cycle of Candida utilis. Can. J.Microbiol. (in preparation).

Novick,A., Szilard,L.:

Description of the chemostat. Science 112:715, 1950

Phillips,K.L.:

Determination of oxygen and carbon dioxide using a paramagnetic oxygen analyzer. Biotechnol.Bioeng. 5:9, 1963.

Powell,E.O.:

Growth rate and generation time of bacteria, with special reference to continuous culture. J. gen. Microbiol. 15:492, 1958.

Powell,E.O.:

A outline of the pattern of bacterial generation times. J.gen. Microbiol.18:382, 1958.

83

Discussion

<u>Nečinová</u>: Did you follow the synthesis or induction od some enzymes in the course of the cell-cycle?

<u>Dawson</u>: Yes.Intracellular and extracellular production of enzymes have been investigated in bacteria. Both types are found to vary during the cell cycle and in post cycle periode.

<u>Schrauwen</u>: Dr.Dawson, you stated the productivity of biomass per unit fermentor volume as being higher in your method than in conventional continuous culture.Can you illustrate this? In continuous culture D = 1.5 x doubling time, in phased culture D = doubling time.

In your explanation you take the fermentor volume as compared with continuous culture as V = D, but when changing medium you need a vessel of 2V capacity so the productivity on unit of volume of the vessel needed is actually in your technique.

<u>Dawson</u>: Question (1): Correct; the time required to harvest one volume of culture in the chemostat is approximately one and a half times the doubling time of the cells, in phased culture only the doubling time is needed. Question (2): Not necessarily so. At the end of the cycle (1/2)V of culture can be removed to the second stage, and then (1/2)V portions of fresh medium can be dosed into each stage to proceed into the next cycle. At the end of this cycle-stage two harvest is removed and the process repeats.

<u>von Meyenburg</u>: I admire Professor Dawson's attempt to find a method to study synchronous populations in continuous culture. But I have to remark on some grave disadvantages of the method described. At each phasing cycle the concentration of the "limiting" substance changes from high to low values; thus the conditions in the medium are not constant over the budding cycle of the phased culture and can certainly give rise to artefacts. I think we have to look for a method to study synchronous populations under constant conditions, e.g. after the phasing cycles to put the phased culture in a second fermentor with a steady flow of substrate according to the generation time chosen.

<u>Dawson</u>: I agree that during the cycle of phased growth the environmental conditions are changing, and that it would be advantageous to be able to follow the cycle under constant conditions We are currently investigating such possibilities, including the approach that you suggest.

In regard to artefact production, I do not think that this is any greater in phased growth than in chemostat or batch culture. I think that it is important to recognize that in the various methods of batch, continuous and phased culture, it is the microbe reaction to its environment that is being examined, and from this we hope to learn something about how the cell operates. The respective techniques help us to appreciate the varying ability of the organisms to grow and develop: for example, the batch culture outlines the range of metabolic activity, the chemostat the particular activity characteristic of a fixed growth rate and phased culture the finer details of this activity.

84

<u>Holme:</u> (1) How is the population density regulated at different concentrations of the limiting factor in continuous-phased culture? (2) Synchrony in the continuous phased culture depends on specific nutrient limitation - might this be a limitation of the method?

<u>Dawson:</u> Question 1. As in the chemostat - by the amount of nutrient limiting to growth present in the medium, the concentration of that nutrient factor in the culture likewise controls the doubling time (i.e. the growth rate). As we have seen, the cell population obtained in continuous phased growth on a given medium depends upon the doubling time: at fast growth rates the cell population is smaller than that which exists at slow growth rates. Consequently upon dosing medium to the respective cultures, the amount of nutrient per cell available in fast growing cultures will be larger than that for the cells growing with a longer doubling time. Question 2. No more than in a chemostat.

$\mathcal{26}$

Reprinted from *European J. Biochem.*, **10**(3), 511–514 (1969)

Synchronization of *Escherichia coli* in a Chemostat by Periodic Phosphate Feeding

B. C. Goodwin

School of Biological Sciences, University of Sussex

(Received January 24/June 25, 1969)

A method is described of producing and maintaining division synchrony of bacterial cells growing in a chemostat. The method depends upon adding the limiting nutrient, phosphate, to the culture periodically, the period being equal to the generation time so that there is one addition per cell generation. Several enzymes were measured and shown to oscillate once during the cell cycle. Each enzyme had a characteristic phase relationship to cell division.

The chemostat has been a very useful tool for the study of microbial physiology since it was first described by Monod [1] and by Novick and Szilard [2]. This paper describes a simple technique for producing and maintaining in a chemostat bacterial cultures which are synchronized with respect to cell division, thus extending the experimental range of the apparatus for studies on the growth and division cycle. The use of this technique for the analysis of the dynamic behaviour of molecular control processes in the growth cycle of *Escherichia coli* will be described in following papers.

MATERIALS AND METHODS

Organisms

E. coli B/1, a non-filamentous strain, was used in all the experiments reported here.

Enzyme Assays

Enzymes were assayed by standard procedures: aspartate transcarbamylase and ornithine transcarbamylase by the method of Gerhardt and Pardee [3]; lactic dehydrogenase by the method of Wroblewski and LaDu [4]; and alkaline phosphatase by the method of Torriani [5]. Bacteria were toluenized by shaking with toluene at 37° for 15 min.

Growth and Cell Counts

Growth of the bacterial cultures was measured by absorbance at 540 nm in an EEL photometer. An absorbance of 0.2 corresponds to about 2×10^8 cells/ml

Enzymes. Aspartate transcarbamylase or carbamoyl-phosphate:L-aspartate carbamoyl transferase (EC 2.1.3.2); ornithine transcarbamylase or carbamoylphosphate:L-ornithine carbamoyltransferase (EC 2.1.3.3); lactic dehydrogenase or L-lactate:NAD oxidoreductase (EC 1.1.1.27); alkaline phosphatase or orthophosphoric monoester phosphohydrolase (EC 3.1.3.1).

and a mass of about 0.35 mg dry weight of bacteria/ml. Cell counts were made with a Coulter Counter Model "A" using a 30 μ orifice tube.

Culture Apparatus and Synchronization Procedure

The chemostat used had a culture volume of 240 ml. The temperature of the culture was maintained at 37°. It was aerated and mixed by vigorous stirring with a magnetic stirrer, a stream of sterile air passing through the culture vessel at a fixed rate.

The basic medium used was M9 minimal salts medium with 0.1 M Tris buffer replacing phosphate. The pH was adjusted to 7.4 with HCl. This medium was pumped continuously into the culture, the dilution rate being adjusted to give a particular cell generation time, say 80 min. Once every 80 min a small pulse of KH_2PO_4 was pumped into the culture, a pair of Crouzet timing clocks being used to control the periodicity and duration of the phosphate pulse. Good division synchrony was obtained when the pulse delivered was enough to make the culture 0.18 mM in phosphate. The duration of the pulse was set at 3 min. The interval between pulses was always set equal to the calculated mean generation time of the culture, as determined from the dilution rate by the relation

$$T = \frac{\ln 2}{D}$$

where T = generation time and D = dilution rate. The limiting nutrient was phosphate, and cell division synchrony was produced by the periodic supply of this limiting growth factor.

RESULTS

The behaviour of the continuously synchronized culture with respect to a number of different variables is shown in Fig. 1. All variables are periodic, as might be anticipated from the fact that the chemical

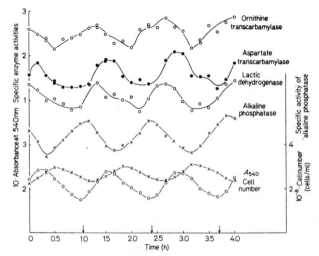

Fig. 1. *Variations in cell number, turbidity (A$_{540}$), and specific activities of four enzymes in a continuously synchronized chemostat culture of* E. coli *B. The limiting nutrient, phosphate, was added to the culture in a brief pulse at the arrows shown on the abscissa, the amount added increasing the phosphate concentration by 0.18 mM*

environment of the cells is itself periodic. The lowest curve shows how cell number varies, the count per ml rising rapidly in coincidence with the beginning of the phosphate pulse, which is marked by an arrow on the abscissa. Since there is continuous cell washout, cell numbers do not double in the chemostat. Cell synchrony is between 60 and 70%, with a mean of 64%; *i. e.*, all cells complete division in about 36% of the generation time, which in this experiment was 80 min.

The curve for the absorbance at 540 nm, which is primarily a measure of cell mass/unit volume in the culture and not of cell number, shows that there is a lag of about 30 min after the phosphate pulse before the growth rate of the cells reaches its maximum, after which it declines continuously. Cells grow at a positive rate throughout the cycle, as will be shown below.

Enzyme activities are represented on this graph in the units: activity of enzyme per ml of culture divided by absorbance at 540 nm; *i. e.*, as specific activities. If the differential rate of enzyme synthesis is constant, the specific activity will be constant; otherwise the specific activity will rise or fall accordingly as the rate of enzyme synthesis is greater than or less than the growth rate.

Alkaline phosphatase is known to vary inversely with the phosphate concentration, which acts as an inhibitor and repressor in *E. coli* [5]. The shape of the curve for this enzyme shows that, after the pulse, the rate of synthesis of the enzyme is decreased and the specific activity consequently drops. The synthetic rate rises again when the phosphate concentration in the cells drops below a value which has been estimated by Torriani [5] to be 10 µM. This occurs about the same time as the growth rate, measured by absorbance at 540 nm, begins to decrease due to phosphate limitation. The other enzymes shown are lactic dehydrogenase, aspartate transcarbamylase and ornithine transcarbamylase. Each of these has a well-defined periodicity in specific activity, and a definite phase relationship to the phosphate pulse. Thus it is clear that they are not all responding to phosphate variations in the same way.

It is useful in analyzing the behaviour of cells in a chemostat to be able to calculate from observed quantities such as turbidity, enzyme activity per unit of culture, *etc.*, the actual rates of variation in these quantities in the cells themselves, correcting for loss due to washout. The equation for the rate of change in turbidity (cell mass) in a chemostat is:

$$\frac{dx}{dt} = \mu x - Dx \qquad (1)$$

where x is the turbidity, μ is the growth rate of the cell, and D the dilution rate [1]. The quantity μ is given by

$$\mu = \frac{1}{x}\frac{dx}{dt} + D. \qquad (2)$$

In a batch culture, writing X as turbidity and μ as growth rate, the exponential law of growth gives

$$\mu = \frac{1}{X}\frac{dX}{dt}. \qquad (3)$$

Eliminating μ from (2) and (3) gives the relation between turbidity in the chemostat and turbidity in an equivalent batch culture:

$$\frac{1}{X}\frac{dX}{dt} = \frac{1}{x}\frac{dx}{dt} + D$$

whence

$$\frac{X}{X_0} = \frac{x}{x_0}\,e^{Dt} \tag{4}$$

where X_0 and x_0 are the values of the variables at $t = 0$. In general one will take $X_0 = x_0$ in the conversion. Knowing x, x_0, and D enables us to calculate X. This calculation was applied to the absorbance data for two cell cycle from Fig. 1, with the result shown by the appropriate curve labelled A_{540} in Fig. 2 where a semilog plot is used. It is evident that growth as measured by absorbance at 540 nm in the chemostat is positive throughout the cell cycle under the conditions employed, although there is a periodicity in the growth rate μ which depends upon the periodic changes in phosphate concentration.

A similar conversion applied to cell number and aspartate transcarbamylase activity gives the other curves shown in Fig. 2.

In Fig. 3 are shown the activities of the enzymes ornithine transcarbamylase and aspartate transcarbamylase per ml of culture rather than per unit volume of cells. In these units there is no longer any periodicity in ornithine transcarbamylase, but a very clear one in aspartate transcarbamylase. When enzyme activities are calculated on the basis of cell numbers, the results are as shown in Fig. 4. Here the periodicity in ornithine transcarbamylase is marked, whereas aspartate transcarbamylase is weakly periodic.

DISCUSSION

The method of cell synchronization described in this paper is essentially similar to those used to produce and maintain division synchrony in cultures of unicellular organisms which can respond to an environmental periodicity, among which are cells with "biological clocks". The commonest periodicity used in the latter case is a light: dark cycle, but temperature cycles are also effective [6]. Chemical periodicities are not normally used to synchronize cell cultures, although the diurnal rhythms of mitosis in the cells of certain metazoan tissues are dependent upon hormonal periodicities [7]. One-cycle synchronization procedures using mitotic inhibitors such as high thymidine concentration [8], or colcemid [9], give good synchrony in mammalian cell cultures. All these synchronization methods depend upon the fact that there are periodic changes in the physiological state of a cell during its growth cycle under the particular growth conditions used. A sufficiently strong single stimulus from the environment can

34*

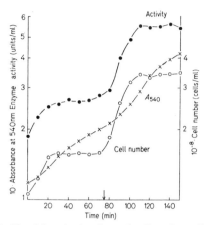

Fig. 2. *Plot of data showing changes in cell number, turbidity (A_{540}), and total activity of aspartate transcarbamylase per ml culture after conversion from chemostat to equivalent batch culture behaviour using Equation (4) of the text*

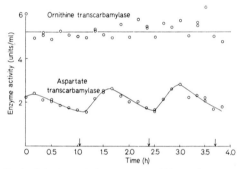

Fig. 3. *Behaviour of the enzymes ornithine and aspartate transcarbamylase expressed as activities per ml of culture, using the same data as that for Fig. 1*

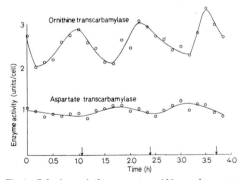

Fig. 4. *Behaviour of the enzymes ornithine and aspartate transcarbamylase expressed as activities per cell, using the same data as that for Fig. 1*

produce a considerable measure of synchrony providing the stimulus is specific for a particular phase of the cell cycle.

Phosphate deprivation would not be expected to affect particular biosynthetic processes in the bacterial cell cycle with any high degree of specificity. Thus it is not surprising that several cycles of phosphate starvation are required to generate cell synchrony in a culture. However, the fact that cell divisions became synchronized by exposing the culture to a periodic phosphate environment means that the rate of phosphate utilization in single cells must be periodic also, under the particular conditions of growth used, the minimum rate of utilization occurring just before cell division when the environmental phosphate concentration is also a minimum. If there were no internal cellular periodicity of phosphate utilization there would be no phase of the cell cycle with which the environmental periodicity could get "aligned" or "entrained". Because the internal periodicity of phosphate utilization has a particular temporal relationship to cell division, the environmental periodicity produces division synchrony when each cell is entrained to it. This does not imply that the rate of phosphate utilization during the bacterial cell cycle is necessarily periodic when the environmental phosphate concentration is constant. Possibly it is the environmental periodicity that generates the cellular periodicity in phosphate utilization.

The data of Fig. 3 and 4 raise a question regarding the correct units to use in studies of enzyme activities in cells. The experiments reported here are concerned with the dynamics of enzyme control processes in bacteria, so it is desirable to use variables which accurately reflect the operation of intracellular control circuits. If one uses the wrong variables, one sees the wrong dynamics. For example, it was concluded from Fig. 1, that the effective activity of aspartate transcarbamylase, measured by its specific activity, varies in a definitely periodic manner during the cell cycle. Then any process controlled by aspartate transcarbamylase activity, such as rate of pyrimidine production, will also vary periodically during the growth cycle. However, if the effective activity of aspartate transcarbamylase is determined by the number of molecules of enzyme per cell rather than per unit cell volume, then the result should be as shown in Fig. 4, where there is a very weak periodicity in aspartate transcarbamylase. A control mechanism regulating the number of aspartate transcarbamylase molecules per cell would then be holding this number relatively constant. To do this, the mechanism must allow rapid synthesis of enzyme

molecules as the cell divides and then stop its production after division is complete. Such a mechanism would have to operate by means of a trigger which was connected with the process of cell division. It is perfectly possible that this type of control process operates in cells, but it is not the type one normally thinks of in terms of feedback control by end-product metabolites. The latter type of biochemical control circuit operates on the basis of concentrations, to which specific activities are taken to be proportional. Hence the use of these units in this study, especially for enzymes like alkaline phosphatase and aspartate transcarbamylase which are known to be under feedback repression control in *E. coli*.

The physiological state of the bacteria growing in the periodic environment described in this study is obviously different from that of cells growing in the normal steady conditions of the chemostat, or in batch cultures. However, their condition is perfectly well-defined and can be used as a reference state for studies on the dynamic behaviour of control circuits in relation to the growth and division cycle of bacteria. Such studies will be reported in subsequent papers.

I would like to acknowledge the excellent technical assistance of Miss Jacqueline Hall and the very valuable advice on continuous culture techniques given by Professor J. Postgate. The research was supported by a grant from the Medical Research Council.

REFERENCES

1. Monod, J., *Ann. Inst. Pasteur*, 79 (1950) 390.
2. Novick, A., and Szilard, L., *Science*, 112 (1950) 715.
3. Gerhardt, J. C., and Pardee, A. B., *J. Biol. Chem.* 237 (1962) 891.
4. Wroblewski, F., and LaDu, B. N., *Proc. Soc. Exptl. Biol. Med.* 90 (1955) 210.
5. Torriani, A., *Biochim. Biophys. Acta*, 38 (1960) 460.
6. Bünning, E., *The Physiological Clock*, Academic Press, New York 1964.
7. Bullough, W. S., and Lawrence, E. B., *Proc. Roy. Soc. B.* 154 (1961) 540.
8. Galavasi, G., Schvenk, H., and Bootsma, D., *Exptl. Cell Res.* 41 (1966) 428.
9. Stubblefield, E., and Klevecz, R., *Exptl. Cell Res.* 40 (1966) 660.
10. Gorini, L., and Gundersen, W., *Proc. Natl. Acad. Sci. U. S.* 47 (1961) 961.
11. Yates, R. A., and Pardee, A. B., *J. Biol. Chem.* 227 (1957) 677.
12. Kuempel, P. L., Masters, M., and Pardee, A. B., *Biochim. Biophys. Res. (Commun.)* 18 (1965) 858.

B. C. Goodwin
School of Biological Sciences, University of Sussex
Brighton (Sussex), Great Britain

Editor's Comments on Papers 27 and 28

27 **Dean and Hinshelwood:** *Integration of Cell Reactions*

28 **Pirt:** *Review of Growth, Function, and Regulation in Bacterial Cells*

These contributions indicate just how far the limits of study for microbial growth have extended. In the mathematical area the development has spread beyond the scope of papers [A. G. Fredrickson, D. Ramkrishna, and H. M. Tsuchiya, *Math. Biosci.*, **1**, 327 –374 (1967)] and reviews [P. R. Painter and A. G. Marr, *Ann. Rev. Microbiol.*, **22**, 519– 548 (1968)] and now extends outside these to the realm of monographs (B. C. Goodwin, *Temporal Organization in Cells*, Academic Press, Inc., New York, 1963) and books (A. C. R. Dean and Cyril Hinshelwood, *Growth, Function and Regulation in Bacterial Cells*, Oxford University Press, New York, 1966), which obviously lie outside the confines of this volume.

Reprinted from *Nature*, **199**(4888), 7–11 (1963)

INTEGRATION OF CELL REACTIONS

By Dr. A. C. R. DEAN and Sir CYRIL HINSHELWOOD, O.M., F.R.S.

Department of Physical Chemistry, University of Oxford

Introduction

THAT living matter reproduces itself by some kind of copying process has been commonly realized for a long time. That this process depends essentially on conformity of new material with existing spacings in molecular patterns was expressed in the various analogies with crystal growth or template action and in the correspondences pointed out by Astbury[1] between various distances revealed in X-ray analysis. That the sequences of monomer units in proteins, nucleic acids and polysaccharides allow an almost infinite variation in possible patterns has long been obvious. All these ideas have in more recent times been succinctly stated in terms of information' and 'coding'[2].

Modern X-ray analysis has focused attention on the helical structure of DNA (deoxyribose nucleic acid), on the great importance of base-pairing in the formation of doubly stranded DNA as a major element in replication, and also on the configuration of protein molecules[3-6]. The ultra-centrifuge has revealed the large range of particle sizes and molecular weights characteristic of the macromolecular cell components, and the use of isotopic tracers has revealed the lability of some of these[7,8]. The correspondence between relatively labile low molecular weight RNA (ribose nucleic acid) and the DNA of an invading phage has led to the conception of 'messenger' RNA, an idea which has been supported by the claim that the low-molecular-weight RNA formed during periods of rapid adjustment of cells to new conditions differs in composition from the normal[8,9].

A further elaboration of the conception of information and coding is represented by the 'repressor', a substance which is supposed to prevent an existing gene from manifesting its effect[10]. The action of the repressor itself is counteracted by the appropriate 'inducer', which allows the repressed gene to express itself in the formation of specific enzymes, previously thought to be called into being by a rather more direct and positive action of the inducer. The essential basis of the repressor idea is that certain substrates or products of an enzyme X may inhibit, not so much this enzyme itself (an action which would be entirely analogous to well-known phenomena of heterogeneous catalysis[11]) as the first member of a whole enzyme sequence by which X is ultimately synthesized[10].

All these various lines of approach represent developments in the quantitative side of biology and as such are highly characteristic of the modern trend. Valuable and indeed spectacular as some of them are, they are in many respects unco-ordinated, and do not yet give a truly quantitative treatment of the cell as a working unit. It seems to us that there is still lacking a proper appreciation of what might perhaps be called the 'total integration of cell function', a principle which we feel is unlikely not to be observed by Nature after countless ages of evolution.

Some Difficulties

The way in which some of the present ideas about cell function seem to us to need enlarging and co-ordinating may first be illustrated by a few examples.

(a) Transfer RNA of a given type is supposed to combine specifically with a single amino-acid which it presently guides into the correct position in a peptide chain. Since some twenty amino-acids may be involved, since there are only four possible structures for the terminal base in the RNA, and since the influence of side-chains in organic reactions is seldom transmitted very far from the reactive centre, the specificity can scarcely be explained on purely intramolecular structural grounds. Indeed, the assumption is generally made that the amino-acid–RNA combination is mediated by a specific enzyme[12,13]. If it is, then the specificity of the enzyme protein is playing a no less important part than that of the RNA, and it appears arbitrary to rule that one is more fundamental than the other. Moreover, the general control of all the individual steps involved in the complex polymerization processes to form the macromolecular cell components can scarcely be casually dismissed as operated by 'enzyme reactions' as though the mechanism of each enzymatic stage did not itself present in one form or another the same major problems of specificity, and conformity to an existing pattern. Familiarity with enzyme reactions may breed contempt, or at least a tendency to regard the name as providing an explanation of the mechanism and a failure to realize that activating enzymes are, in the words of Zamecnik[12], "unusual catalysts deserving of more attention than they currently receive".

When RNA and DNA are both built up under the influence of specific enzymes, are we really justified in saying that DNA can affect the cytoplasm and not vice versa ? Without the cytoplasmic constituents, the right DNA could not be made at all.

(b) The idea of repressor action would, it seems to us, benefit from incorporation into a wider setting. At the moment we have the picture of one gene responsible for the formation of an enzyme which is inhibited by a repressor formed under the influence of a second gene. The effect of the repressor can be neutralized by the action of an inducer which the cell may never have encountered before. Natural selection seems to have allowed a good deal of unnecessary activity to escape elimination for a long time. Inhibition of the actual activity of an enzyme, either by a substrate or by a product of its own working, would be perfectly understandable on the analogy of what is frequently observed with inorganic catalysts. Catalytic action depends on some degree of affinity between the catalyst and the substrate or the product, and this fact means that excess of either may block the surface (for example, SO_3 poisons platinum in the reaction of SO_2 with O_2). 'Repression', however, is assumed to affect not the working of the enzyme itself, but that of quite different enzymes preceding it in a synthetic chain. The normal principles of catalysis suggest no reason for this. The difficulty would no longer exist if the actual functioning of the enzyme were in fact impaired and if this impairment reacted back on the synthesis of its own precursors through a cycle of interdependent reactions in the way which will be dealt with later. Thus here, too, the disregard of the integrated function of the cell leaves unsolved a problem which might in the end prove not to be real. Evolution has in all probability ensured that synthesis and function are seldom completely uncoupled.

(c) One may well feel some degree of uneasiness about the failure to qualify statements such as the common one that chloramphenicol specifically inhibits protein synthesis. Protein synthesis is a complex affair with many stages. If protein is not formed, no enzymatic activity or anything dependent on it can continue for long. All that can be said is that the ratio nucleic acid/protein continues to rise for a short time during the unco-ordinated events following the interruption by this drug of normal cell function.

Interplay of Cell Components

Protein cannot be formed without the mediation of RNA. Neither can RNA nor DNA be synthesized without the necessary specific enzymes. Cells cannot function steadily without periodic division, and cell division is not possible without the production of a cell wall. All these constituents are clearly in mutual dependence.

There may well be a sense in which the DNA may be given the most fundamental role. We may say that an imposed mutation affecting DNA soon transmits its consequences to the other constituents, while changes chemically induced in these will soon disappear provided the DNA code remains intact. Even if this is accepted as beyond all doubt, the working of the cell as a machine cannot possibly be understood except in terms of the interplay and coupling of all its parts. Work on isolated cell reactions or components may yield valuable information; but this can never be properly assembled until the laws of integration of cell function are applied.

Although the contrary might sometimes seem to be tacitly implied, the self-replication of DNA in the absence of other cell components and in particular of enzymes has never been observed. The DNA is, moreover, never continuously replicated except in the intact functioning cell. In cell-free systems its synthesis may proceed to a limited extent, but must then stop.

Since the cell cannot continue to function without periodically dividing, the steady formation of wall material is necessary. The structure of the cell wall, with its complex pattern of polysaccharide, muco- and lipo-protein, is highly specific, and indeed probably as specific as that of proteins and nucleic acids. It determines essential permeability relations, and is probably the actual seat of certain enzyme activities. Moreover, the enzymes and nucleic acids of the interior cannot be correctly built up unless the cell wall makes the proper provision for the access of the raw materials. Nor can the wall be maintained and expanded without the appropriate activity of the nuclear and cytoplasmic components. Once again we have the picture of total integration.

Something in a way equivalent to a code resides in the secondary structures constituted by the foldings and spatial distributions of the macromolecular components of the cell. Geometrical factors directly affect the diffusion paths of the many labile intermediate metabolites, and thus may determine the ratio of alternative reactions, and may even affect the probability of incorporation of a given type of unit into a growing macromolecular chain. The equilibrium folding of proteins may, as has been suggested[15], be pre-determined by the amino-acid sequence; but equilibrium does not normally prevail in biological systems. In the last resort the primary DNA code may be able to impose its message on everything else, but it would be rash to reject all consideration of the possibility that an established geometry of diffusion paths, controlling the synthesis of enzymes, polysaccharides and nucleic acids alike, can have no influence at all on its own perpetuation, or in any event on the speed of its own modification. Once again it would seem wiser to remember the principle of total integration.

Some Sources of Misconception

The true perspective of cell phenomena is sometimes disturbed by the way in which experiments are planned. In crossing and recombination tests a given character is often recorded as positive or negative on the result of an observation made at a single fixed time such as 24 or 48 h. The consequence is to impose an artificial discontinuity which the phenomena do not inherently possess. The appearance of sharp distinctions instead of gradations tends to focus attention on structural differences rather than on continuously varying processes and so gives a false picture of the true relationships. The distinction between constitutive and inducible enzymes[16] is, for example, probably much too sharply drawn. The less information there is, the more clearly defined a classification may appear. Suppose, for example, we have three related enzyme activities all quantitatively and continuously variable. Crude criteria of assessment can easily lead to a matrix of possibilities ranging from + + + to − − − with all other combinations of + and − between. This can be interpreted in terms of single and separate genes. Since the activities are related, they may not separate easily, and a further conclusion may be drawn about the contiguity of the genes on a chromosome. The arbitrary nature of this could easily escape attention.

Complaints have been voiced that genetic data not conforming to clear-cut or preconceived patterns are often discarded[17,18]. Indeed, it must be said that the propounders of genetic theories themselves subject one another to a good deal of criticism, which, taken as a whole, suggests the need for a wider view of the entire field.

Lederberg[19] has pointed out that streptomycin gives exceptionally clear-cut results in the distinction between sensitive and resistant types and has directed attention to the fact that in this respect it is 'unusually suitable'. But the distinction he makes is seldom borne in mind by those who quote and base wide generalizations on the results obtained with this drug (which in fact there are good reasons for regarding as quite untypical[20]).

Another experimental need to which insufficient attention may be paid is that of ensuring a proper steady-state relation between the organisms under test and their environment. A strain of bacteria is sometimes maintained on a nutrient agar slope or in broth for long periods, and then in a matter of hours or minutes after transfer to a defined artificial medium it may be tested for growth rate or enzyme activity, or it may be crossed with another strain as far removed as itself from equilibrium with the environment. Such procedures may give very misleading results unless it is borne in mind that at a first transfer to a new medium the growth rate is low, the enzyme activities are abnormal, and the nucleic acid and protein contents out of balance.

Even when thoroughly acclimatized cultures of known history are used, variations in nucleic acids and proteins occur during the growth cycle[21,22] (a completely steady state being reached only in continuous culture[22-27]) so that snap tests taken at a fixed time may be completely unrepresentative.

Even if minutely standardized procedures are laid down, it may still be true that the results are an unspecified function of the experimental conditions and will be difficult to assess unless this function is itself investigated.

The present-day picture tends to be fragmented, static, and in some respects artificially simplified. The cell is often spoken of as though it were an unco-ordinated collection of mechanisms. 'Regulator genes' are spoken of as being switched on and off; but what are regulator genes and repressors if not part of the expression of the total quantitative cell balance ?

The structural code, at whatever level it may be inscribed in the cell, controls the whole set of potentialities. The expression and balance of these are governed by certain general principles which cannot be ignored without resulting confusion.

A number of communications on the kinetics of cell and tissue growth have been published at different times[28-35]; but it may be said on the whole that their influence needs re-asserting.

Integration of Cell Processes

The 'principle of total integration' is capable of mathematical expression. If the rate of formation of a nucleic acid, A_1, depends on an enzyme reaction, we should have in the simplest case $dA_1/dt = k_1 P_1$, where P_1 is the concentration of the appropriate enzyme. Conversely if the protein formation is guided by a nucleic acid template, we should have another equation such as $dP_1/dt = k_1^1 A_2$. The general principle is unaffected by the consideration that the individual equations are not necessarily of the simple linear form given in the example. Having regard to the active part that the cell wall plays in permeability phenomena as well as being the seat of certain enzyme activities, the polysaccharide, mucopeptide and lipoprotein constituents can all be brought into the scheme.

Now since the cell reproduces itself as a whole, and no part can continuously function without the others, it is reasonable to express the integration principle by closing the set of equations according to a scheme of which the following is the simplest example:

$$dX_1/dt = k_1 X_2, \; dX_2/dt = k_2 X_3, \; dX_3/dt = k_3 X_4 \ldots$$
$$\ldots \; dX_n/dt = k_n X_1$$

Great elaboration of such schemes, with complicated branchings, is, of course, possible, without change in the basic principle[29].

A second principle, without which no satisfactory theory of cell function is possible at all, is that of the preservation of physico-chemical balance. Inflow of nutrients and outflow of waste products in general can only maintain a proper balance within certain limits of cell size. Hence the necessity for division. Control of division must rest on a third principle, namely, that of some degree of coupling between the act of division and the process of growth. Many factors may be involved in this[36-40], but one simple overall formulation of the condition would be that division occurred when the amount of some key component in the cell reached a critical value[28-30,37]. More elaborate specifications could be laid down without changing the essential need for some such coupling condition. These could bring in reducing potential, size of nuclear particles and so on. The main point is, however, that without some attempt at a formulation no theory can have contact with reality.

Systems of equations based on the foregoing principles have already been examined. They are, it should be emphasized, in no way in conflict with the idea of genetic codes or any other legitimate conceptions of genetics. They lead to some general conclusions which the theory of cell function cannot properly ignore.

The cell emerges as a system which automatically adapts itself to give the optimum growth rate in a given medium. In so doing it changes the proportions of its components, and when it is transferred from one medium to a fresh one it may abandon one pattern of chemical reactions for another. In the process it brings itself into an optimum relation with its environment[29,30].

It is legitimate enough for particular purposes to talk of genes, operons, cistrons, repressors, messenger and transfer RNA, ribosomes, and so on, as parts of a fragmented and often virtually static system; but it is only when they are assembled into a machine that they can possibly make complete sense. When the assembly is ignored, the need will often arise to explain each new fact by a fresh *ad hoc* hypothesis.

It is the actual coupling of the assembly which is responsible for the very important cell property of adaptability to optimum growth, and for the existence of whole continuous ranges of properties rather than discrete differences only. In drug resistance phenomena, for example, although the resistant forms may on occasion show a single degree of response whatever the concentration of drug to which they have been exposed (selection

of pre-existent types)[20], they more often display a continuous spectrum of degrees of resistance[30].

We shall not re-capitulate here all the experimental evidence for direct adaptation[30,41-48] nor the objections to the fluctuation test[49] which have been claimed to disprove it. The isolated demonstration (by replica plating) of pre-existent forms resistant to bacteriophage and to relatively low concentrations of streptomycin has, as mentioned here, led to the quite illegitimate assumption that the result is general.

Drabble and Hinshelwood[20] showed that adaptation and mutation occur together in the case of streptomycin resistance, and Kilkenny and Hinshelwood[50] showed that even yeast cells which segregated in a Mendelian manner still underwent direct adaptation (to galactose) after separation.

The examination of the growth equations shows, furthermore, that in a given environment the establishment of a true steady state may be fast or slow[25]. Correspondingly, adaptive changes may be rapidly or slowly reversible. This result corresponds with experimental observations that adaptations are sometimes easily reversible and sometimes persistently held[30], the notion of stable heredity with unicellular organisms being one of degree only. The argument based on stability that adaptive changes are purely selective cannot be consistently upheld. Cells with greatly changed characteristics, relating, for example, to fermentative and respiratory properties and to drug resistance, can be obtained under conditions which preclude any important degree of selection of special types of individual[47,48,51-56]. The importance of enzymatic differences in a directed antigenic transformation is stressed by Austin, Widmayer and Walker[57] and the interplay of microbe and environment in clinical practice is discussed by McDermott[58].

Those who, rightly, lay great stress on the principle of natural selection sometimes fail nevertheless to understand that the selection of an appropriate reaction pattern by self-replicating matter is just as effective and, therefore, may be just as important as the selection of special individuals. A preference could only be justified by some kind of exaltation of structure over function, a distinction which Nature is unlikely to observe.

Unbalanced Growth

A good deal of importance is attached in the literature to observations about the differential synthesis of cell components, for example, uneven replication of DNA at different phases of individual cell cycles. The observations on this seem to have been made entirely with cells not in dynamic equilibrium with their environment. The cells have, for example, been deprived of special growth requirements such as thymine[59], very recently transferred to a new medium[59], or subjected to temperature shock[60]. In these circumstances, the principles we have outlined here would predict very severe disturbances of the co-ordination between the synthesis of different components. To take the most highly simplified example possible, suppose:

$$dX/dt = \alpha Y$$
$$dY/dt = \beta X$$

In dynamic equilibrium X and Y will settle down to average values of X_0 and Y_0 per cell. Suppose we transfer to a medium where the X component (say an enzyme) functions poorly, then the rate of formation of Y is reduced very greatly (possibly to zero), while the formation of X for some time can continue at the original rate. It would be wrong to infer from this that the replication of X necessarily and naturally is a temporal antecedent of that of Y. When continuous growth is allowed again, the proportions will eventually settle down to a balanced ratio. How contingent the whole phenomenon must be is shown by the fact that disturbances in relative synthesis rates characterizing growth of 'synchronous' cultures has proved to be a function of the way in which the synchronization is effected[27,61,62].

Significance has been attached to the observation that in some synchronized cells mass may increase linearly instead of exponentially with time. Quite apart from the question whether the distinction can be accurately established for the small changes in mass involved, any cell thrown right out of balance can show anomalous rates of increase for any components. With a more carefully planned technique of synchronization (depending only on the selection by filtration of appropriately sized cells and not subjection to chemical or temperature shock) synchronized cells can be obtained in which RNA, DNA and total protein all increase exponentially and maintain a constant ratio over the whole cell cycle[27,61].

Experiments in which the formation of protein or nucleic acid, or the course of specific enzyme activities, is followed for short periods after disturbances such as the addition of drugs are extremely difficult to interpret in the absence of much more complete studies. If a motor-car factory were partially destroyed by a bomb, finished cars might continue to be delivered for a short time, or alternatively individual components might pile up. But conclusions about the sequence of the manufacturing process could be drawn only by the careful co-ordination of various kinds of information.

Conclusion

Although some of the ideas about cell integration put forward in this article have received a good deal of opposition they are supported by much that can be found in the literature.

A very positive attitude is adopted not only by Graham Cannon[63] but also by Uvarov[64], who says : "In spite of a number of cases when environmentally induced variations have been shown to be, at least temporarily, inheritable, there is a reluctance to accept any evidence that might appear to favour the much maligned Lamarckian principle" . . . (and) . . . "this should encourage more entomologists to study their problems without being afraid of infringing some outdated canons of text-book genetics". Ephrussi[65] has remarked: "The second thing we should avoid is basing the distinction of mechanisms on nuclear or cytoplasmic localization. In my opinion this has been a major source of confusion in the past, and it is not going to be so easy to avoid it in the future because we have all been trained to regard the problem of differentiation as a nucleus/cytoplasm dilemma". Danielli[66], commenting on this, refers to: "evidence of either nuclear or cytoplasmic control, or both, in relation to structure and function". Our belief is that the relation needs to be made more explicit.

A common attitude is to admit the possibility of direct adaptive changes in principle but to ignore, dismiss or minimize the significance of the evidence that they occur[67-70]. When this is done the picture of cell function is left very incomplete.

Some authors speak occasionally of specifically induced mutations[71]. They accept the directing influence of the environment on what must, according to the accepted view, be the DNA code. The change in the DNA must involve enzyme action at some stage, so that the distinction between this kind of mutation and an adaptation becomes very indefinite. In this connexion we may refer to the gradual stabilization of induced enzyme adaptation which is observed with bacterial cultures[72,73] and also point out again the arbitrary nature of the distinction between inducible and constitutive enzymes.

Biologists concerned with the differentiation of tissues in complex organisms seem to take a wider view of the potentialities of the cell. Weiss[74], for example, says that: "genic replication, while seemingly the initiating step in the growth process, is followed by chains of events the nature of which is determined by the specific constitution of the (extragenic) rest of the cell". He also says[75] : "In short, the story of the molecular control of cellular activities is bound to remain fragmentary and incomplete unless it is matched by knowledge of what makes a cell the unit that it is, namely, the cellular control of molecular activities". Sonneborn[76] refers to: "successive cycles of differentiation of cytoplasm by nucleus and then of nucleus by cytoplasm". Chantrenne[77] says: "For if the cytoplasm in bacterial systems appears ancillary in regard of the all-mighty genome, embryologists view the cells and embryos as a much more involved system in which the cytoplasm exerts a very strong control on the genome". These authors are concerned with systems much more complex than those in which the kinetic implications can be easily treated but their basic ideas certainly seem to agree with the concept of cell integration.

Perhaps, if all that has been said in this article is conceded, its relevance to the discussion of other modern work may still be questioned. Our contention is that intensive work on individual fragments of a complex machine is not only incomplete but also may be misleading unless attention is paid to the way in which the parts fit together. Coding, information, messengers and carriers are all perfectly valid conceptions: so also are various kinds of enzyme-mediated condensation reaction. But these terms have in themselves little explanatory value, and present-day ideas will only achieve maximum fruitfulness when so combined that they are not left simply to describe bits of a machine without anything to make it work. Sometimes the claim is made that machines can only be understood in terms of their parts. The counterclaim can also be advanced that to understand how the parts fit together the machine must be observed in action. Relatively simple phenomena do reveal the machine in action. Apart from adaptive responses, to which reference has already been made, we have the variation of enzyme contents of cells according to need, enzyme changes compensating for the effects of adverse pH, the rise and fall of stores of cell components, and the whole rhythmic variation of cell composition during the growth cycle. One last example may be cited. When cells are multiplying two tendencies may be observed. On the one hand, the DNA content per unit of mass tends to be constant under optimum conditions. On the other hand, the cell tends to divide so as to maintain a constant amount of DNA per cell[21]. When nutrients become deficient a compromise has to be struck: there is an actual increase in the DNA per unit of mass; but a more economical sharing out so that the cells become much smaller and the amount of DNA per cell drops[22]. Facts such as these can only find an explanation in terms of total integration.

It must be emphasized again in conclusion that the ideas we have expressed are in no way in conflict with the conception of the genetic code, nor do they ignore the principle of mutation and selection. Cells in the course of evolution must have profitably exploited the variability of reaction patterns and the consequences of complete integration. It is those who affirm the contrary who fail to give the natural selection principle its due.

[1] Astbury, W. T., *Symp. Soc. Exp. Biol.*, **1**, 66 (1951).

[2] Crick, F. H. C., *Discovery* (March 1962).

[3] Watson, J. D., and Crick, F. H. C., *Nature*, **171**, 737 (1953).

[4] Watson, J. D., and Crick, F. H. C., *Cold Spring Harbor Symp. Quant. Biol.*, **18**, 123 (1953).

[5] Kendrew, J. C., *Pontif. Acad. Sci. Scripta varia*, 109 (1961).

[6] Perutz, M. F., *Pontif. Acad. Sci. Scripta varia*, 217 (1961).

[7] See for refs. *Microsomal Particles and Protein Synthesis*, edit. by Roberts, R. B. (Pergamon Press, New York, 1958).

[8] See for refs. Spiegelman, S., *Fed. Proc.*, **22**, 36 (1963).

[9] Brenner, S., Jacob, F., and Meselson, M., *Nature*, **190**, 576 (1961).

[10] Monod, J., and Jacob, F., *Cold Spring Harbor Symp. Quant. Biol.*, **26**, 389 (1961).

[11] Schwab, G. M., Noller, H., and Block, J., *Handbuch der Katalyse*, 160 (Springer-Verlag, Wien, 1957).

[12] Zamecnik, P. C., *Pontif. Acad. Sci. Scripta varia*, 431 (1961).

[13] Zamecnik, P. C., *Biochem. J.*, **85**, 257 (1962).

[14] See for refs. Rogers, H. J., *Ciba Found. Study Group No. 13, Resistance of Bacteria to the Penicillins*, edit. by de Reuck, A. V. S., and Cameron, Margaret P. (Churchill, London, 1962).

[15] Afinsen, C. B., *Pontif. Acad. Sci. Scripta varia*, 6 (1961).

[16] See Jacob, F., and Wollman, E. L., *Sexuality and the Genetics of Bacteria*, 261 (Academic Press, New York, 1961).

[17] Lindegren, C. C., *The Yeast Cell* (Educational Publishers, Inc., Saint Louis, 1949).

[18] Ravin, A. W., *Symp. Soc. Gen. Microbiol.*, **3**, 46 (1953).

[19] Lederberg, J., *J. Bact.*, **59**, 211 (1950).

[20] Drabble, W. T., and Hinshelwood, Sir Cyril, *Proc. Roy. Soc.*, B, **154**, 449 (1961). Drabble, W. T., *ibid.*, **154**, 571 (1961).

[21] Dean, A. C. R., and Hinshelwood, Sir Cyril, *Proc. Roy. Soc.*, B, **151**, 348 (1960).

[22] Dean, A. C. R., *Proc. Roy. Soc.*, B, **155**, 580, 589 (1962).

[23] Monod, J., *Ann. Inst. Pasteur*, **79**, 390 (1950).

[24] Powell, E. O., *J. Gen. Microbiol.*, **15**, 492 (1956).

[25] Moser, H., *The Dynamics of Bacterial Populations Maintained in the Chemostat*, Carnegie Institution of Washington Pub. 614 (Washington, D.C., 1958).

[26] Maaløe, O., *Symp. Soc. Gen. Microbiol.*, **10**, 272 (1960).

[27] Herbert, D., *Symp. Soc. Gen. Microbiol.*, **11**, 391 (1961).

[28] Hinshelwood, C. N., *The Chemical Kinetics of the Bacterial Cell* (Clarendon Press, Oxford, 1946).

[29] Hinshelwood, Sir Cyril, *J. Chem. Soc.*, 745 (1952); 1304, 1947 (1953).

[30] Dean, A. C. R., and Hinshelwood, Sir Cyril, *Progr. Biophys. Biophys. Chem.*, **5**, 1 (1955).

[31] Nedler, J. A., *Biometrics*, **17**, 220 (1961).

[32] Perret, C. J., *J. Gen. Microbiol.*, **22**, 589 (1960).

[33] Perret, C. J., and Levey, H. C., *J. Theor. Biol.*, **1**, 542 (1961).

[34] Kacser, H., *Symp. Soc. Exp. Biol.*, **14**, 13 (1960).

[35] Waddington, C. H., *Colston Papers*, **7**, 105 (1954).

[36] Swann, M. M., *Cancer Res.*, **17**, 727 (1957).

[37] Scherbaum, O., *Exp. Cell. Res.*, **13**, 11 (1957).

[38] Plesner, P., *Cold Spring Harbor Symp. Quant. Biol.*, **26**, 159 (1961).

[39] Mazia, D., *Harvey Lectures*, **53**, 130 (1957–58).

[40] Stern, H., *Developing Cell Systems*, edit. by Rudnick, Dorothea, 135 (Ronald Press Co., New York, 1960).

[41] Dean, A. C. R., and Hinshelwood, Sir Cyril, *Ciba Found. Symp. Drug Resistance in Micro-organisms*, edit. by Wolstenholme, G. E. W., and O'Connor, Cecilia M., 4 (Churchill, London, 1957).

[42] Dean, A. C. R., and Hinshelwood, Sir Cyril, *Ciba Found. Symp. Cell Metabolism*, edit. by Wolstenholme, G. E. W., and O'Connor, Cecilia M., 311 (Churchill, London, 1959).

[43] Dean, A. C. R., *Proc. Linnean Soc.*, **169**, 45 (1958).

[44] Dean, A. C. R., and Hinshelwood, Sir Cyril, *Proc. Roy. Soc.*, B, **151**, 435 (1960).

[45] McCarthy, B. J., and Hinshelwood, Sir Cyril, *Proc. Roy. Soc.*, B, **153**, 339 (1960).

[46] Midgley, J. E. M., *Proc. Roy. Soc.*, B, **153**, 250 (1960).

[47] Dean, A. C. R., *Proc. Roy. Soc.*, B, **153**, 329 (1960).

[48] Dean, A. C. R., and Broadbridge, P. H., *Proc. Roy. Soc.*, B (in the press).

[49] Dean, A. C. R., and Hinshelwood, Sir Cyril, *Proc. Roy. Soc.*, B, **139**, 236 (1952); **140**, 339 (1952).

[50] Kilkenny, B. C., and Hinshelwood, Sir Cyril, *Proc. Roy. Soc.*, B, **139**, 575; **140**, 352 (1952).

[51] Baskett, A. C., and Hinshelwood, Sir Cyril, *Proc. Roy. Soc.* B, **139**, 58 (1951).

[52] Dean, A. C. R., *Proc. Roy. Soc.*, B, **147**, 247 (1957).

[53] Shrift, A., Nevyas, Joann, and Turndorf, Sietske, *Plant Physiol.*, **36**, 502 (1961).

[54] Nagai, S., and Nagai, H., *Naturwiss.*, 441 (1958).

[55] Kossikov, K. V., *Ciba Found. Symp. Drug Resistance in Micro-organisms*, edit. by Wolstenholme, G. E. W., and O'Connor, Cecilia M., 102 (Churchill, London, 1957).

[56] Kossikov, K. V., *Fifth Intern. Congr. Biochem.: Section Medical Genetics* (1961).

[57] Austin, Mary L., Widmayer, Dorothea, and Walker, Lola M., *Physiol. Zool.*, **29**, 261 (1956).

[58] McDermott, W., *Yale J. Biol. Med.*, **30**, 257 (1958).

[59] See for refs. Maaløe, O., *Cold Spring Harbor Symp. Quant. Biol.*, **26**, 45 (1961).

[60] See for refs. Campbell, A., *Bact. Rev.*, **21**, 263 (1957).

[61] Abbo, F. E., and Pardee, A. B., *Biochim. Biophys. Acta*, **39**, 47 (1960).

[62] McFall, Elizabeth, and Stent, G. S., *Biochim. Biophys. Acta*, **34**, 580 (1959).

[63] Cannon, H. G., *Proc. Linnean Soc.*, **169**, 41 (1958).

[64] Uvarov, B. P., presidential address, *Proc. Roy. Entomol. Soc. Lond.*, **25**, 5 (1961).

[65] Ephrussi, B., *J. Cell. Comp. Physiol.*, **52** (Supp. 1), 35 (1958).

[66] Danielli, J. F., *J. Cell Comp. Physiol.*, **52** (Supp. 1), 51 (1958).

[67] Bryson, V., and Szybalski, W., *Origins of Resistance to Toxic Agents*, edit. by Sevag, M. G., Reid, R. D., and Reynolds, O. E. (Academic Press, New York, 1955).

[68] Bryson, V., and Szybalski, W., *Adv. Genetics*, **7**, 1 (1955).

[69] Cavalli-Sforza, L. L., and Lederberg, J., *Sixth Intern. Cong. Microbiol. Symp. Growth Inhibition and Chemotherapy*, 108 (Istituto Superiore di Sanità, Rome, 1953).

[70] See, for example, discussion in *Ciba Found. Symp. Drug. Resistance in Micro-organisms*, edit. by Wolstenholme, G. E. W., and O'Connor, Cecilia M., 62 (Churchill, London, 1957).

[71] Lindegren, C. C., and Pittman, D. D., *J. Gen. Microbiol.*, **9**, 494 (1953).

[72] Hinshelwood, Sir Cyril, *Chem. and Indust.*, 1050 (1961).

[73] Lindegren, C. C., *Nature*, **189**, 959 (1961).

[74] Weiss, P., *The Hypophyseal Growth Hormone, Nature and Actions*, edit. by Smith, R. W., Gaebler, O. H., and Long, C. N. H., 3 (McGraw-Hill, New York, 1955).

[75] Weiss, P., *The Molecular Control of Cellular Activity*, edit. by Allen, J. M., 1 (McGraw-Hill, New York, 1932).

[76] Sonneborn, T. M., *Proc. U.S. Nat. Acad. Sci.*, **46**, 149 (1960).

[77] Chantrenne, H., *Nature*, **197**, 27 (1963).

28

Reprinted from *Sci. Progr.*, **55**(3), 489–492 (1967)

Growth, Function and Regulation in Bacterial Cells: A Review

S. J. PIRT

Growth, function and regulation in bacterial cells

By A. C. R. DEAN and SIR CYRIL HINSHELWOOD. *pp. xvi + 439. Clarendon Press: Oxford University Press, 1966. 84s.*

The principles of growth

Quantitative studies on the growth of protists and cells began to make substantial progress only about 20 years ago. Previous studies on growth were largely confined to the qualitative description of the nutrition, the fate of substrates and the external morphology of cells. A contribution which greatly influenced this re-orientation towards quantitative growth studies was Sir Cyril Hinshelwood's book on *The Chemical Kinetics of the Bacterial Cell*. The successor to this book by Dean and Hinselwood may

therefore be anticipated with great interest. It is a re-synthesis of the concepts of growth kinetics developed by Hinshelwood and his colleagues and takes into account the prolific amount of investigations which they have made in the field since the first book. The subject is a difficult and consequently a controversial one. Although the first glimmer of light on the mechanism of growth has appeared, there is a lack of balance in opinions about the control of growth. Dean and Hinshelwood's book should help to restore the balance.

The kinetics of growth developed by Hinshelwood and the experimental tests have lead him to challenge some firm beliefs about the mechanism of growth and adaptation of the organism to its environment. A lively controversy has

397

developed which has stimulated more rigorous thinking and experiment in the field. Unfortunately the argument has been marred by some misrepresentation of Hinshelwood's view, and both sides have been partisan in their approach. The main issue in the controversy has been—to what extent must one attribute adaptation of the cell to spontaneous mutation and selection of a genetically different organism rather than to non-genetic or phenotypic change. Hinshelwood realized the potential of bacteria for studying the kinetic problems of cell growth because of the relative ease and rapidity of reproduction of these organisms. The Hinshelwood school therefore took certain common bacterial types and studied particularly their mode of adaptation to grow in the presence of certain inhibitors of growth and to utilize new substrates. Simultaneously much work was being done by molecular biologists on the more material aspects of enzyme synthesis and its relation to gene (deoxyribonucleic acid) action. There has been a gulf between these two schools which has not helped their mutual understanding. The monograph of Dean and Hinshelwood puts the two approaches in perspective and provides links between them.

The central theme of the new book is this, Given a certain hereditary constitution in the cell, how far can we account for the properties of the cell by the kinetics of growth? The conclusion which the kinetics lead to is that, in a growth medium the balance of constituents in the cell, including nucleic acids, proteins, polysaccharides and lipids, must change in response to environmental changes so as to maintain the maximum growth rate. This may be interpreted to mean that the cell modifies the relative amounts of its constituents, that is, its phenotypic composition, in response to need. In contrast the molecular biologists tend to attribute control of the cell entirely to gene enzyme interaction and to deny the concept of phenotypic response to need. The argument is a fascinating one and must be capable of resolution by appropriate experiments.

The experimental approaches of the two opposing schools are very different in character: the molecular biologists devote attention to studying the material nature, activity and amount of enzymes and nucleic acids in cells, particularly in the period immediately after a change in the conditions and up to a few generations after the event. In contrast, Hinshelwood's school has studied the long term effect following prolonged growth ('training') up to 2000 generations (one subculture is about ten generations) in the new environment. It seems possible therefore that the phenomena most prominent in each approach are not the same. It may be concluded that in the short term experiment one only observes the initial transient states whereas in the long term one observes the final steady state. To this point consideration must be given in the future. It was also found that the 'training' or adaptation to a new growth condition could be achieved much more rapidly ('accelerated training') by gradually changing to a new environment whilst the culture was growing, rather making the change while the organisms were in the lag period before growth. Accelerated training is of particular significance because it shows that the time taken for adaptation could be too short to allow selective growth of a few mutants as a possible explanation. Opponents of the concept of phenotypic adaptation to drug resistance have questioned the interpretation of 'accelerated training' on the basis of the criticism of the experimental technique advanced by Sinai and Yudkin. However, as Dean and Hinshelwood point out, further work has rebutted this criticism.

Another distinctive feature of the experimental work of the Hinshelwood school is the significance which they attach to the duration of the lag period

490

before growth begins in a bacterial culture. Their analysis of the factors which influence the lag is the most authoritative and complete study of the phenomenon. Subtle mathematical analysis is used to relate duration of lag to the phenomenon of adaptation; one feels at times that the analysis is too subtle and attaches more than due significance to it. But, as an example of the application of mathematical logic to a biological problem, the study is of great interest.

Of the many striking chapters in the book, by far the most important is that on 'total integration' of cell functions. Here we have the basic theorems of the kinetics of cell growth, that is, the laws by which the cell can take in materials, produce more cells, and react to the environment. The basic kinetics can be referred to the two simple models (A and B) illustrated.

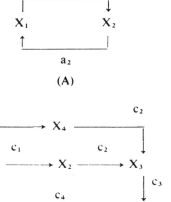

(A)

(B)

In (A) the elements X_1 and X_2 could be enzymes each providing an amino acid (a_1 or a_2) which is required in synthesis of the other enzyme. Thus production of enzyme X_2 is dependent on the activity of enzyme X_1 and vice versa. Hence the model represents *interdependent* elements

in the cell structure. The arrow indicates the interdependence, it does not mean that enzyme X_1 synthesizes enzyme X_2 and so on. The elements X_1, X_2, etc., could be any structural element which has a function in the cell. It might for instance be a lipid which controls the access of a substrate to an enzyme. A more complex system of dependences is shown in model (B). Here X_1, X_2, etc., are the structural elements and c_1, c_2, etc., are the diffusible metabolites formed by the preceding enzyme and incorporated in the succeeding one in the sequence. Model (B) introduces branches and alternative sequences. The model therefore represents the cell as a branched network of elements, each of which is dependent on other units and each of which to some extent controls the formation of the next in the sequence. Analysis of the kinetics of growth of such a system shows that eventually all elements in the system will increase in amount at the same exponential rate, that is according to the autocatalytic law which is typical of growth. Also it is found that in a particular constant environment the amounts of the elements present in the cell must settle down to constant proportions which allow the maximum growth rate. It follows that, if the environment changes, the proportion of the enzymes, nucleic acid and other constituents will change so as to maintain the maximum growth rate. Where alternative sequences of enzymes exist it can be shown that one pathway may decay and give place to the other. Such a process will simulate what molecular biologists term 'repression' and 'induction' of enzyme synthesis. These predictions and many others follow from what Dean and Hinshelwood have termed the 'network theorem'. The kinetics of growth therefore lead to the conclusion that all elements in the cell in some measure contribute to the control not just the genetic material or DNA.

There is now a growing body of evidence in favour of this view of the cell.

491

Much of the experimental work of the Hinshelwood school supports it. Continuous flow culture (chemostat) studies are beginning to provide powerful evidence in support. For instance, Tempest & Herbert (*J. gen. Microbiol.* **41**, 143, 1965) showed by chemostat studies that the amount of so-called 'constitutive' enzymes, that is, enzymes supposed always present in the cell in constant amount, do in fact vary widely in amount with change in growth rate of the cell.

At many points the book shows the relevance of the kinetics to fundamental biological problems: for example, the way in which drug addiction may be caused in man. The authors are well aware of the way in which the kinetics and the 'spatial map' or ultrastructure of the cell are interdependent. It is shown by an elegant argument that the history of a cell might make a phenotypic change irreversible and therefore appear to be a genetic change. The development of 'communities of cells' or 'tissues' is discussed in a thoughtful, fresh manner.

To conclude—this monograph represents the contribution of kinetic studies to the physiology of growth. The overriding impression is one of the great significance of total integration of cell functions. We have to regard the cell as a highly organized structure to account for its complex functions, in particular its ability to reproduce itself. The great merit of the total integration hypothesis is that it accounts for the overall control. The idea of the interdependence of cell structural elements upon each other's functions seems inescapable. Molecular biology appears to overlook the basic effects of these interdependences and to be solely concerned with operation of special mechanisms like the action of repressor proteins which inhibit gene action, or the inhibition of a reaction sequence by the end product of the sequence. The total integration hypothesis does not exclude more subtle refinements such as the repressor proteins being superimposed on the basic kinetic mechanism.

The book introduces salutary criticism of some of the accepted bases of genetics and molecular biology, for example, the Luria and Delbrück fluctuation test or the interpretation of the varying rates of development of colonies on nutrient agar plates, important in distinction between phenotypic and genotypic variation. Occasionally the authors' scepticism seems carried to undue lengths, for instance in questioning the existence of repressor proteins or the visual evidence for the circular bacterial chromosome obtained by Cairns. However, the presentation is exceptionally well-balanced. The style is such as to hold one's attention to the subject matter with ease. The mathematics required is an elementary acquaintance with the calculus and algebra, about the equivalent of G.C.E. 'A' level which should be within the competence of the majority of biologists. The book leaves the impression that we are on the verge of mathematical prediction of cell behaviour. The matter could hardly be more important since it deals with the fundamental basis of life, the control and exploitation of the micro-organisms, the differentiation of function in cells and even the development of the mind.

492

Editor's Comments on Paper 29

29 Brock: *Microbial Growth Rates in Nature*

The study of microbial growth has previously been limited, almost entirely, to the study of pure cultures, being largely concerned with changes taking place in the populations of individual species. There is a need now to understand the interplay between individuals and populations of different species, of their relative increases or growth, and of the influence of environment on these, not only in the laboratory but, more especially, in nature. This is virtually an unexplored area, and in this review Brock outlines some of the pioneering incursions being made into it.

$\mathcal{29}$

Reprinted from *Bacteriol. Rev.*, **35**(1), 39–58 (1971)

Microbial Growth Rates in Nature

THOMAS D. BROCK

Department of Microbiology, Indiana University, Bloomington, Indiana 47401

INTRODUCTION

The survival of a species in a natural habitat depends ultimately on its ability to grow at a rate sufficient to balance death due to predation, parasitism, or natural causes. Although many aspects of the ecology of a species may be deducible from behavior of the species in culture, it seems evident that at some stage of the study the investigator must return to nature and study the behavior of the population in its natural habitat. This seems even more important now than it was several decades ago because we are now more aware of the tremendous physiological versatility of microbes as evidenced by the existence of biochemical feedback loops, inducible and repressible enzymes, etc. What a species can do in a culture medium is not necessarily what it is doing in its natural habitat. Environmental factors of all kinds may differ drastically in nature from what they are in culture, and they may differ in ways we may not even perceive. This is especially true when we consider that in nature microbes live in microenvironments which may differ widely from the macroenvironment which we are capable of measuring with ordinary instruments and chemical procedures. For instance, nutrient quality and quantity, pH, and osmotic pressure are among the factors which can differ between micro-

environment and macroenvironment. Hence, it is difficult (if not impossible) to simulate precisely the physicochemical conditions of the natural environment in the laboratory. Furthermore, in nature an organism is always faced with competition from other organisms, difficult to duplicate in the laboratory.

It might seem virtually impossible to measure microbial growth rates in nature, as a result of the small size of microbes and the fact that many species live together. It is the purpose of this review to show that natural microbial growth rates can be measured by a variety of methods adaptable to different kinds of habitats or different kinds of organisms. The review also considers some of the misconceptions about the measurement of growth rates and evaluates some of the pitfalls to successful application of these methods. An attempt is made, when possible, to compare natural growth rates with those of the same species under presumably optimal conditions in the laboratory.

Commitment to Nature

There is a natural tendency for the microbial ecologist to take the easy way out and stay in the laboratory. All sorts of laboratory devices have been constructed which give the illusion of duplicating natural conditions: chemostats, turbidostats, temperature-gradient blocks, soil perfusion columns. Although these devices permit interesting experiments, they do not replace study of organisms in nature. At some stage, studies in nature must be conducted, preferably before the investigator's ideas about what microorganisms *might* be doing in nature become too firmly fixed. Once some idea of what organisms *are* doing in nature is obtained, ecologically relevant laboratory study of organisms can be conducted in a more meaningful way. Ideally, one should move back and forth from nature to laboratory rather frequently, always checking natural observations with laboratory study and vice versa.

What is Nature?

Any microbial habitat which is not completely controlled can be called a natural habitat. A microbial habitat may be quite tiny. The mammalian intestinal tract, for instance, is not one but many microbial habitats. A single soil crumb may be the home of several kinds of organisms, each in its own niche, and one niche may differ chemically or physically from other niches on the crumb. Except in extreme environments in which selective conditions are so rigorous that populations of single organisms may attain macroscopic dimensions, microbial niches are microscopic ones, and the microscope is an instrument indispensable for this exploration.

POPULATION PROBLEMS

A microbial population is any group of cells of one or more types which can be defined in terms of its extent in space and time. A population may be, for instance, a microbial colony which has a fixed location in space, a species of alga which is distributed widely throughout the plankton of a lake but defined in terms of its characteristic morphology, or the sum total of cells in the rumen, which may be regarded as a single population defined in terms of location.

In terms of growth kinetics, two types of populations can be easily recognized: the steady-state population and the exponentially growing population. The latter, not common in nature, is a closed population of which the members are all growing continuously with no gain from or loss of cells to the environment. The steady-state population may also contain cells which are growing continuously, but cells are being lost from the population at the same rate they are being added. There are, of course, many gradations between exponential and steady state. Thinking about the steady state makes us consider the whole problem of cell migration, which

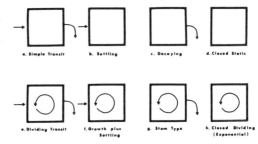

FIG. 1. *Diagrammatic representation of possible cell populations. Arrows outside the boxes indicate cell migration; arrows inside the boxes refer to cell divisions taking place within the system. Based on Cleaver (15).*

can play a significant role in changes in population number. (In fact, one of the most important distinctions between a natural and a laboratory population is that in the former migration can occur.) Examples of different kinds of populations, based on the ideas of Gilbert and Lajtha (22), are shown in Fig. 1.

The upper part of Fig. 1 portrays situations in which no cell division is occurring, and change occurs only due to input and output. An example of a simple transit population (Fig. 1a) is the microflora of the air, or the drift phytoplankton being swept down a river from one lake to another. An example of a settling population (Fig. 1b) is the mud surface of a lake, on which algae may be deposited, or the rectum of a mammal, which is the repository of vast microbial populations from the large intestine. An example of a decaying population (Fig. 1c) is the population in the mammalian intestinal tract after oral antibiotic therapy, in which washout but not growth can occur. Another example of a decaying population is the darkened hotspring algal mats described by Brock and Brock (11). An example of a closed static population (Fig. 1d) is a dry sample of soil held in the laboratory.

The lower part of Fig. 1 portrays situations in which cell division occurs as well. An example of a dividing transit population (Fig. 1e) is the bacterial flora at a given location in the mammalian intestine, or an algal component able to divide in a moving stream. Of course, virtually any natural dividing population will experience some immigration and emigration, and hence could be assigned formally in this category, although from a practical standpoint it is only when transit is the major component of population change that it becomes relevant to recognize the situation. Depending on the relative rates of immigration and emigration, the dividing transit population (Fig. 1e) grades into the other three types of growing populations (Fig. 1f, 1g, and 1h). An example of growth plus settling (Fig. 1f) might be the bottom flora of a shallow lake, where an algal species lives both in the water and on the bottom. A *Sphaerotilus* bloom is an example of a stem-type population, to borrow a term from the mammalian cell biologist. In this case (Fig. 1g), immigration is unimportant and emigration by sloughing or release of swarmers is common. The closed dividing population (Fig. 1h) is the familiar exponentially dividing population of the microbial physiologist, and is found only transiently in nature.

It is important to emphasize that a steady-state population as defined by an ecologist is quite different from that defined by a microbial physiologist. The physiologist thinks of the steady state from the viewpoint of the individual cell and defines it as the situation in which "the distribution of each intensive random variable (e.g., cell age or cell protein) does not depend on the time when the sample is chosen" (45), a situation which exists most commonly when the population is in exponential growth. To the ecologist, concerned with the population rather than the individual, exponential growth is anything but a steady state.

ASSESSMENT OF MICROBIAL NUMBERS OR MASS

In any study of microbial growth rates, it is necessary to assess population numbers. The phycologist or protozoologist does this by performing direct microscopic counts of the organisms in a sample from the habitat. If the organisms vary widely in volume, he will not make a simple count, but rather an estimate of the volume of cellular material of a given species. Procedures for such estimates are well established (43, 58). The mycologist may use similar methods (46). Because of the small size of bacteria, the bacteriologist has used direct microscopic counts rather infrequently (16), relying instead on viable counts. Although such methods are useful for physiological and genetic studies in the laboratory on single species with plating efficiencies that approach 100%, they are usually unsuited for ecological investigations. Many studies (e.g., 32, 48) have shown that the viable count is always much lower than the direct microscopic count. Perfil'ev and Gabe (48) give ratios varying from 13,000:1 to 737,000:1. Although low ratios could be due to faulty plating techniques, even under the best conditions at the most about 10% of the bacteria in a natural population may be counted. The reason, of course, that viable counts are done in spite of this serious limitation is that it is often virtually impossible to recognize a bacterial species microscopically, and at least when one has a colony on an agar plate one can subculture it and proceed to an identification. Another reason is that direct microscopic counts are difficult and tedious. Finally, viable counts permit assessment of populations whose densities are too low for measurement by more direct methods.

It should be clearly recognized, however, that viable counts cannot reveal how rapidly organisms are growing in nature. (For an exception to this statement, see the discussion below of Meynell's work.) It is merely necessary to note that a bacterial spore, which contributes nothing to the function of the ecosystem at the time of sampling, will produce a colony on an agar plate. Viable counts are useful in studies of microbial dispersal, and it is in this way that they are best used in ecological studies. The tracing of a source of pollution from a viable count of *Escherichia coli* in a water supply is an example of this kind of study.

If we cannot assess bacterial numbers in a natural habitat by viable counts, how then are we to do it? First, direct microscopic counts are often possible when the bacterium in question is large or morphologically distinct. Examples of this approach are the enumeration of *Chromatium* in lakes (35) or of *Leucothrix mucor* attached to seaweeds (10). Second, identification may be possible at the microscopic level by the use of fluorescent-antibody procedures (26, 52). Third, we may be dealing with habitats in which only one or a restricted number of kinds of organisms are present, so that species identification is no problem (2). Admittedly there are cases in which none of these situations obtains, but new methods can probably be devised.

El-Shazly and Hungate (18) obtained a measure of the microbial mass in a rumen fermentation by measuring the rate of gas production of samples of rumen contents containing a rate-saturating quantity of fermentable substrate, so that gas production was a function only of population size. The absolute population size could not be measured, but *relative* changes in population density could be. A precisely analogous method can be used by those working with photosynthetic organisms. The rate of photosynthesis (measured with $^{14}CO_2$) can be measured under conditions of saturating light intensity. Changes in population density will then be reflected by changes in the rate of $^{14}CO_2$ uptake.

A cell constituent which may be of considerable value in determining relative population densities of actively metabolizing microbes is adenosine triphosphate (ATP) (29). Sensitive techniques are available for its measurement, and ATP concentration should be proportional to biomass. Of course, only in natural systems where microbes constitute the majority of the biomass could ATP determinations be used to estimate the relative population density of microbes.

Chlorophyll has been used frequently as an index of algal biomass, but is not a reliable one because chlorophyll content per cell may vary widely as a function of light intensity (12). Epply (19) described a method for measuring standing crop by extrapolating the photosynthesis-rate curve back to zero time.

Estimating population size of filamentous organisms is more complex, but at least the temptation to use viable counting procedures does not arise. The direct way to measure lengths of filaments is to enlarge the field of view either by photography, projection, or drawing with camera lucida, and then to measure the lengths of filaments with a ruler or map reader. This approach has been used by soil mycologists (46). A simpler way, and one which is probably just as accurate, is to use a stereological technique. A grid is thrown across the field and the number of intersections of filaments with lines of the grid is counted. Olson (43) described five different stereological procedures, each suitable to a particular system or providing a particular degree of accuracy. We have used one of his methods for assessing aggregate lengths of filamentous bacteria on artificial substrates with considerable success (3), and we can recommend it for convenience, rapidity, and relative accuracy.

DEFINING BOUNDARIES OF THE HABITAT

Before a study is begun, the boundaries of the habitat must be defined. Initially one must decide on the scale of the investigation. For study of microbial growth in soil, for instance, a dimension as large as a field or as small as a soil crumb could be selected. For a lake, the whole body of water could be investigated, or any fraction of this body, such as the planktonic regions, the littoral, the thermocline, or the bottom sediments. Even in a habitat as sharply bounded as the rumen, it might in some cases be useful to consider subsections (e.g., rumen wall, food particles) in which particular species or processes might be concentrated.

For study of the growth rate of a single species, it is usually preferable to examine the smallest habitat encompassing the boundaries or the field of influence of the organism. Thus *L. mucor* lives as an epiphyte of benthic algae on the littoral of sea coasts (6). It would be unprofitable to study the whole vast littoral region rather than the specific seaweeds with which *L. mucor* is associated.

It is essential to know the type of distribution which the organism has throughout the habitat. Animal and plant ecologists (27, 38) use the terms "dispersion" or "pattern" when referring to this aspect. If a population is randomly dispersed, its distribution will follow a Poisson distribution. Random dispersion is rare in natural populations, the species more frequently being patchy or clumped. In such cases, the variance and the mean are not equal, as in the Poisson, and departures from randomness can be detected by deviations of the distribution from a Poisson series. Greig-Smith (27) gives a variety of statistical tests which can be used to detect such differences and to express them quantitatively. In general, all these methods use data derived by counting the number of individuals in a series of quadrats.

A quadrat is a real or imaginary area of known size which is thrown across the field. Despite its name, a quadrat need not be square but can have any convenient shape; for example, the area encompassed by a microscope field is a quadrat. It is obvious that the quadrat size must be large enough so that distributions obtained are not biased by quadrat size. This point, discussed in detail by Kershaw (34), is illustrated in Fig. 2. It can be seen that if the quadrat size is either too small or too large, patchiness will be missed. The ideal quadrat size for detecting patchiness would have an area approximately equal to the area of the clump, since some quadrats would contain large numbers of organisms and others would be virtually devoid of organisms.

To assess growth rates over the whole extent of the habitat, it is important to use quadrat sizes large enough to eliminate changes in number caused by patchiness. At the same time, very large quadrat sizes make the work of counting unnecessarily tedious.

It is essential to study habitats in which the organism of interest is actually growing. Many organisms are carried by streams or currents far from the site where they reproduce. For instance, *Staphylococcus aureus* grows in the kidney, forming so-called foci of infection. Cells from these foci are sloughed into the blood and carried to distant parts of the body. It is easy to assess staphylococcal numbers in the blood stream, but it would be incorrect to use changes in these numbers as a measure of staphylococcal growth rates in the kidney. In algal blooms, the cells often float to the surface and are carried by wind and current to quiet backwaters where they accumulate in vast numbers. The factors that con-

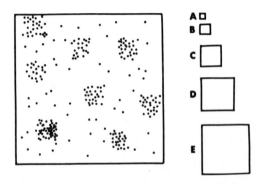

FIG. 2. *Relationship between quadrat size and the detection of nonrandom distribution. If the population were sampled with quadrats A or B, only slight nonrandomness would be seen, with C very marked nonrandomness; with D and E the distribution would appear progressively more random. Thus, the most marked demonstration of nonrandomness would be with a quadrat area about equal to the area of the clump. Based on Kershaw (34).*

trol algal blooms can not be ascertained by studying algae in such backwaters rather than in the part of the water where the algae are actually growing.

ARTIFICIAL SUBSTRATES

The use of artificial substrates greatly simplifies many studies in microbial ecology. The substrate which has been most commonly used in both aquatic and soil microbiology is the glass microscope slide (14, 53). The main advantage is ease of microscopic examination, but an additional advantage for rate studies is that one knows precisely the time microbial growth *could* commence. For studies on colonization and succession artificial substrates are ideal. The substrate may be both physically and chemically analogous to many natural substrates such as rocks and minerals. A wide variety of materials (e.g., plastics) other than glass can be used (53). If necessary, even opaque materials can be used in conjunction with incident fluorescence microscopy. In fouling studies, the artificial substrate is at the same time the habitat of interest.

In soil or other heterogeneous systems, microscope slides are difficult to use because (i) they cannot be removed and replaced without disturbing the environment; (ii) the opacity of the particulate material makes microscopy difficult; and (iii) it is difficult to make microscopic observations of living organisms, so that subsequent isolation of an organism of interest is usually not possible. Because of these and other difficulties, Perfil'ev and Gabe (48) devised special flat-glass

microcapillaries which could be inserted into the habitat for colonization and then removed and examined periodically (Fig. 3). These flat capillaries, called by them "pedoscopes" (if used in soil) or "peloscopes" (if used in mud), are far superior to round capillaries for microscopy, and have been used in the Soviet Union for many years. Their use will no doubt become more general as a result of the appearance of English translations of Perfil'ev's two books (47, 48). Through the use of these capillaries, Russian workers have discovered and studied new bacteria that live at the mud-water interface: *Lieskeella, Dictyobacter, Cyclobacter, Trigonobacter, Metallogenium,* and others. (The taxonomic validity of all of these entities is not accepted by all bacterial taxonomists.) In addition, new bacterial genera have been described from soil and water. Further, with the use of microcapillaries, Perfil'ev has been able to observe the microzonation which develops at the mud-water interface. For instance, in the mud of Lake Khepo-Yarvi, eight microzones developed over a vertical distance of 2 mm. From the surface down, these were photosynthetic zone (diatoms), iron oxidation zone (*Ochrobium* and *Gallionella*), predatory bacterial zone (*Dictyobacter*), *Azotobacter* zone, filamentous bacterial zone, *Cyclobacter* zone, *Lieskeella* zone, and sulfide oxidation zone (*Thiospira*). Below the eighth zone, conditions became highly reducing, as shown by the presence of FeS.

The value of flat-glass capillaries for many ecological studies is clear. The book by Perfil'ev and Gabe (48) should be consulted for details of methods and results, as well as for techniques used to actually construct the capillaries. In addition, this book has a detailed discussion of how capillaries can be used in continuous-flow systems and with micromanipulators.

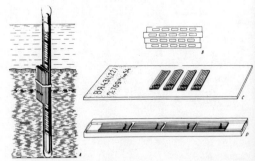

FIG. 3. *Capillary peloscope of Perfil'ev and Gabe (48). A. Capillary peloscope inserted at the mud-water interface. B. Cross section of typical capillaries. C. Capillaries mounted on a microscope slide for observation. D. Capillary cells in a protective channel with fixing fluid.*

USE OF THE MICROSCOPE IN NATURE

Two approaches are possible. One can take the microscope to the field—in the sense of actually submerging it in the environment—or one can bring the field to the microscope. The first approach is fraught with difficulties and has been used only rarely. Staley (54) immersed a phase microscope in a small pond in Michigan and used it to study algal growth. This technical tour de force may serve as an example of the great difficulties with which the microbial ecologist is faced. In the first place, even with a submerged microscope a glass slide was necessary and served as an artificial surface. The slide was immersed in the water near the microscope and was transferred periodically to the microscope stage for observation and photography. To permit periodic viewing without diving, the microscope was placed at the surface of the water, with the eyepieces emerging from the pond. Staley watched and periodically photographed algae and bacteria which appeared on the slide, and thus determined the manner and rate of growth. Although in principle the same or a modified technique could be used in more rigorous environments such as deep waters, rivers, the ocean, perhaps even hot springs, in practice the difficulties arising are probably too great in terms of the information gained. A much simpler approach is to bring the submerged glass slide from the field to the microscope and to observe and photograph it by using a water-immersion lens. This was the approach used by Bott and Brock (3) to study bacterial growth rates in a small greenhouse pond. Microscope slides which were marked with a diamond pencil at a number of locations were submerged in the water. At intervals, the slides were brought to the surface and carried submerged to the microscope in glass petri dishes. Locations containing organisms were photographed and the slides were then returned to the pond. The temperature was not altered significantly and the slides were out of the pond for just a few minutes. Since at least 5 mm of water was present over the surface of the slide, presumably the organisms, 0.5 to 1.0 μm in thickness, did not become disturbed. An alternate method would be to use the microcapillaries of Perfil'ev and Gabe (48), discussed earlier.

In soil, experiments of this type are considerably more difficult due to the nature of the substrate. A single slide cannot be removed and reinserted periodically without disturbance. Microscopy of undisturbed soil particles is even more difficult. Recently, Casida (13) made a useful advance towards direct microscopy of undisturbed soil

particles. The procedure involves the use of incident illumination of the soil particle by light which passes through the objective in reversed fashion. The bacterial cell so illuminated diffracts the light, and light of some wavelengths returns through the objective to the ocular. Various colors are seen, the cell thus being differentiated from the dark background of the soil. However, actively growing bacteria do not exhibit this diffraction phenomenon, so that the technique permits observation in soil only of those cells which are in a dormant state. Microorganisms larger than bacteria cannot easily be seen either. Casida concludes that the technique will permit continuous observation of the growth, activities, and interactions of soil microorganisms, although if only dormant cells are observed it is not clear how one can study growth in situ; no growth rates have as yet been presented. If nothing else, the report of this technique may perhaps encourage further study and experimentation with the light microscope, the most powerful tool of the microbial ecologist.

MEASUREMENT OF GROWTH RATE OF SINGLE ORGANISMS

The most direct method of measuring growth rate is to measure the size or mass of a single organism at various times. This method, which is relatively easy to use in plant and animal ecology, is difficult to use in microbial ecology for several reasons. (i) It is often difficult or impossible to return periodically to precisely the same organism. (ii) Only nondestructive measurements can be used, such as measurement of size. Mass, the most important ecological parameter, may not relate too precisely to size. Interference microscopy does permit an estimate of cell mass, but is difficult to apply routinely or to very small cells. (iii) The procedure is tedious, especially with slow-growing organisms, as one must wait for long periods of time to obtain data on even a single cell. During the wait, the microscope may be tied up and cannot be used for other things. (iv) Because only a single organism is studied at a time, the data obtained may not be representative of the population as a whole.

There has been a considerable amount of work involving measurements of growth rates of single cells (usually bacteria or yeasts) in pure culture. This work, most recently reviewed by Painter and Marr (45), provides some clues concerning the growth of microbes in nature. Painter and Marr discuss some of the difficulties of accurately measuring growth rates of individual cells microscopically from changes in cell size. For cells the size of bacteria, the error of measurement is such that distinction between linear and exponential growth cannot readily be made. The consensus seems to be that the individual bacterial cell may grow exponentially, whereas the individual eucaryotic cell does not. For most ecological work, the distinction may not be too important. What is needed is merely a measure of how long it takes for a young cell to increase in size and to divide.

In his studies with a submerged microscope, Staley (54) observed the growth of *Chlorella* in a small Michigan pond. During the day a single cell increased in size, and at night this enlarged cell divided into four daughter cells. This is identical with the behavior of *Chlorella* cultures growing in alternating light and dark in the laboratory (56). Not all of the cells observed by Staley managed to complete division in 24 hr. In certain cases, if a cell did not complete division during the first night it continued to enlarge during the second day and then completed division the following night.

Measurement of growth rates of filaments involves the same principle as that of unicells, but microscopic recognition of the growing element is easier. No study of *direct* measurement of filament growth in nature has come to my attention, although much work on growth of single fungus filaments in laboratory culture has been done (50).

GROWTH RATES OF DEVELOPING MICROCOLONIES

The growth rate of microorganisms that form microcolonies in nature can be estimated by counting the number of cells per microcolony at different time intervals. Counting numbers is easier than measuring the sizes of single cells; hence, it is more convenient for routine work. When evaluating counts of microcolonies, one should keep in mind that immigration or emigration of cells could have occurred during the intervals between counts.

Bott and Brock (3) determined growth rates of aquatic bacteria developing at marked locations on microscope slides immersed in a small pond. The changes observed during the growth of a single microcolony are shown in Fig. 4. Considerable heterogeneity in developing microcolonies was observed. Some colonies never got beyond the two- to four-cell stage. Others developed into larger aggregates, after which most of the cells in the colony lysed. In others, the cells spread out after the colonies reached the 16- to 32-cell stage, making identification of the colony difficult. Only a few colonies developed to really large size (e.g., 128 cells or larger). Presumably, many of the organisms which initiated colonies on the

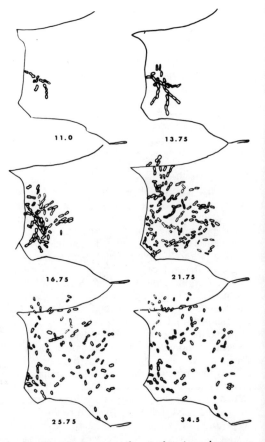

FIG. 4. *Development of a single microcolony on a glass slide immersed in a small pool. Time is expressed in hours after immersion. Drawings from photomicrographs of Bott and Brock (3).*

slide were not well adapted to a sessile existence, and they succumbed after utilizing organic material adsorbed to the slide in their immediate environment. However, even the organisms which were well adapted (as shown by their formation of larger colonies) never completely dominated the habitat because grazing organisms such as protozoa eventually moved in and consumed them. Doubling times obtained by counting on the photographs the numbers of cells per microcolony were in the range of 2 to 10 hr.

In this study (3), both immigration and emigration were observed. Not infrequently, a cell would undergo division and one of the daughter cells would swim away; the remaining cell would remain attached to the glass and undergo further division. This phenomenon is common in stalked bacteria such as *Caulobacter* (49), but was also seen for bacteria without obvious stalks. The technique of Helmstetter (28) for obtaining growth-synchronized populations of bacteria depends on the fact that the parents remain adherent to a support and the just-formed daughters are washed free.

However, in flowing-water habitats the organisms are usually types in which both products of division adhere tightly to the substrate, so that emigration is of little consequence; the evolutionary advantage of adhering tightly to the substrate in this situation is obvious.

In estimating growth rates, immigration is probably a lesser source of error than emigration; if microcolonies are widely spaced, the probability that an immigrant will settle within a colony is fairly low. Two biological factors which could affect immigration are antibiosis, which could reduce it, and chemotaxis, which could increase it. A technique involving ultraviolet radiation for quantifying immigration rates is described below.

For routine work, it would be desirable to be able to quantify growth rates of microcolonies without the complications involved in removing the slides, locating and photographing the colonies, and replacing the slides, especially when all of these operations must be done quickly and without disturbance to the organisms. To obviate these problems, Bott and Brock (3) immersed a series of slides at a single location and removed two or three at different time periods. On each slide, the cell count in a large number of microcolonies was made, and the average number of cells per microcolony was calculated for that time period. By plotting numbers of cells per microcolony at different immersion times, the growth rate could be calculated. Assuming that the organisms in the developing microcolonies are sufficiently distinct so that the microcolonies arising from cells belonging to the same species can be recognized microscopically, the lumping of data from different microcolonies is valid as a measure of the growth rate of a single organism. If, however, such recognition is not possible, the growth rates obtained represent the average growth rate for the aggregate of microbial species on the slide. For some ecological work, this latter information is still useful.

In many aquatic habitats, bacteria do not develop on the slides as discrete microcolonies but are scattered more or less randomly. This distribution might arise through growth if daughter cells moved away from mothers before settling down. On the other hand, it could arise as a result of immigration in the absence of growth. To distinguish growth from immigration, Bott and Brock (2, 3) employed ultraviolet radiation. Slides which were immersed were irradiated with a germicidal ultraviolet source at regular intervals during the experiment, the intervals calculated to

be about equal to the generation time of the population. Thus, newly attached immigrants were killed before they divided. The rate of increase of cell numbers on unirradiated slides is a function of both immigration and growth, whereas on the irradiated slides it is a function only of immigration.

Several conclusions could be drawn from these irradiation experiments. Microcolonies never developed on irradiated slides, showing that microcolonies arose as a result of growth. In most habitats, immigration is quantitatively unimportant. The rate of increase in cell number, even if distribution on the slides is essentially random, is caused primarily by growth on the slides. Therefore, if the immigration rate is not subtracted no serious error will be introduced in calculating growth rates. It should be noted that the habitats studied were relatively unpolluted and microbial drift through the water was low. In habitats in which drift is high, immigration may be more significant.

Modifications of the ultraviolet radiation technique can be envisaged which would make it adaptable to other habitats. In soil, which ultraviolet radiation would not penetrate, gamma radiation might be substituted. In some other habitats, germicidal chemicals might be used. The only requirement is that the agent in question not cause cells to be liberated from the surface, since if that occurred the immigration rate would be underestimated. Bott and Brock (*unpublished data*) tested a number of chemicals, including ethyl alcohol, formaldehyde, mercuric bichloride, and hydrochloric acid, but these proved unsatisfactory either because they modified the surface in such a way that attachment was inhibited or they caused detachment.

GROWTH RATES OF FILAMENTOUS ORGANISMS

Various methods (43) are available for quantifying filamentous organisms and can be used in conjunction with slide immersion and ultraviolet radiation to measure growth rate. Representative data on the growth rate of *Sphaerotilus* on glass slides (4) are given in Fig. 5. The glass slides were colonized initially by unicellular swarmer cells. In the ultraviolet-irradiated slides no *Sphaerotilus* filaments appeared, whereas on the unirradiated slides colonization by swarmers was followed by outgrowth of *Sphaerotilus* filaments. Growth rates for filamentous bacteria living in hot springs have also been measured in this way (Bott and Brock, *unpublished data*). In most of these habitats, both unicellular and filamentous bacteria developed on the same slides, and the doubling times of both kinds of bacteria were fairly similar.

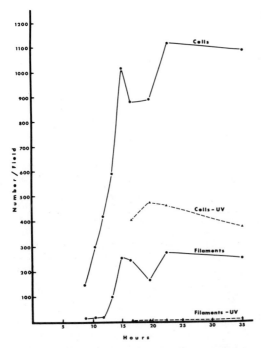

FIG. 5. *Quantitative counts of total Sphaerotilus cells and of filaments on glass slides immersed in a small stream. On the ultraviolet-irradiated slides, single cells never develop into filaments. From Bott and Brock (4).*

When expressing the doubling time of the filaments, the value given is the rate of increase in the aggregate length of filaments per unit area. This could be due either to increase in lengths of existing filaments or to the increase in number of attaching filaments. Again, studies using ultraviolet radiation permit these alternatives to be distinguished. In all cases studied, growth was due primarily to increase in length of filaments. On irradiated slides only short filaments were observed. On unirradiated slides, the lengths of filaments got progressively longer as immersion time increased. The rate of increase of length of filaments was exponential over two decades, suggesting that growth was occurring along the whole lengths of the filaments, rather than only at the tips as it does in filamentous fungi. If the latter growth habit occurred, rate of increase should have been as the square rather than the exponential.

METHODS BASED ON ANALYSIS OF THE CELL-DIVISION CYCLE

The typical division cycle of a eucaryotic cell is given in Fig. 6. In this cycle, mitosis can be recognized morphologically and deoxyribonucleic acid (DNA) synthesis can be recognized auto-

radiographically by the use of radioactively labeled thymidine. In many cell types, the time of mitosis represents a constant fraction of the total cell-division cycle. If this is true and if the time occupied by mitosis is known, the division rate of a population can be calculated by a formula derived originally by Crick (*cited in* 30), time of mitosis/total time of cell-division cycle $\times$ $\log_e 2 = \log_e (1 + 2R/1 + R)$, where R is the fraction of cells in mitosis. If interphase is relatively long and R is thus small, this formula reduces to time of mitosis/total time of cell-division cycle $= 1.44R$. The use of this formula assumes that the rate of entry of cells into mitosis is uniform.

This approach was used by Warner (57) to estimate division rate of the protozoan *Entodinium* in the rumen. The time for mitosis was determined by microscopic observation of entodinia using a warm stage; a value of about 15 min was obtained. It was assumed that this value holds for all cells of that species in the population. The fraction of cells in mitosis (r) was then counted at different times and the time for cell division was thus obtained. The minimum mean doubling time found was 5.5 hr, which corresponds to a maximum division rate of about four generations per day. However, dividing forms were much more common at night than during the day. The true division time of the population can be calculated by averaging the various division times over the 24-hr period; it is 15 hr (*see* 31). The maximal growth rate obtained for entodinia in culture was one division every 2 to 4 days, suggesting that conditions in culture are far from optimal for multiplication.

Labeled Thymidine Methods

The duration of the cell cycle can also be estimated by labeling with radioactive (^3H or ^{14}C)

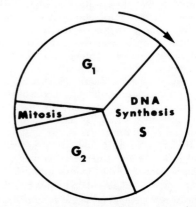

FIG. 6. *Cell cycle in an eucaryote. Arrow indicates the direction in which cells progress in the cycle.*

thymidine. The problem of determining growth rate of mammalian cell population in vivo by thymidine labeling is, in principle, the same as the problem of determining growth rate of a microbial population in nature, and the book by Cleaver (15) provides a detailed summary of the concepts involved.

If a population is labeled with a brief pulse of tritiated thymidine, only those cells in S phase (i.e., engaged in DNA synthesis) will be labeled. After continued growth of the population, the labeled cells will pass through mitosis as a wave, and this fraction can be estimated by preparing autoradiograms at different times and counting the fraction of labeled cells which are in mitosis. At least in mammalian cells, mitotic figures can be easily recognized. The results to be expected from the ideal case and the results obtained in an actual example are given in Fig. 7. Initially, there are no labeled mitoses until a time equal to G_2 has passed, after which the fraction of labeled mitoses rises to 1.0 in a period equal to the duration of mitosis (M). The labeled fraction remains at 1.0 for a period equal to the duration of the S phase minus the duration of mitosis (S − M), and subsequently falls to zero. The second wave of labeled mitoses appears after a further time equal to $G_1 + G_2$. It is thus possible to calculate the duration of the various phases from the time periods exhibited by the graph. Although the shape of the experimental curve does not precisely follow the ideal, it is close enough for most practical purposes.

The limitations of any technique involving the use of radioactive thymidine should be kept in mind. It must be known that the population in question assimilates this compound and incorporates it only into DNA. If there is a large pool of nonradioactive thymidine, this will affect the kinetics of labeling; consequently, it is essential at the end of the pulse to add an excess of nonradioactive thymidine, which dilutes any labeled material remaining in the pool. It is essential to know that the tritiated thymidine itself does not alter the cell cycle (e.g., through radiation damage). The labeling time must be kept short, but it must be long enough to permit labeling sufficient for detection on the autoradiograms. The technique can be applied only to systems in which the number of dividing cells is sufficiently high that the fraction of labeled mitoses can be determined accurately.

If pulse labeling, which often requires manipulation of the culture, is not possible, continuous labeling can be used. With this method, the isotope is added once and maintained in excess throughout the labeling period. To determine the

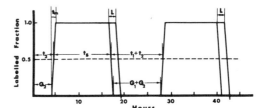

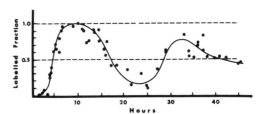

FIG. 7. *Fraction of labeled mitoses as a function of time after a pulse of* ^3H-*thymidine. Upper: ideal case with no intercell variation. Abbreviations: L, duration of labeling;* t_m, *time for mitosis;* t_s, *duration of S phase;* t_1, *duration of G, plus one-half of* T_m; t_2 *duration of* G_2 *plus one-half of* t_m; G_1 *and* G_2 *as on Fig. 6. Bottom: actual data for mouse L cells.* G_1, *5.7 hr; S, 12.2 hr;* G_2, *4.4 hr; M, 0.9 hr; T (total cell cycle time), 23.2 hr. Taken from Fig. 4.5 of Cleaver (15).*

parameters of the cell cycle with this method, three functions must be determined: the fraction of labeled mitoses, the fraction of labeled cells, and the average grain count per cell. A typical continuous-labeling experiment is shown in Fig. 8. The upper curve, showing the rate of increase in labeled mitoses, is used to calculate G_2, as in the pulse-labeling technique. The fraction of labeled cells increases linearly (middle curve) as a result of the entry of cells into the S period. With populations in exponential growth, there are more young cells than old; hence the rate of entry of cells into S is greater than the rate of cell division. For this reason, the rate of increase in labeled cells is faster than the rate of cell division and cannot be used directly as a measure of division rate. The labeled fraction will reach 1.0 at a time equal to $G_1 + G_2$, and since G_2 was estimated already G_1 can now be calculated by using the data obtained in the middle curve. The maximum amount of radioactive thymidine which a cell can incorporate in one cycle (estimated by the average grain count) corresponds to a single complete round of DNA synthesis. Since this takes the whole S phase to complete, the time it takes for the average grain count per cell to reach a constant level corresponds approximately to the S period (lower part of Fig. 8). [In actual fact, this procedure slightly overestimates the duration of the S period; the complications are discussed by Cleaver (15).]

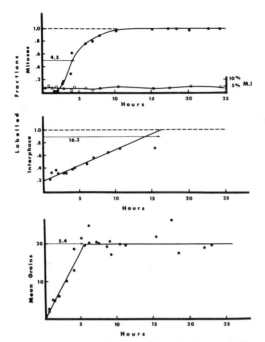

Fig. 8. *Continuous-labeling experiment with ^3H-thymidine and human skin epithelium cells. The upper curve is used to determine t_2 (4.5 hr) in the same manner as in Fig. 7. The middle curve is used to estimate $t_1 + t_2$ (16.3 hr); thus both t_1 and t_2 can be calculated. S (5.4 hr) is estimated from the lower curve. The other parameters can then be calculated. Taken from Figure 4.6 of Cleaver (15).*

An alternative method of measuring the rate of entry of cells into the S phase involves a double labeling technique, a pulse of tritiated thymidine being used for the first labeling followed after a defined time interval by a pulse of ^{14}C-thymidine. Autoradiograms can be prepared by using a double-emulsion technique which permits counting separately radioactivity due to tritium or carbon-14. The interval between exposures to the two labels should be less than $G_2 + M$ to ensure that labeled cells do not divide during the period of labeling (36).

If it is possible to determine the DNA content per cell by microspectrophotometry of Feulgen-stained autoradiograms, the relative duration of the four phases of the cell cycle can be determined more directly, since cells in G_2 will have twice the DNA content of cells in G_1. The population need be labeled only for a brief period before fixation so as to label the S-phase cells. By microspectrophotometry, the unlabeled cells can be divided into G_1 and G_2; the mitotic cells are detected cytologically, and the S phase cells by autoradiog-

raphy. In this way, the fraction of the cells in each of the four phases is estimated and the *relative* durations of the four phases can be determined (*see* p. 123 of reference 15 for details). To convert these relative values into absolute ones, it is necessary to determine by other methods the duration of one of the stages of the cell cycle in absolute terms. The advantage of this method is that it requires only brief labeling and only a single sample of the population. The disadvantage is that it requires microspectrophotometry, which is laborious and can be performed only on fairly large cells.

Although the methods described above have been applied extensively to mammalian systems, they have been used only rarely with eucaryotic microorganisms, usually with pure cultures [*see* Cleaver (15) for references]. However, the potential value of these methods for microbial ecology is clearly so great that their description here appears justified.

GROWTH RATES OF PROCARYOTES AS MEASURED BY THYMIDINE AUTORADIOGRAPHY

The measurement of the growth rates of procaryotes presents some new complications because procaryotes do not exhibit the typical mitotic cell-division cycle described above for eucaryotes. In the bacteria which have been studied, DNA synthesis in exponential populations occurs throughout virtually the whole of the cell cycle, being interrupted only briefly at the time of cell division (37). Consequently, if a pulse of tritiated thymidine is given to a growing bacterial population, virtually all cells become labeled to some extent, in contrast to the situation in eucaryotes in which only cells in S phase get labeled at all, and these heavily. A second problem is that since procaryotes do not divide by mitosis, it is impossible to evaluate microscopically the fraction of cells in division. Last, because of the small size of most procaryotes and the low DNA content per cell, grain counting on autoradiograms is difficult; even heavily labeled cells have only a few silver grains over them.

If grain counting is possible, growth rate can be estimated by a continuous-labeling method. Referring to Fig. 8, it can be seen that the time required for the mean grain count per cell to reach saturation is approximately equal to the length of the S phase, and if the S phase occupies 90 to 97% of the cell cycle (the situation in *E. coli*), the length of the cell cycle (virtually the same as the doubling time) can be estimated.

It would be desirable to have methods which did not require grain counting. An approach I used (10) was to measure the rate of accumulation of labeled cells in the population during continuous labeling. The marine bacterium *L. mucor* was used in this study. This is a large, filamentous epiphyte of seaweeds which is readily recognized in natural material. Since the filaments project perpendicularly from the surface of algal fronds, microscopy and preparation of autoradiograms is relatively easy. In growing laboratory cultures, the rate of accumulation of radioactive cells is linear with time at least until 80% of the cells are radioactive. (Some cells never become radioactive even after long time periods, so that if incubation with tritiated thymidine is continued indefinitely the rate of accumulation of radioactive cells falls off.) The rate of accumulation of radioactive cells is proportional to the doubling time of the culture, and about 1% of the cells get labeled in 0.002 generation. This relationship, determined for a pure culture, could then be used to estimate doubling time in nature (10). With this technique, one assumes that the rate of accumulation of radioactive cells is linear and begins without a lag, just as it does in pure cultures, and that nonradioactive thymidine, which might dilute the radioactive material, is absent from seawater. It is desirable to analyze samples incubated at several different times to ensure that the rate of incorporation is indeed linear in nature.

This technique requires a study with laboratory cultures growing at known rates so that the field data can be converted into real doubling times, and it requires the assumption that there is nothing intrinsically different about the way cultures and natural samples incorporate tritiated thymidine.

For determinations of relative rather than absolute growth rates, preliminary studies with laboratory cultures are not necessary. From the labeling rates of different natural populations it is possible to deduce which populations are growing fastest. For some studies, such relative rates suffice. However, to determine the contribution of a particular organism to food chains, absolute rates are essential.

Ecological Approximations for Labeling Experiments

The labeling of DNA is a particularly valuable technique for the study of growth rates because DNA is the only macromolecule in the cell which does not turn over and because synthesis does not occur in nongrowing cells. In exponentially growing populations, in which turnover is negligible, the rate of synthesis of any macromolecule could be measured because all materials are synthesized at the same rate. Since even in steady-state populations turnover and resting synthesis may be low, the measurement of the rate of synthesis of some macromolecule other than DNA might be acceptable as a first approximation for ecological work.

A point which needs to be emphasized is that in any labeling experiment the time of incubation with the isotope must be kept short (i) to prevent any secondary changes from taking place during the time the population is confined to the bottle, (ii) so that the concentration of isotope always remains in excess, and (iii) so that the experiment is terminated before any recycling of the label has occurred.

Rate of Appearance of Unlabeled Cells

Another radioautographic method which has some advantages over those described above involves measuring the rate at which unlabeled cells appear in a population which is fully labeled and then placed in medium which lacks label. Because of the semiconservative replication of DNA, after one cell division the specific radioactivity of each of the two offspring will be halved; consequently, one can measure growth rate by measuring the time taken for the grain count to decrease by half. The advantage of this method is that the labeled population can be placed under completely natural conditions, eliminating effects of confinement to bottles or to effects of the isotope itself. The disadvantages are that the population must be removed from its natural environment for labeling and that after several cell divisions the mean grain count will be reduced to such low levels that an accurate count will not be possible. Thus, the method seems best adapted for studying the behavior of species after introduction into foreign environments. This method apparently has not been used in ecology, although it is widely used by mammalian cell biologists (15). The genetic method of Meynell (discussed below) is formally equivalent to this isotope-dilution method.

In bacteria or other small organisms in which grain counting is difficult, a modification of this method might be to measure the rate of appearance of unlabeled cells in a fully labeled population. After the first cell division, all cells will still be labeled, but the labeled fraction will decrease by half after the second and subsequent cell divisions (Fig. 9). By counting the labeled and unlabeled cells after various periods of time, the doubling time of the population can be calculated.

Although for both of these methods a fully labeled population is desirable, the methods can be less accurately applied to partially labeled populations. This would be of considerable value

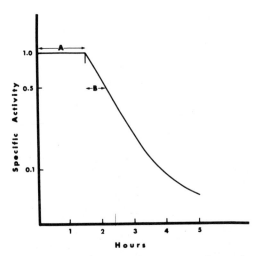

FIG. 9. *Change in specific radioactivity of a population as a function of time after labeling. "A" is the time that elapses before labeled cells pass through the nondividing compartments and begin to decrease in radioactivity. "B" is the time for the specific radioactivity to decrease by one-half, which is equivalent to the doubling time. After Fig. 7.5 of Cleaver (15).*

if full labeling required excessively high isotope concentrations or long incubation times.

GENETIC METHODS

An interesting method using genetically marked cells was devised by Meynell (40, 41) to measure the growth rates of bacteria in the animal body. As Meynell notes, bacterial growth in vivo is usually much slower than it is in laboratory cultures. This could be either because the animal is an unfavorable habitat for bacterial growth or because bacteria are being killed in vivo so that the division rate is greater than the rate of increase of the viable count. These two possibilities can be distinguished only if a method is available to measure bacterial growth when some cells are dying. The technique which Meynell used was to introduce a nonreplicating genetic marker into the bacteria before inoculation into the animal. As shown in Fig. 10, at each cell division the fraction of the population which contains the label will be halved. One determines the total population and the population with the genetic label at different time periods; from the rate at which the marker is lost, the doubling time can then be calculated. This method is formally analogous to the isotope dilution method described above, but it is simpler because it eliminates autoradiography and direct microscopy, the labeled and unlabeled populations being assessed by plating procedures.

In his first work, Meynell (40) made use of the fact that when a lysogenic bacterium is superinfected with an appropriate mutant of its prophage, the mutant enters the bacterium but does not replicate. Its presence can be assessed after induction by plating on an indicator bacterium which is specific for the superinfecting phage. Preliminary experiments showed that the marked population did behave as predicted during cell division up until the 10th generation, after which the proportion of marked bacteria decreased at a rate less than predicted, probably because about 0.1% of the original bacteria were lysogenized by the superinfecting phage so that replication could occur. Mice were inoculated intravenously with *E. coli* K-12 lysogenized with phage lambda *b* and superinfected with lambda *hc*. Within 30 min, the bacteria had cleared from the blood and were lodged in the liver and spleen. Assessment of total viable counts showed that the viable count dropped about three decades in the first 8 hr and did not decrease farther over the next 48 hr. The proportion of labeled organisms remained constant over the duration of the experiment, showing that no replication occurred.

In a second study (42), Meynell applied this technique to a study of the replication, killing, and excretion in the feces of *Salmonella typhimurium* which had been inoculated into the mouse orally. The genetic label was a histidine gene introduced in the abortively transduced state. Since the gene in an abortive transductant does not replicate, it is passed to only one of the daughter offspring at division. The proportion of cells in the population containing the gene is determined by plating on medium without histidine and scoring for the tiny colonies formed by the abortive transductants. Since excretion was a

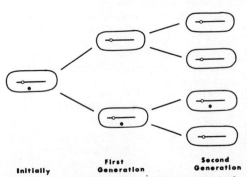

FIG. 10. *Distribution of a nonreplicating genetic element among dividing bacteria. The closed circle is the nonreplicating element; the open circle and line represent the replicating genome. At each generation, the fraction of cells containing the marker decreases by one-half. Based on Meynell (40).*

major factor in loss of viable cells from the intestine, the ability to distinguish excretion from death is clearly of value. In most cases, the bacteria underwent only a few divisions in the intestine. If the mice had been treated with streptomycin to eliminate normal intestinal bacteria before inoculation (the infecting strain used being streptomycin resistant), cell division was more rapid. The results showed that the normal mouse intestine contained a mechanism which was bacteriostatic and weakly bactericidal to *S. typhimurium*.

In neither of the two systems described above did much bacterial division occur. In another study, Maw and Meynell (39) studied the growth in the mouse of *S. typhimurium*, a natural pathogen of this animal. The marker was a superinfecting mutant of phage P22. After intravenous inoculation the organism became lodged in the spleen. The viable count doubled every 24 hr, whereas the true doubling time as determined from rate of dilution of the genetic marker was 8 to 10 hr. This should be compared with the minimum division rate observed in the test tube of 0.5 hr. The division rate in the spleen was at best only 5 to 10% of the maximum observed in vitro, and the death rate in the spleen was quite low. These results show that the maximum potential growth rate is not reached in vivo even by an organism which is highly pathogenic.

Despite the fact that Meynell's method is restricted to organisms for which the relevant genetic combinations can be constructed, it is clearly an interesting one with considerable promise. It should be useful in many cases in which it is of value to measure the division rate of organisms newly introduced into a habitat. Its use in studies of host-parasite relations may give us new insights into mechanisms involved in the establishment of colonizing species.

METHODS ESPECIALLY SUITED TO STEADY-STATE POPULATIONS

A system in steady state is in a time-independent condition in which production and consumption of each element of the system are exactly balanced, the concentrations of all elements within the system remaining constant even though there is continual change. Steady-state microbial populations are found in a number of natural habitats: the rumen (31), the intestinal tract (21), hot springs (11), and perhaps even infected animals (23, 24). In the steady state there is no change in population size, even though cell division is taking place. Because of this, it is impossible to measure growth rates from changes in absolute numbers, and one must seek more ingenious methods.

As I have pointed out elsewhere (8), an important item in characterizing a specific steady-state system is its time constant. Although fluctuations in growth rate may occur over a short period of time, on a long view of the system these fluctuations can be ignored. Thus, in the system of decomposition which occurs on the deciduous-forest floor, there is a great change in nutrient input in the autumn when leaves fall, but on an annual basis the forest floor ecosystem is in steady state because no net accumulation of leaves occurs (44). In the rumen, great changes occur daily as a result of feeding, but in the long run the system is constant (31).

The chemostat is an excellent laboratory model of a steady-state population (5), and it is instructive to consider how one measures growth rate in a chemostat, the principles developed there being readily adapted to natural situations. Measurement of growth rate requires a knowledge of the volume of the system (v) and its flow rate (f). The turnover time is v/f and the doubling time is equal to the turnover time.

Let us consider how the principle of the chemostat can be applied to a natural system. In a well-mixed system such as the rumen, the problem reduces to one of determining the rate at which material passes through, that is, the flow rate f. Hungate (31) discussed this problem in detail. If a pulse of an inert material (the marker) is given at zero time and if mixing is completed quickly, the rate of change in concentration of this inert material will give a measure of the turnover time. Because of mixing of the marker with unmarked contents and with material entering later, not all the marker leaves during one turnover time, even though during this period the total material leaving equals the total entering. The half-time, that is, the time at which the concentration of marker reduces to one-half the initial, is equal to 0.69 of the turnover time. Suitable markers for use in the rumen include nondigestible materials from the feed such as lignin or silica, or artificial markers such as iron oxide, chromic oxide, polyethylene glycol, rubber, etc.

If we know the turnover time (flow rate), and we know that the microbial population is in steady state, we know the doubling time of the population. To calculate the *productivity* of the population, that is, the number of new cells produced per day, we must also know the absolute population size, which can be determined by direct microscopic count. In the sheep rumen, there are about 10^{10} bacteria per ml and since the average volume of the sheep rumen is 5 liters, the total population is about 5×10^{13} bacteria. If the turnover time were 24 hr, 5×10^{13} bacteria would be produced per day.

In this discussion it has been assumed that bacteria disappear from the system only by passing out of it. If predation or lysis occurs within the system, the estimate of productivity could be low. Further, the estimate is for the rumen as a whole and assumes that it is a completely mixed system. It is quite possible that there are microhabitats within the rumen (e.g., the surfaces of food particles) where growth is rapid and other habitats with reduced ability to support microbial growth, so that the technique does not necessarily measure the maximum potential growth rate possible for rumen microorganisms. If for some reason knowledge of this potential were needed, one of the techniques described earlier in this paper, such as measurement of the rate of development of microcolonies, would be necessary. Although this has not been specifically done for the rumen, it could probably be done relatively easily (R. Hungate, *personal communication*).

If measurement of flow rate is not possible, growth rate in a steady-state system can be determined from the rate at which organisms leave the system, assuming that all organisms produced within the system are excreted intact and can be collected and counted. If the total number of organisms in the system is also known, the growth rate can be calculated. With this technique, neither the volume nor the flow rate needs to be measured, but the system must be completely mixed. This method was used by Gibbons and Kapsimalis (21) to measure the growth rate of the total intestinal microflora of hamsters, guinea pigs, and mice. The animals were placed in individual cages containing coarse-screen bottoms, and the fecal pellets which dropped through were collected. Twice a day for 3 days the pellets were collected, weighed, and homogenized in a diluent, after which direct microscopic counts were performed. Estimates of the total quantity of bacteria excreted per day were made by multiplying the number of bacteria per gram of feces by the average number of grams of feces excreted per day. After the excretion rate was determined, the animals were sacrificed and the bacterial count of the entire intestinal canal was determined. The average number of generations per day was then calculated. The doubling times of the intestinal bacteria calculated in this way were fairly low, varying from less than one to about six doublings per day. The intestinal flora of mice seemed to grow faster than those of hamster and guinea pig, but for all three species of animals the rates were much lower than those of typical intestinal bacteria in laboratory culture.

One possible limitation of this experiment is that bacteria might be destroyed within the intestine by predation, parasitism, or lysis. To check this, Gibbons and Kapsimalis (21) repeated their experiments with gnotobiotic mice inoculated with a pure culture of *E. coli*. Even under these presumably ideal conditions, *E. coli* grew at average rate of only 1.2 divisions per day. One explanation for this observation may be that environmental conditions (e.g., nutrient availability, *p*H, presence of natural inhibitors) may be inherently unfavorable for more rapid growth. Another possibility is that in certain microenvironments within the intestinal tract the growth rate is much higher than for the intestinal tract as a whole. It seems to me that the latter explanation has considerable merit, since the intestinal tract is highly heterogeneous both physically and chemically, and it is well established (51) that high bacterial populations live attached to the walls of the intestinal tract. If all the bacteria are growing in microenvironments which occupy only 10% of the total intestinal volume and after being sloughed into the lumen grow no further, the growth rate in these microenvironments would be 10 times that of the intestinal tract as a whole. Furthermore, in these experiments the assessment of total number of cells per intestinal tract included both growing and nongrowing cells. If all the cells were produced in 10% of the habitat and after entering the rest of the habitat grew no further, they would still be counted in the assessment of total numbers and would enter into the overall calculation. For instance, if the total bacteria excreted were $1,000 \times 10^9$ and if the count for the whole intestine were $1,000 \times 10^9$ but only 10% of the cells were from growing habitats, then the growth rate would be $1,000/100 = 10$ generations per day rather than 1. This serious error could be eliminated if only the cells in growing habitats were counted, although such counts could be done only if it were known where growth was taking place. Autoradiographic methods might provide a solution to this last problem.

Another approach to the measurement of steady-state growth was taken by Brock and Brock (11) for use in their studies on thermophilic algae in hot-spring drainways. It involved measurement of the algal wash-out rate after growth was prevented by darkening the system. Imagine a chemostat operating under steady-state conditions but with unknown volume and flow rate. In principle, the generation time could still be determined if growth were completely inhibited in a manner which did not otherwise alter the system. Wash out would still continue and the wash-out rate could be estimated by performing cell counts at various intervals of time. Since

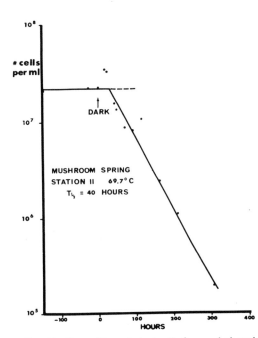

FIG. 11. *Rate of loss of algal cells from a darkened portion of a hot-spring algal mat. From Brock and Brock (11).*

growth balances wash out, the growth rate could then be determined.

This approach was relatively easy to use in hot springs because the effluent channels provide flow rate and temperature characteristics that are relatively constant over extended periods of time and because the algal population which develops reaches a steady state when the thickness of the mat is such that self shading occurs. Water flowing across the surface of the mat causes erosion of the cells through mechanical action, and the lost cells are replaced by newly grown ones. In areas of laminar water flow, a uniform mat develops from which replicate samples can be taken at different time periods. In studies at Mushroom Spring extending over 3 years (7, 11, 12), population density at selected stations remained fairly constant. However, if the system was darkened the algae disappeared quickly and were virtually absent within 2 to 3 weeks.

The results of a typical darkening experiment

are given in Fig. 11. It can be seen that within 2 days after darkening the population density begins to decrease and that the rate of loss is exponential over at least two decades. In the example of Fig. 11, the half-time of the rate of loss was 40 hr. If the assumptions made above are valid, the doubling time of the population in the steady state was 40 hr.

During the first day or two after darkening there is virtually no loss of cells. Perhaps this lag reflects normal adaptation of the algae to alternating day and night, wherein during the day products of photosynthesis are stored for use as energy sources during the night (12). Further, although cell division may occur only during the night, as in *Chlorella* (54), this does not affect the calculation of the doubling time of the *population*. Finally, for the technique to work the mechanism by which the cells are lost from the system does not really matter; erosion, lysis, or predation could be responsible. (In a system in which cells die but do not disappear, viable counting or autoradiography might be used to measure loss rate.)

In the hot-spring system it was possible to show directly that cell loss was due to erosion in the following way (*unpublished data*). Temperatures around 70 C are near the upper limit for blue-green algal growth, so that water flowing over the mats comes from alga-free areas. Large volumes of water were collected just downstream from stations where darkening experiments had previously been done (the mats being first allowed to return to the steady state), and these water samples were passed through membrane filters to remove the algal cells. The algae were then counted on the membrane filters by using vertical illumination on a fluorescence microscope, the algae being recognized because of the intense red autofluorescence imparted by chlorophyll and phycocyanin. In this way, the density of algal cells in water which had flowed over the mat could be calculated. This calculated erosion rate was similar to the loss rate measured in the darkening experiments, thus providing evidence that cell loss in the steady state is due to erosion rather than to lysis or predation.

Another factor to be taken into consideration is the error which might arise if cells from water settled under the darkened area. If the mat under study receives water from an upper algae-containing area, since such water contains algae, the loss rate would be underestimated if settling occurred. This point was checked (*unpublished data*) by placing glass slides containing depression wells on top of portions of algal mats under the black covers. Some algal cells did settle onto the slides and, at intervals, slides were removed and algal densities were quantified. The study showed that settling rate was two to three decades lower than wash-out rate, so that this factor was not a material source of error. Apparently, once the cells are taken into the rapidly flowing water their chances of settling are poor until the water has reached the foot of the spring, at which flow rate is much slower. In other habitats, settling may be more significant; thus, controls for settling should always be used.

It should be emphasized that the steady-state darkening technique measures only the growth rate of the population as a whole. It is possible (in fact, likely) that growth rates of different subsets of the population may vary quite markedly from this average value. In these compact algal mats in which self-shading is considerable, cells in the deeper parts of the mat receive much less light than do surface cells. Quantitative grain counts of autoradiographs of mats labeled with ^{14}C from CO_2 showed a progressive decrease in label from the surface to the bottom of the mats, which were 5 mm thick (12). Thus, although in the study presented in Fig. 11 the population as a whole was doubling every 40 hr, the cells at the surface were photosynthesizing faster and were probably developing faster than those at the bottom. This is, of course, the same problem which we discussed earlier in relation to measurement of growth rate in the rumen and intestinal tract. In terms of what the organisms are doing to the ecosystem, it is the growth rate of the whole population which is of interest. If one wishes to determine the fastest growth rate which an organism is capable of realizing in nature, another technique will be necessary.

The measurement of wash-out rates in growth-inhibited, steady-state populations can be applied with suitable modifications to other systems, both photosynthetic and heterotrophic. With photosynthetic systems, removal of the energy source without disturbing the system is relatively easy and could be applied to other algal habitats such as rivers, soils, and rocks. (The darkening of a whole lake would probably be beyond the capabilities of most research studies.) It is important to know in such studies that the factors affecting wash out are not themselves altered by darkening. For instance, if loss of algae was due not to erosion but to animal grazing, it would be necessary to ascertain that the animals involved were not themselves affected (physiologically or behaviorally) by darkening.

With heterotrophic systems, the inhibition of growth would probably require the use of an antibiotic which was specific for the organism of interest but did not affect other activities of the system. For studies of bacterial growth in higher animals, a whole array of suitable antibiotics is already available. For instance, to study the growth rates of intestinal organisms, antibiotics such as streptomycin and neomycin, which do not pass through the intestinal wall into the blood stream, could be easily used. It is well known that administration of such antibiotics orally results in inhibition of growth and elimination of intestinal bacteria. Although these antibiotics have been used both experimentally and clinically to sterilize the intestine, they have not been used in quantitative measurements of the rate of wash out of bacteria from the intestines.

One study involving the growth of *Mycobacterium tuberculosis* in the lungs of mice is relevant to this discussion. Gray and Cheers (24) showed that after an initial establishment period which took 6 weeks, the *M. tuberculosis* population reached a steady-state level in the lung which was maintained for at least the next 12 weeks. If infected mice were treated with drugs (pyrazinamide plus isoniazid) after the steady state had become established, there was a prompt exponential drop in numbers with a half-time of less than a week. This interesting experiment was done for another purpose than measuring growth rate, and because only viable counts were made it is not clear whether loss is due to lethal effects of the drugs or to destruction of cells by the animal after growth had been inhibited. In the latter event, these data could be used to calculate the steady-state growth rate.

Even if a population is not in steady state, growth rate might in principle still be measured. In a habitat in which absolute cell numbers are increasing, losses will still occur through predation, lysis, grazing, or wash out. Under these conditions, loss rates could be measured by a growth-inhibition procedure for samples taken at different times, and the loss rates obtained at each time could then be added to the observed rate of increase at each time, thus making possible a calculation of the actual growth rate. Admittedly, this technique is laborious.

A stable population size does not necessarily indicate a steady state in which growth is occurring. The organism may be in a habitat in which grazing, lysis, and wash-out losses are essentially zero but in which conditions are unfavorable for growth. A situation of this kind was found by W. N. Doemel (Ph.D. Thesis, Indiana University, 1970) in his studies on the growth rate of the eucaryotic alga *Cyanidium caldarium* in acid hot springs. Artificial channels placed in the flowing water provided fresh habitat for colonization and growth by *C. caldarium*. After an extended exponential increase in cell numbers, the population reached a constant level. Darkening of a portion of the channel at this time did not result in immediate wash out, although colonization did not occur on a fresh darkened channel. The constant population level reached by the organism thus resembles a stationary phase rather than a steady state. The probable explanation is that in its niche there are rarely predatory or grazing organisms, and, because the cells are quite dense and the flow in the acid streams is not very turbulent, erosion of cells from the mat by the mechanical action of the water is minimal.

Another method for measuring growth rate in the steady state is to inhibit wash out, grazing, or other losses and to measure the rate at which the population whose density was kept in check by one or more of these factors increases after the check has been removed. This presupposes that the limiting factor can be eliminated without altering the system in other ways. Gambaryan (20) used this method to measure generation times of bacteria in muds of Lake Sevan (U.S.S.R.). In this habitat, bacterial growth took place primarily in the liquid phase, and protozoa or other animals feeding on the bacteria could be eliminated by passing the muddy water through filter paper. Controls showed that these filtrates did not differ significantly in physiochemical respects from the original muddy water. To obtain an estimate of the generation time in mud, bottom samples were filtered and the filtrate was divided into two samples; one of these was used to obtain the initial bacterial count and the other was placed in a test tube which was sealed and immersed in the lake at the sampling site. After 24 hr, the test tube was removed, samples of the initial and final samples were passed through membrane filters, the organisms were stained on the filters, and microscopic counts were performed. Generation time was calculated from the formula: generation time = (time of incubation) (log 2) divided by (log final count minus log initial count). In various parts of the lake, generation times of 10 to 281 hr were determined. Unfortunately, controls to show that the population density was being kept in check by protozoa and other animals which consume bacteria were not included.

El-Shazly and Hungate (18) used an analogous approach to measure net growth rate of rumen microorganisms. The population densities of samples of rumen material were assessed immediately after removal and again after incubation in vitro for 1 hr under simulated rumen conditions. The relative population densities were estimated by making maximal fermentation rate measurements; thus, rates of change in population size could be determined. It must be assumed that the only role of the glass bottle is to confine the growing population and prevent wash out and that secondary changes do *not* take place during the 1-hr incubation. To check this, studies were also carried out on samples incubated in dialysis bags in the rumen of a fistulated animal; the data obtained were similar to those obtained in vitro. It must also be assumed that losses due to protozoal grazing are negligible. In these studies, the net growth values per hour ranged from −7 to +27% of the population, but even in a single animal marked differences were observed over a diurnal feeding cycle. Net growth was smallest (or even negative) just before the morning feeding, when bacterial nutrients were probably limiting; the highest growth rate was found several hours after feeding. By adding growth rates obtained over the diurnal cycle, a daily growth rate (turnover time) of 1.92 was obtained, in good agreement with estimates made from calculations of wash-out rates from use of nonmetabolizable tracers to measure dilution rate.

Experiments of this type could, in principle, be done in any habitat in which losses due to grazing are low or in which grazers could be eliminated. However, since incubation time must be kept very short to avoid secondary changes in the bottle, accurate methods for assessing population size are necessary to permit detection of small differences. An incubation time less than one-tenth the doubling time is probably desirable. In phytoplankton studies, population densities can be assessed very accurately by use of $^{14}CO_2$, making it possible to use this method fairly easily; Eppley (19) has applied it with some modifications to measurement of growth rates of phytoplankton living in the sea off La Jolla, Calif. (The study was actually designed to measure the standing crop of photosynthetically active phytoplankton, but it determines growth rate as well.) Samples of seawater were filtered through 150-μm netting to remove larger animals and were then incubated on shipboard under in situ light and temperature conditions with $^{14}CO_2$. Samples were taken at 24 and 48 hr, and the assimilated radioactivity was determined by membrane filtration. From changes in rate of ^{14}C uptake after 24 and 48 hr, the specific growth rate (k) was calculated from the equation $k = 1/\text{days}$ ln (48 hr-uptake − 24-hr uptake/24-hr uptake). Apparent growth rates of about one to two doublings per day were obtained. The incubation times which Eppley used were long, but were so chosen to avoid problems associated with diurnal periodicity in algal photosynthesis. Shorter incubation times could be used if high specific radioactivity $^{14}CO_2$ was used and if periodicity was obviated by incubating under saturating light to obtain the potential maximum photosynthesis rate. With such modifications, the method becomes precisely analogous to that of El-Shazly and Hungate (18).

INOCULUM SIZE: RATE OF EFFECT

An interesting technique quite different from any of the above has been used to estimate the growth rate of the spirochete *Treponema pallidum* in syphilitic lesions in the rabbit (17). An average incubation time of 17 days is required for a demonstrable lesion after the intradermal injection of 500 organisms; and the time required for the development of a lesion is reduced by 4 days for each 10-fold increase in inoculum size. Since a 10-fold increase in number is equivalent to 3.3 generations, the generation time of the spirochete in vivo can be estimated as 4 days/3.3 generations or 1.2 days (29 hr) per generation. It is assumed that each inoculated cell is capable ultimately of producing a lesion.

With modifications, this technique might be applicable to various other situations (in, for example, nodulation rate by rhizobia or the rate of establishment of a component of the intestinal flora).

CONCLUDING STATEMENT

A wide variety of methods for measuring microbial growth rates in nature is available. At least one of the methods described should be applicable to any specific ecological situation.

Is the result obtained worth the effort? This can be answered only in the context of a particular study. In a large ecosystems study, the unavailability of information on microbial growth rates would be a serious lack. In a simple study of the physiological ecology of a single organism, the knowledge of growth rate may be a useful parameter for the interpretation of the response of an organism to environmental variables. In our own studies on the ecology of *L. mucor* (33) and thermophilic blue-green algae (11), a knowledge of growth rates provided an important insight into how these organisms succeed in their natural habitats. Such information is of considerable value in interpretation of evolutionary processes, and it significantly supplements the conclusions drawn from studies of the same organisms in laboratory systems. Indeed, without such studies in nature, the laboratory studies are not too useful.

ACKNOWLEDGMENTS

This investigation was supported by research grants GB-7815 and GB-19138 from the National Science Foundation and by research contract C00-1804-22 from the Atomic Energy Commission.

LITERATURE CITED

1. Allen, O. N. 1957. Experiments in soil bacteriology, 3rd ed. Burgess Publishing Co., Minneapolis.
2. Bott, T. L., and T. D. Brock. 1969. Bacterial growth rates above 90 C in Yellowstone hot springs. Science 164:1411–1412.
3. Bott, T. L., and T. D. Brock. 1970. Growth and metabolism of periphytic bacteria: methodology. Limnol. Oceanogr. 15:333–342.
4. Bott, T. L., and T. D. Brock. 1970. Growth rate of *Sphaerotilus* in a thermally polluted environment. Appl. Microbiol. 19:100–102.
5. Brock, T. D. 1966. Principles of microbial ecology. Prentice-Hall, Inc. Englewood Cliffs, N.J.
6. Brock, T. D. 1966. The habitat of *Leucothrix mucer*, a widespread marine microorganism. Limnol. Oceanogr. 11:303–307.
7. Brock, T. D. 1967. Relationship between standing crop and primary productivity along a hot spring thermal gradient. Ecology 48:566–571.
8. Brock, T. D. 1967. The ecosystem and the steady state. Bioscience 17:166–169.
9. Brock, T. D. 1967. Mode of filamentous growth of *Leucothrix mucor* in pure culture and in nature, as studied by tritiated thymidine autoradiography. J. Bacteriol. 93:985–990.
10. Brock, T. D. 1967. Bacterial growth rate in the sea: direct analysis by thymidine autoradiography. Science 155:81–83.
11. Brock, T. D., and M. L. Brock. 1968. Measurement of steady-state growth rates of a thermophilic alga directly in nature. J. Bacteriol. 95:811–815.
12. Brock, T. D., and M. L. Brock. 1969. Effect of light intensity on photosynthesis by thermal algae adapted to natural and reduced sunlight. Limnol. Oceanogr. 14:334–341.
13. Casida, L. E., Jr. 1969. Observation of microorganisms in soil and other natural habitats. Appl. Microbiol. 18:1065–1071.
14. Cholodny, N. 1930. Ueber eine neue Methode zur Untersuchung der Bodenmikroflora. Arch. Mikrobiol. 1:620–652.
15. Cleaver, J. E. 1967. Thymidine metabolism and cell kinetics. North-Holland Publishing Co., Amsterdam.
16. Collins, V. G., and C. Kipling. 1957. The enumeration of waterborn bacteria by a new direct count method. J. Appl. Bacteriol. 20:257–264.
17. Davis, B. D., R. Dulbecco, H. N. Eisen, H. S. Ginsberg, and W. B. Wood. 1968. Microbiology. Hoeber Medical Division, New York.
18. El-Shazly, K., and R. E. Hungate. 1965. Fermentation capacity as a measure of net growth of rumen microorganisms. Appl. Microbiol. 13:62–69.
19. Epply, R. W. 1968. An incubation method for estimating the carbon content of phytoplankton in natural samples. Limnol. Oceanogr. 13:574–582.
20. Gambaryan, M. E. 1966. Method of determining the generation time of microorganisms in benthic sediments. Microbiology 34:939–943.
21. Gibbons, R. J., and B. Kapsimalis. 1967. Estimates of the overall rate of growth of the intestinal microflora of hamsters, guinea pigs, and mice. J. Bacteriol. 93:510–512.
22. Gilbert, C. W., and L. G. Lajtha. 1965. Cellular radiation biology. The Williams & Wilkins Co., Baltimore.
23. Gray, D. F., and C. Cheers. 1967. The steady state in cellular immunity. Aust. J. Exp. Biol. Med. 45:407–416.
24. Gray, D. F., and C. Cheers. 1967. The steady state in cellular immunity. Aust. J. Exp. Biol. Med. 45:417–426.
25. Gray, T. R. G. 1967. Stereoscan electron microscopy of soil microorganisms. Science 155:1668–1670.
26. Gray, T. R. G., P. Baxby, I. R. Hill, and M. Goodfellow.

1968. Direct observation of bacteria in soil. *In* T. R. G. Gray and D. Parkinson (ed.), The ecology of soil bacteria. Liverpool University Press, Liverpool.

27. Greig-Smith, P. 1964. Quantitative plant ecology, 2nd ed. Butterworths, London.

28. Helmstetter, C. E. 1969. Sequence of bacterial reproduction. Annu. Rev. Microbiol. 23:223–238.

29. Holm-Hansen, O., and C. R. Booth. 1966. The measurement of adenosine triphosphate in the ocean and its ecological significance. Limnol. Oceanogr. 11:510–519.

30. Hughes, A. F. 1952. The mitotic cycle. Butterworths, London.

31. Hungate, R. E. 1966. The rumen and its microbes. Academic Press Inc., New York.

32. Jannasch, H. W., and G. E. Jones. 1959. Bacterial populations in sea water as determined by different methods of enumeration. Limnol. Oceanogr. 4:128–139.

33. Kelly, M. T., and T. D. Brock. 1969. Physiological ecology of *Leucothrix mucor*. J. Gen. Microbiol. 59:153–162.

34. Kershaw, K. A. 1964. Quantitative and dynamic ecology. Edward Arnold, London.

35. Kuznetsov, S. I. 1959. Die Rolle der Mikroorganismen im Stoffkreislauf der Seen. Deutscher Verlag der Wissenschaften, Berlin.

36. Lala, P. K. 1968. Measurement of *S* period in growing cell populations by a graphic analysis of double labeling with ^{3}H- and ^{14}C-thymidine. Exp. Cell Res. 50:459–463.

37. Maaloe, O., and N. E. Kjeldgaard. 1966. Control of macromolecular synthesis. W. A. Benjamin, New York.

38. Macfadyen, A. 1963. Animal ecology, 2d ed. Isaac Pitman, London.

39. Maw, J., and G. G. Meynell. 1968. The true division and death rates of *Salmonella typhimurium* in the mouse spleen determined with superinfecting phage P22. Brit. J. Exp. Pathol. 49:597–613.

40. Meynell, G. G. 1959. Use of superinfecting phage for estimating the division rate of lysogenic bacteria in infected animals. J. Gen. Microbiol. 21:421–437.

41. Meynell, G. G. 1968. A new look at infection. New Sci. 40: 360–361.

42. Meynell, G. G., and T. V. Subbaiah. 1963. Antibacterial mechanisms of the mouse gut. Brit. J. Exp. Pathol. 44:197–208.

43. Olson, F. C. W. 1950. Quantitative estimates of filamentous algae. Trans. Amer. Microsc. Soc. 59:272–279.

44. Ovington, J. D. 1962. Quantitative ecology and the woodland ecosystem concept. Advan. Ecol. Res. 1:103–192.

45. Painter, P. R., and A. G. Marr. 1968. Mathematics of microbial populations. Annu. Rev. Microbiol. 22:519–548.

46. Parkinson, D., G. S. Taylor, and R. Pearson. 1963. Studies on fungi in the root region. I. The development of fungi on young roots. Plant Soil 19:332–349.

47. Perfil'ev, B. V., D. R. Gabe, A. M. Gal'perina, V. A. Rabinovich, A. A. Sapotnitskii, E. E. Sherman, and E. P. Troshanov. 1965. Applied capillary microscopy; the role of microorganisms in the formation of iron-manganese deposits. Consultants Bureau, New York.

48. Perfil'ev, B. V., and D. R. Gabe. 1969. Capillary methods of investigating micro-organisms. Oliver and Boyd, Edinburgh.

49. Poindexter, J. S. 1964. Biological properties and classification of the *Caulobacter* group. Bacteriol. Rev. 28:231–295.

50. Robertson, N. F. 1959. Experimental control of hyphal branching and branch form in hyphomycetous fungi. J. Linn. Soc. London Bot. 46:207–211.

51. Savage, D. C., R. Dubos, and R. W. Schaedler. 1968. The gastrointestinal epithelium and its autochthonous bacterial flora. J. Exp. Med. 127:67–76.

52. Schmidt, E. L., R. O. Bankole, and B. B. Bohlool. 1968. Fluorescent-antibody approach to study of rhizobia in soil. J. Bacteriol. 95:1987–1992.

53. Sladecekova, A. 1962. Limnological investigation methods for the periphyton ("Aufwuchs") community. Bot. Rev. 28:286–350.

54. Staley, J. T. 1970. *In situ* observations on the growth of algae using an immersed microscope. J. Phycol., *in press*.

55. Strugger, S. 1948. Fluorescence microscope examination of bacteria in soil. Can. J. Res. (section C) 26:188–193.

56. Tamiya, H. 1966. Synchronous cultures of algae. Annu. Rev. Plant Physiol. 17:1–26.

57. Warner, A. C. I. 1962. Some factors influencing the rumen microbial population. J. Gen. Microbiol. 28:129–146.

58. Welch, P. S. 1948. Limnological methods. Blakiston Co. Philadelphia.

Author Citation Index

Subject Index

Abnormal growth, 312, 379
Absolute cell count, 90
Acceleration phase
  negative, 26, 37–38
  positive, 26, 28–36, 51–58, 74
Acetic acid, 81, 98, 200, 224, 230, 288
Actinomycetes, 280
Activity functionals, 336
Adaptation, 82–83, 104–105, 394–395, 398
Adaptive changes, 395
Adenosine triphosphate, 404–405
Aeration, effects of, 77, 192 (*see also*
  Oxygen)
Aerobacter aerogenes, 121–122, 190, 207, 209,
  214, 216, 247, 309, 327–328, 331
Aerobacter cloacae, 168–169, 179–189
Aerobic growth, 171, 230, 256, 289, 316
Affinity constant, 149 (*see also* Saturation
  constant)
Age, 200, 352–353
  of cell, 217
  of culture, 74, 217
  terminology, 1
Alcohol, 275, 276
Algae
  Chlorella pyrenoidosa, 230, 263–273, 277,
    408, 417
  Cyanidium caldarium, 418–419
Amino acid pool, 222
Amino acids, 80, 113–114, 205, 226, 311
  casamino acids, 114
Aminopterin, 340
Ammonia, 69, 75, 95, 179, 200, 205–206,
  209, 224–226, 256, 331 (*see also*
  Nitrogen)
Anaerobic growth, 171, 230, 239, 250, 256,
  289

Analysis
  chemical, 115, 163, 193, 204, 294 (*see
    also specific substance*)
  empirical aspects, 310, 339
  experimental (*see* Methods, experimental)
  growth curves, 73, 183, 184
  mathematical, 30–34, 230, 241
  microbiological assay, 76, 98
  microscope, 10–15, 127
  Monod growth, 87
  protein, 310–311, 339
  theoretical, of continuous culture,
    230–262
Animal cells, 340
Animalcules, 1
Anthrone, 409
Antibiotics, 103, 230, 232, 276, 280, 285,
  392–395
  neomycin, 311
  penicillin, 190–202
Antigens, 205, 394
Antiseptics, 2 (*see also* Disinfectants)
Apparatus
  (*see also* Bactogen; Chemostat;
    Turbidostat)
  for continuous culture, 161–163,
    166–167, 192, 178–179, 265–268,
    295–296, 309
  operation, 178–179, 192, 265–268,
    269–279, 364
  for synchrony, 262–263, 348, 360, 379
Apparatus effect, 187
Arginine, 114, 331
Artifacts, 4, 7, 280, 339, 385
Aspergillus niger, 75, 311
Assays
  chemical, 2

429